T0240108

Accelerator-Driven System at Kyoto University Critical Assembly

Cheol Ho Pyeon
Editor

Accelerator-Driven System at Kyoto University Critical Assembly

 Springer

Editor
Cheol Ho Pyeon
Institute for Integral Radiation and Nuclear Science
Kyoto University
Osaka, Japan

ISBN 978-981-16-0346-4   ISBN 978-981-16-0344-0   (eBook)
https://doi.org/10.1007/978-981-16-0344-0

Cover Illustration: ©KURNS

This Springer imprint is published by the registered company Springer Nature Singapore Pte Ltd.
The registered company address is: 152 Beach Road, #21-01/04 Gateway East, Singapore 189721, Singapore

Preface

A great deal of interest has been shown in the accelerator-driven system (ADS), to produce energy and transmute radioactive wastes in a clean and safe way, from the viewpoint of a promising and innovative technology. ADS is considered an interestingly hybrid system comprising of a reactor and an accelerator, and many experts of reactor physics and nuclear data have dedicated their important time and a huge effort to the implementation of a new system of nuclear transmutation by ADS, through numerical analyses by stochastic and deterministic approaches.

The few introductory books on ADS present a rather narrow view of numerical simulations by concentrating only on the behaviors of the neutrons under the existence of an external neutron source such as the accelerator. Meanwhile, neutron characteristics of ADS cannot be separated from other neutronics and experimental aspects, including accelerator, radiation detection, and nuclear data, since the validity and verification of numerical simulations, and the accuracy of experimental results are confirmed by integrating with the relevant fields.

The authors have attempted to write the results and discussions of experimental analyses accommodated to the needs of the scientists and researchers in our ADS community. In particular, the authors have tried to introduce the researchers to measurement methodologies and numerical simulations of reactor physics parameters in a subcritical reactor, including neutron spectrum, subcritical multiplication factor, subcriticality, prompt neutron decay constant, and effective delayed neutron decay constant. Nearly all of the illustrations have been specifically devised for this book to facilitate insight into various aspects of statics and kinetics in a subcritical core. Profound works have been written on the subjects of subcriticality by Kengo Hashimoto and of effective delayed neutron fractions by Masao Yamanaka.

The authors have engaged in feasibility studies of ADS since 2003 with the combined use of the Kyoto University Critical Assembly (KUCA) core and two accelerators (pulsed-neutron generator: 14 MeV neutrons and fixed-field alternating gradient accelerator: 100 MeV proton accelerator). ADS feasibility has been mainly examined by statics and kinetics experiments carried out at KUCA and numerical simulations by the Monte Carlo calculations with major nuclear data libraries. This work is based on a lot of invaluable reactor operations and important experiences in the KUCA core by the authors, research staff, and students over 15 years.

The main objective of this book is to lead to a considerable emphasis on actual data of the ADS experiments, showing mainly outstanding features of the KUCA facilities, including two external neutron sources (14 MeV neutrons and 100 MeV protons) and a variety of nuclear fuel materials (highly enriched uranium, thorium, and natural uranium) and reflector materials (polyethylene, graphite, beryllium, aluminum, iron, lead and bismuth), which play an important role in determining neutron spectrum of the KUCA core.

The book has been organized into three parts with eight chapters and Appendix. The authors present, firstly, kinetics parameters: subcriticality and delayed neutron decay constant covering studies of measurement methodologies and correction factors, in Chaps. 2–4. Secondly, the authors include static parameters: reaction rates, nuclear transmutation of minor actinide (Np-237 and Am-241), neutronics of Pb-Bi (Pb and Bi isotopes) that is a coolant material of actual ADS experimental facilities, and sensitivity and uncertainty of criticality in the KUCA core, dedicating studies of neutron spectrum, nuclear transmutation, sensitivity, and uncertainty, in Chaps. 5–8, respectively. Thirdly, the authors provide all experimental data of ADS carried out in the KUCA core so that many researchers should easily access, compiling in Appendix.

The authors would like to acknowledge the assistance of a group of truly exceptional former students of Kyoto University, including Yoshiyuki Hirano, Hervault Morgan, Hiroshi Shiga, Takahiro Yagi, Hesham Shahbunder, Hak Sung Kim, Yuki Takemoto, Tetsushi Azuma, Hiroyuki Nakano, Kiichi Sukawa, Yoshimasa Yamaguchi, Atsushi Fujimoto, and Makoto Ito, and former other research staff, including Dr. Jae Yong Lim, Dr. Kyeong Won Jang, Dr. Thanh Mai Vu, and Dr. Song Hyun Kim, carrying out successfully all the ADS experiments at KUCA and demonstrating interestingly the experimental and numerical results.

It is particularly important to acknowledge the invaluable comments and fruitful discussions with respect to a series of the ADS experiments at KUCA, of the members of the ADS Coordinated (Collaborative) Research Projects (CRP) organized by the International Atomic Energy Agency (IAEA) in 2004 through 2019, including Dr. Alberto Talamo (Argonne National Laboratory: ANL, US), Dr. Mario Carta (Energia Nucleare ed Energie Alternative: ENEA, Italy), Prof. Piero Ravetto (Politecnico di Torino, Italy), Dr. Fabrizio Gabrielli (Karlsruhe Institute of Technology, Germany), Prof. Chang Hyo Kim (Seoul National University: SNU, Korea), Prof. Mate Szieberth (Budapest University of Technology and Economics, Hungary), Prof. Hyung Jin Shim (SNU, Korea), Dr. Alexander Stanculescu (Idaho National Laboratory, US), and Dr. Yousry Gohar (ANL, US).

The authors also wish to express our gratitude to Ms. Maki Nakatani and Ms. Mari Yamahana of KURNS for their efforts in helping prepare the various drafts and manuscripts.

The authors are thankful to Springer Nature for having the patience for the completion of the manuscript and their responses all throughout its preparation for over a year.

Osaka, Japan
December 2020

Prof. Cheol Ho Pyeon

Contents

Contributors

Kengo Hashimoto Atomic Energy Research Institute, Kindai University, Osaka, Japan

Cheol Ho Pyeon Institute for Integrated Radiation and Nuclear Science, Kyoto University, Osaka, Japan

Masao Yamanaka Institute for Integrated Radiation and Nuclear Science, Kyoto University, Osaka, Japan

Chapter 1
Introduction

Cheol Ho Pyeon

Abstract At the Kyoto University Critical Assembly (KUCA), the accelerator-driven system (ADS) is composed of a solid-moderated and solid-reflected core (A-core) and a pulsed-neutron generator (14 MeV neutrons) or the fixed-filed alternating gradient (FFAG) accelerator (100 MeV protons). At KUCA, two external neutron sources, including 14 MeV neutrons and 100 MeV protons, are separately injected into the A-core, and employed for carrying out the ADS experiments. With the combined use of the A-core and two external neutron sources, basic and feasibility studies of ADS have been engaged in the examination of neutronics of ADS, through the measurements of statics and kinetics parameters of reactor physics, including subcritical multiplication factor, subcriticality, prompt neutron decay constant, effective delayed neutron fraction, neutron spectrum, and reaction rates.

Keywords KUCA · ADS · Pulsed-neutron generator · FFAG accelerator

1.1 Kyoto University Critical Assembly

1.1.1 KUCA Facility

The Kyoto University Critical Assembly (KUCA [1]; Fig. 1.1) is a multi-core type critical assembly developed by Kyoto University, Japan, as a facility that can be used by researchers from all the universities in Japan to carry out studies in the field of reactor physics. KUCA was established in 1974, as one of the main facilities of the Research Reactor Institute, Kyoto University (KURRI; currently, Institute for Integrated Radiation and Nuclear Science: KURNS), located at Kumatori-cho, Sennan-gun, Osaka, Japan.

KUCA is a multi-core-type critical assembly consisting of two solid-moderated cores (A- and B-cores) and one light-water-moderated core (C-core). Then, a single

C. H. Pyeon (✉)
Institute for Integrated Radiation and Nuclear Science, Kyoto University, Osaka, Japan
e-mail: pyeon.cheolho.4z@kyoto-u.ac.jp

© The Author(s) 2021
C. H. Pyeon (ed.), *Accelerator-Driven System at Kyoto University Critical Assembly*,
https://doi.org/10.1007/978-981-16-0344-0_1

1

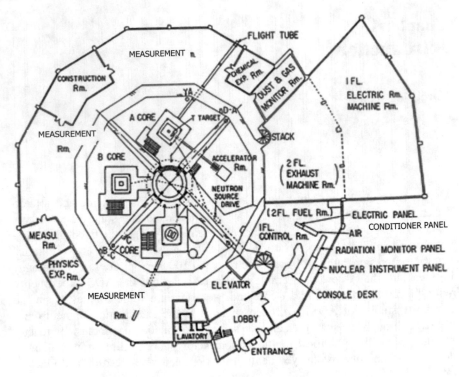

Fig. 1.1 Horizontal cross section of the KUCA building (Ref. [1])

core only can attain the critical state at a given time because the assembly is equipped with a single control mechanism that prevents the simultaneous operation of multiple cores. Users can select the core that is the most appropriate for their experiments. Also, a pulsed-neutron source is also installed, which can be used in combination with the A-core.

Owing to the compatibility of KUCA, a wide variety of research and education has been performed at the facility. Furthermore, KUCA has been used for the following research and education activities:

- New reactor concepts
- Thorium-fueled reactors
- Fusion-fission hybrid systems
- Subcritical systems
- Accelerator-driven system (ADS)
- Neutron characteristics of minor actinide
- Experimental education course for students (Ref. [1]).

Experimental and numerical studies on the validation and verification of nuclear data and nuclear calculation codes and on the development of new detector systems are also conducted.

Fig. 1.2 Solid-moderated and -reflected core (A-core) (Ref. [1])

1.1.2 Solid-Moderated and Solid-Reflected Cores

Two solid-moderated cores (A-core; Fig. 1.2 and B-core) are installed at KUCA. Squared-shaped coupon-type uranium fuel plates (93 wt% enriched) of 2″ length, 2″ breadth, and 1/16″ thick covered with a thin plastic coating are used as the fuel material. Solid moderator materials, including polyethylene and graphite, are combined with highly-enriched uranium (HEU), thorium, and natural uranium plates to form the fuel elements. Polyethylene, graphite, beryllium, aluminum, iron, lead, and bismuth are used as the reflector elements.

A wide variety of neutron spectra could be achieved by varying the composition of the fuel and moderator plates in the fuel element, and also by varying the reflector elements.

The A-core can be used in combination with the pulsed-neutron generator, and also, it is used for carrying out studies on ADS and fission-fusion hybrid reactor systems.

1.1.3 Light-Water-Moderated and Light-Water-Reflected Core

A single light-water-moderated core (C-core; Fig. 1.3) is installed at KUCA. Plate-type uranium fuel (93 wt% enriched) of 600 mm length, 62 mm width, and 1.5 mm thick with an aluminum (Al) cladding of 0.5 mm thick are used in the C-core. In addition, there are two types of curved fuel plates: one 93 wt%, and the other 45 wt%

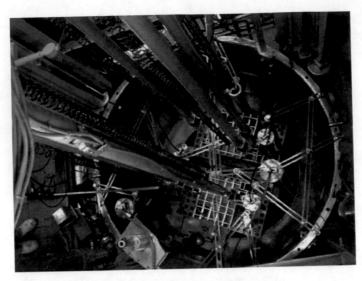

Fig. 1.3 Light-water-moderated and -reflected core (C-core) (Ref. [1])

enriched of 650 mm length and 1.4 mm thick with an Al cladding of 0.45 mm thick; 32 plates with different curvatures and widths are used.

The fuel element is formed by assembling the fuel plates in aluminum fuel frames. The fuel elements are loaded in a core tank of 2,000 mm diameter and 2,000 mm depth and then immersed in light water to form the core. A part of the reflector region can be substituted for a heavy-water reflector. Three types of fuel frames, each having different fuel loading pitch and thus providing different neutron spectra in the core, are available. The core can be separated into two parts with arbitrary gap width, and it is suitable for coupled core and criticality safety studies.

The C-core is used for carrying out a wide variety of basic studies on light-water-moderated systems, including the development of high-flux research reactor, enrichment reduction in a research reactor, criticality safety, and study of coupled core theory.

The C-core is also used for conducting a graduate-level joint reactor laboratory course in affiliation with many Japanese universities in addition to conducting an undergraduate-level reactor laboratory course in affiliation with the Kyoto University. International reactor laboratory courses for students overseas have been also conducted since 2003.

1.1.4 Pulsed-Neutron Generator

A pulsed-neutron generator (Fig. 1.4) is attached to KUCA. Deuteron (D) ion beams are injected onto a tritium (T) target to generate the pulsed neutrons (14 MeV

Fig. 1.4 Pulsed-neutron generator (D-T accelerator) (Ref. [1])

neutrons) through ^{3}H(d, n)^{4}He reactions. The pulsed-neutron generator can be used in combination with the critical assembly (A-core). The system consists of a duoplasmatron-type ion source, a high-voltage generator with a capacity of 300 kV, an acceleration tube, a beam pulsing system, and a tritium target. The main charac-teristics of deuteron ion beams are follows: the acceleration voltage 300 kV at most; the beam current 5 mA at most; the neutron pulsed width ranging between 300 ns and 100 μs; the pulsed repetition rate between 0.1 Hz and 30 kHz (max. duty ratio 1%).

The pulsed-neutron generator has been used, in combination with the A-core, for carrying out basic studies on ADS and fission-fusion hybrid reactor systems.

1.1.5 Fixed-Field Alternating Gradient Accelerator

Fixed-Field Alternating Gradient (FFAG) accelerator, which was originally proposed forty years ago, attracts much attention because of its advantages, such as a large acceptance or a possible fast repetition rate, compared with that for synchrotrons. Furthermore, the operation of an FFAG accelerator is expected to be very stable because no active feedback is required for the acceleration. From these features, FFAG accelerator is considered a good candidate for the proton driver in ADS.

FFAG accelerator is available to combine strong focusing optics like a synchrotron with a fixed-magnetic field like a cyclotron. Unlike a synchrotron, the magnetic field experienced by the particles is designed to vary with radius, rather than time. This naturally leads to the potential to operate at high repetition rates limited only by

Fig. 1.5 FFAG accelerator complex at KURRI (https://www.rri.kyoto-u.ac.jp/facilities/irl)

the available RF system, while strong focusing provides a possibility of maintaining higher intensity beams than in cyclotrons.

A revival in interest since the 1990s has seen a number of FFAGs constructed, including scaling and linear non-scaling variants. However, high bunch charge operation remains to be demonstrated. A collaboration has been formed to use an existing proton FFAG accelerator at KURNS to explore the high-intensity regime in FFAG accelerators.

At KURNS, the FFAG accelerator complex (Fig. 1.5; Refs. [2–4]) was installed in the experimental facility for the basic study of ADS in 2003. The main characteristics of proton beams are follows: the energy 150 MeV at most; the beam current 1 nA at most; the repetition rate 30 Hz; the pulsed width less than 100 ns. Other than ADS experiments, irradiation for the materials, aerosol, and living animals (e.g., rats) are performed for the basic studies in various research fields.

1.2 Accelerator-Driven System

1.2.1 Overview of Research and Development

ADS was first proposed as an energy amplifier system [5] that couples with a high-power accelerator and a thorium sustainable system. Another possible function of ADS was resolving the issue of transmuting minor actinide (MA) and long-lived fission product (LLFP) generated from nuclear power plants. ADS has attracted worldwide attention because of its superior safety characteristics and potential for burning plutonium and nuclear waste. An outstanding advantage of its use is the anticipated absence of reactivity accidents when sufficient subcriticality is ensured. Also, ADS is expected to provide capabilities for power generation, nuclear waste transmutation, and a reliable neutron source for research purposes.

The ADS experimental facilities are being prepared for the investigations of nuclear transmutation of MA and LLFP, as are the Transmutation Experimental Facility (TEF) [6] at the Japan Atomic Energy Agency and the Multi-purpose Hybrid Research Reactor for High-tech Applications (MYRRHA) [7] at SCK/CEN in Belgium. Research activities on ADS involved mainly the experimental feasibility study using critical assemblies and test facilities: MASURCA in France [8–10], YALINA-booster and -thermal in Belarus [11–13], VENUS-F in Belgium [14–17], and KUCA in Japan [18–78]. At these facilities, feasibility studies on ADS have been conducted by combining a reactor core (fast or thermal core) with an external neutron source by the D-T accelerator (14 MeV neutrons) or 100 MeV proton accelerator (KUCA only), through experimental and numerical analyses of reactor physics parameters, including statics parameters: reaction rates, neutron spectrum, and subcritical multiplication factor; kinetics parameters: subcriticality, prompt neutron decay constant, effective delayed neutron fraction, and neutron generation time. Here, to ensure measurement methodologies of the statics and kinetics parameters and confirm numerical precision by stochastic and deterministic calculations, many attempts were made for uniquely developing new-type and high-precision detectors, and interestingly for introducing advanced-numerical approaches, respectively.

1.2.2 Feasibility Study at KUCA

At KURRI, a series of preliminary experiments on the ADS with 14 MeV neutrons was officially launched at KUCA in 2003, with sights on a future plan (Kart & Lab. Project) [79, 80]. The goal of the plan was to establish a next-generation neutron source, as a substitution for the current 5 MW Kyoto University Research Reactor established in 1964, by introducing a synergetic system comprising a research reactor and a particle accelerator. High-energy neutrons generated by the interaction of high-energy proton beams (100 MeV) with heavy metal was expected to be injected into the KUCA core, and finally, the world's first injection [21] of high-energy neutrons obtained by a new accelerator was successfully conducted into the KUCA core in 2009. The new accelerator is called the FFAG accelerator of the synchrotron type developed by the High Energy Accelerator Research Organization in Japan.

Prior to actual ADS experiments with 100 MeV protons, it was requisite to establish measurement techniques for various neutronics parameters and the method for evaluating neutronic properties of the ADS with 100 MeV protons. Uniquely, KUCA has outstanding features of two external neutron sources (14 MeV neutrons and 100 MeV protons) and a variety of neutron spectrum cores, although the KUCA cores provide almost a thermal neutron spectrum. For the accomplishment of research objectives, a series of basic experiments with 14 MeV neutrons obtained by the

D-T accelerator had been carried out using the A-core (Fig. 1.2) and the pulsed-neutron generator (Fig. 1.4) of KUCA. Then, neutron characteristics of statics and kinetics parameters in reactor physics were experimentally investigated through the development of measurement methodologies and numerically examined through the confirmation of calculation precision by the Monte Carlo calculations. After the first injection of 100 MeV protons, the next campaign of ADS experiments was devoted to basic research (feasibility study) on ADS coupling with the FFAG accelerator (a tungsten or lead-bismuth: Pb–Bi target) and the A-core, by varying external neutron source (14 MeV neutrons or 100 MeV protons), neutron spectrum of reactor core combined with nuclear fuel (uranium-235, thourim-232, and natural uranium), moderators (polyethylene and graphite) and reflectors (polyethylene, graphite, beryllium, aluminum, iron, lead, and bismuth), and subcriticality. Significantly, in 2019, the world's first nuclear transmutation of MA by ADS [66] was accomplished at the condition that the spallation neutrons were supplied to a subcritical core through the injection of 100 MeV protons onto a Pb–Bi target, demonstrating fission and capture reactions of neptunium-237 and americium-241.

All experimental data of ADS were compiled as "ADS experimental benchmarks at KUCA," publishing the KURNS technical reports [81–85], and employed as "Coordinated (Collaborative) Research Projects (CRP) of Accelerator-Driven System and Low-Enriched Uranium Cores in ADS" organized by the International Atomic Energy Agency (IAEA). Through the CRP programs of ADS by IAEA ranging between 2007 and 2018, experimental data of ADS at KUCA were shared with all IAEA state members and used for conducting the validation and verification of nuclear calculation codes and major nuclear data libraries.

References

1. Misawa T, Unesaki H, Pyeon CH (2000) Nuclear reactor physics experiments. Kyoto University Press, Kyoto, Japan
2. Planche T, Lagrange JB, Yamakawa E et al (2011) Harmonic number jump acceleration of muon beams in zero-chromatic FFAG rings. Nucl Instrum Methods A 632:7
3. Lagrange JB, Planche T, Yamakawa E et al (2013) Straight scaling FFAG beam line. Nucl Instrum Methods A 691:55
4. Yamakawa E, Uesugi T, Lagrange JB et al (2013) Serpentine acceleration in zero-chromatic FFAG accelerators. Nucl Instrum Methods A 716:46
5. Rubbia C (1995) A high gain energy amplifier operated with fast neutrons. AIP Conf Proc 346:44
6. Tsujimoto K, Sasa T, Nishihara K et al (2004) Neutronics design for lead-bismuth cooled accelerator-driven system for transmutation of minor actinide. J Nucl Sci Technol 41:21
7. Abderrahim HA, D'hondt P (2007) MYRRAHA, a European experimental ADS for R&D applications status at mid-2005 and prospective towards implementation. J Nucl Sci Technol 44:491
8. Soule R, Assal W, Chaussonnet P et al (2004) Neutronic studies in support of accelerator-driven systems: The MUSE experiments in the MASURCA facility. Nucl Sci Eng 148:124
9. Plaschy M, Destouches C, Rimpault G et al (2005) Investigation of ADS-type heterogeneities in the MUSE4 critical configuration. J Nucl Sci Technol 42:779

10. Lebrat JF, Aliberti G, D'Angelo A et al (2008) Global results from deterministic and stochastic analysis of the MUSE-4 experiments on the neutronics of the accelerator-driven systems. Nucl Sci Eng 158:49

11. Persson CM, Fokau A, Serafimovich I et al (2008) Pulsed neutron source measurements in the subcritical ADS experiment YALINA-booster. Ann Nucl Energy 35:2357

12. Tesinsky M, Berglöf C, Bäck T et al (2011) Comparison of calculated and measured reaction rates obtained through foil activation in the subcritical dual spectrum facility YALINA-booster. Ann Nucl Energy 38:1412

13. Talamo A, Gohar Y, Sadovich S et al (2013) Correction factor for the experimental prompt neutron decay constant. Ann Nucl Energy 62:421

14. Uyttenhove W, Baeten P, Van den Eyden G et al (2011) The neutronic design of a critical lead reflected zero-power reference core for on-line subcriticality measurements in accelerator-driven systems. Ann Nucl Energy 38:1519

15. Uyttenhove W, Lathouwers D, Kloosterman JL et al (2014) Methodology for modal analysis at pulsed neutron source experiments in accelerator-driven systems. Ann Nucl Energy 72:286

16. Lecouey JL, Marie N, Ban G et al (2015) Estimate of the reactivity of the VENUS-F subcritical configuration using a Monte Carlo MSM method. Ann Nucl Energy 83:65

17. Marie N, Lecouey JL, Lehaut G et al (2019) Reactivity monitoring of the accelerator driven VENUF-F subcritical reactor with the "current-to-flux" method. Ann Nucl Energy 128:12

18. Pyeon CH, Hirano Y, Misawa T et al (2007) Preliminary experiments on accelerator-driven subcritical reactor with pulsed neutron generator in Kyoto University Critical Assembly. J Nucl Sci Technol 44:1368

19. Pyeon CH, Hervault M, Misawa T et al (2008) Static and kinetic experiments on accelerator-driven system in Kyoto University Critical Assembly. J Nucl Sci Technol 45:1171

20. Pyeon CH, Shiga H, Misawa T et al (2009) Reaction rate analyses for an accelerator-driven system with 14 MeV neutrons in the Kyoto University Critical Assembly. J Nucl Sci Technol 46:965

21. Pyeon CH, Misawa T, Lim JY et al (2009) First injection of spallation neutrons generated by high-energy protons into the Kyoto University Critical Assembly. J Nucl Sci Technol 46:1091

22. Shahbunder H, Pyeon CH, Misawa T et al (2010) Experimental analysis for neutron multiplication by using reaction rate distribution in accelerator-driven system. Ann Nucl Energy 37:592

23. Taninaka H, Hashimoto K, Pyeon CH et al (2010) Determination of lambda-mode eigenvalue separation of a thermal accelerator-driven system from pulsed neutron experiment. J Nucl Sci Technol 47:376

24. Kawaguchi S, Misawa T, Pyeon CH et al (2010) A new experimental correction method for the first-order perturbation approximation on the steady subcritical reactor. J Nucl Sci Technol 47:550

25. Shahbunder H, Pyeon CH, Misawa T et al (2010) Subcritical multiplication factor and source efficiency in accelerator-driven system. Ann Nucl Energy 37:1214

26. Shahbunder H, Pyeon CH, Misawa T et al (2010) Effects of neutron spectrum and external neutron source on neutron multiplication parameters in accelerator-driven system. Ann Nucl Energy 37:1785

27. Pyeon CH, Shiga H, Abe K et al (2010) Reaction rate analysis of nuclear spallation reactions generated by 150, 190 and 235 MeV protons. J Nucl Sci Technol 47:1090

28. Yagi T, Misawa T, Pyeon CH et al (2011) A small high sensitivity neutron detector using a wavelength shifting fiber. Appl Radiat Isot 69:176

29. Taninaka H, Hashimoto K, Pyeon CH et al (2011) Determination of subcritical reactivity of a thermal accelerator-driven system from beam trip and restart experiment. J Nucl Sci Technol 48:873

30. Pyeon CH, Lim JY, Takemoto Y et al (2011) Preliminary study on the thorium-loaded accelerator-driven system with 100 MeV protons at the Kyoto University Critical Assembly. Ann Nucl Energy 38:2298

31. Taninaka H, Miyoshi A, Hashimoto K et al (2011) Feynman-α analysis for a thermal subcritical reactor system driven by an unstable 14 MeV-neutron source. J Nucl Sci Technol 48:1272

32. Pyeon CH, Takemoto Y, Yagi T et al (2012) Accuracy of reaction rates in the accelerator-driven system with 14 MeV neutrons at the Kyoto University Critical Assembly. Ann Nucl Energy 40:229

33. Lim JY, Pyeon CH, Yagi T et al (2012) Subcritical multiplication parameters of the accelerator-driven system with 100 MeV protons at the Kyoto University Critical Assembly. Sci Technol Nucl Install 2012:395878

34. Yagi T, Pyeon CH, Misawa T (2013) Application of wavelength shifting fiber to subcriticality measurements. Appl Radiat Isot 72:11

35. Takahashi Y, Azuma T, Nishio T et al (2013) Conceptual design of multi-targets for accelerator-driven system experiments with 100 MeV protons. Ann Nucl Energy 54:162

36. Pyeon CH, Azuma T, Takemoto Y et al (2013) Experimental analyses of external neutron source generated by 100 MeV protons at the Kyoto University Critical Assembly. Nucl Eng Technol 45:81

37. Sakon A, Hashimoto K, Sugiyama W et al (2013) Power spectral analysis for a thermal subcritical reactor system driven by a pulsed 14 MeV neutron source. J Nucl Sci Technol 50:481

38. Sakon A, Hashimoto K, Maarof MA et al (2014) Measurement of large negative reactivity of an accelerator-driven system in the Kyoto University Critical Assembly. J Nucl Sci Technol 51:116

39. Pyeon CH, Yagi T, Sukawa K et al (2014) Mockup experiments on the thorium-loaded accelerator-driven system in the Kyoto University Critical Assembly. Nucl Sci Eng 177:156

40. Sakon A, Hashimoto K, Sugiyama W et al (2015) Determination of prompt-neutron decay constant from phase shift between beam current and neutron detection signals for an accelerator-driven system in the Kyoto University Critical Assembly. J Nucl Sci Technol 52:204

41. Pyeon CH, Yagi T, Takahashi Y et al (2015) Perspectives of research and development of accelerator-driven system in Kyoto University Research Reactor Institute. Prog Nucl Energy 82:22

42. Pyeon CH, Nakano H, Yamanaka M et al (2015) Neutron characteristics of solid targets in accelerator-driven system with 100 MeV protons at Kyoto University Critical Assembly. Nucl Technol 192:181

43. Pyeon CH, Fujimoto A, Sugawara T et al (2016) Validation of Pb nuclear data by Monte Carlo analyses of sample reactivity experiments at Kyoto University Critical Assembly. J Nucl Sci Technol 53:602

44. Kim WK, Lee HC, Pyeon CH et al (2016) Monte Carlo analysis of the accelerator-driven system at Kyoto University Research Reactor Institute. Nucl Eng Technol 48:304

45. Yamanaka M, Pyeon CH, Yagi T et al (2016) Accuracy of reactor physics parameters in thorium-loaded accelerator-driven system experiments at Kyoto University Critical Assembly. Nucl Sci Eng 183:96

46. Endo T, Yamamoto A, Yagi T et al (2016) Statistical error estimation of the Feynman-α method using the bootstrap method. J Nucl Sci Technol 53:1447

47. Yamanaka M, Pyeon CH, Misawa T (2016) Monte Carlo approach of effective delayed neutron fraction by k-ratio method with external neutron source. Nucl Sci Eng 184:551

48. Dulla S, Hoh SS, Marana G et al (2017) Analysis of KUCA measurements by the reactivity monitoring MAρTA method. Ann Nucl Energy 101:397

49. Yamanaka M, Pyeon CH, Kim SH et al (2017) Effective delayed neutron fraction in accelerator-driven system experiments with 100 MeV protons at Kyoto University Critical Assembly. J Nucl Sci Technol 54:293

50. Iwamoto H, Nishihara K, Yagi T et al (2017) On-line subcriticality measurement using a pulsed spallation neutron source. J Nucl Sci Technol 54:432

51. Pyeon CH, Fujimoto A, Sugawara T et al (2017) Sensitivity and uncertainty analyses of lead sample reactivity experiments at Kyoto University Critical Assembly. Nucl Sci Eng 185:460

52. Pyeon CH, Yamanaka M, Endo T et al (2017) Experimental benchmarks on kinetic parameters in accelerator-driven system with 100 MeV protons at Kyoto University Critical Assembly. Ann Nucl Energy 105:346

53. Van Rooijen WFG, Endo T, Chiba G et al (2017) Analysis of the KUCA ADS benchmarks with diffusion theory. Prog Nucl Energy 101:243

54. Pyeon CH, Yamanaka M, Kim SH et al (2017) Benchmarks of subcriticality in accelerator-driven system experiments at Kyoto University Critical Assembly. Nucl Eng Technol 49:1234

55. Talamo A, Gohar Y, Gabrielli F et al (2017) Coupling Sjostrand and Feynman methods in prompt neutron decay constant analyses. Prog Nucl Energy 101:299

56. Kim SH, Yamanaka M, Pyeon CH (2018) Kinetics solution with iteration scheme of history-based neutron source in accelerator-driven system. Ann Nucl Energy 112:337

57. Pyeon CH, Vu TM, Yamanaka M et al (2018) Reaction rate analyses of accelerator-driven system experiments with 100 MeV protons at Kyoto University Critical Assembly. J Nucl Sci Technol 55:190

58. Endo T, Chiba G, Van Rooijen WFG et al (2018) Experimental analysis and uncertainty quantification using random sampling technique for ADS experiments at KUCA. J Nucl Sci Technol 55:450

59. Kong CD, Choe JW, Yum SP et al (2018) Application of advanced Rossi-alpha technique to reactivity measurements at Kyoto University Critical Assembly. Ann Nucl Energy 118:92

60. Pyeon CH, Yamanaka M, Oizumi A et al (2018) Experimental analyses of bismuth sample reactivity worth at Kyoto University Critical Assembly. J Nucl Sci Technol 55:1324

61. Yamanaka M, Jang KW, Shin SH et al (2019) Proton beam characteristics with wavelength shifting fiber detector at Kyoto University Critical Assembly. Jpn J Appl Phys 58:036002

62. Endo T, Yamamoto A, Yamanaka M et al (2019) Experimental validation of unique combination numbers for the third- and fourth-order neutron correlation factors in the zero power reactor noise. J Nucl Sci Technol 56:322

63. Talamo A, Gohar Y, Yamamoto T et al (2019) Calculation of the prompt neutron decay constant from the cross and auto power spectrum densities of pulse mode detectors. Ann Nucl Energy 131:138

64. Aizawa N, Yamanaka M, Iwasaki T et al (2019) Effect of neutron spectrum on subcritical multiplication factor in accelerator-driven system. Prog Nucl Energy 116:158

65. Pyeon CH, Yamanaka M, Sano T et al (2019) Integral experiments on critical irradiation of ^{237}Np and ^{241}Am foils at Kyoto University Critical Assembly. Nucl Sci Eng 193:1023

66. Pyeon CH, Yamanaka M, Oizumi A et al (2019) First nuclear transmutation of ^{237}Np and ^{241}Am by accelerator-driven system at Kyoto University Critical Assembly. J Nucl Sci Technol 56:684

67. Katano R, Yamanaka M, Pyeon CH (2019) Application of linear combination method to pulsed-neutron source measurement at Kyoto University Critical Assembly. Nucl Sci Eng 193:1394

68. Pyeon CH, Talamo A, Fukushima M (2020) Special issue on accelerator-driven system benchmarks at Kyoto University Critical Assembly. J Nucl Sci Technol 57:133

69. Watanabe K, Endo T, Yamanaka M et al (2020) Real-time subcriticality monitoring system based on a highly sensitive optical fiber detector in an accelerator-driven system at the Kyoto University Critical Assembly. J Nucl Sci Technol 57:136

70. Talamo A, Gohar Y, Yamanaka M et al (2020) Calculation of the prompt neutron decay constant of the KUCA facility configurations driven by a californium of spallation external neutron sources. J Nucl Sci Technol 57:145

71. Talamo A, Gohar Y, Yamanaka M et al (2020) Paralyzable and non-paralyzable dead-time corrections for the neutron detectors of the KUCA facility using external neutron sources. J Nucl Sci Technol 57:157

72. Katano R, Yamanaka M, Pyeon CH (2020) Measurement of prompt neutron decay constant with spallation neutrons at Kyoto University Critical Assembly using linear combination method. J Nucl Sci Technol 57:169

73. Shim HJ, Kim DH, Yamanaka M et al (2020) Estimation of kinetics parameters by Monte Carlo fixed-source calculations for accelerator-driven system. J Nucl Sci Technol 57:177

74. Endo T, Watanabe K, Chiba G et al (2020) Nuclear data-induced uncertainty quantification of prompt neutron decay constant based on perturbation theory for ADS experiments at KUCA. J Nucl Sci Technol 57:196
75. Yamanaka M, Pyeon CH, Endo T et al (2020) Experimental analyses of β_{eff}/Λ in accelerator-driven system at Kyoto University Critical Assembly. J Nucl Sci Technol 57:205
76. Yamanaka M, Watanabe K, Pyeon CH (2020) Subcriticality estimation by extended Kalman filter technique in transient experiment with external neutron source at Kyoto University Critical Assembly. Eur Phys J Plus 135:256
77. Pyeon CH, Yamanaka M, Lee B (2020) Reaction rate analyses of high-energy neutrons by injection of 100 MeV protons onto lead-bismuth target. Ann Nucl Energy 144:107498
78. Pyeon CH, Yamanaka M, Endo T et al (2020) Neutron generation time in highly-enriched uranium core at Kyoto University Critical Assembly. Nucl Sci Eng. https://doi.org/10.1080/00295639.2020.1774230
79. Shiroya S, Unesaki H, Kawase Y et al (2000) Accelerator driven subcritical system as a future neutron source in Kyoto University Research Reactor Institute (KURRI)—basic study on neutron multiplication in the accelerator driven subcritical reactor. Prog Nucl Energy 37:357
80. Shiroya S, Yamamoto A, Shin K et al (2002) Basic study on accelerator driven subcritical reactor in Kyoto University Research Reactor Institute (KURRI). Prog Nucl Energy 40:489
81. Pyeon CH (2012) Experimental benchmarks for accelerator-driven system (ADS) at Kyoto University Critical Assembly. KURRI-TR-444
82. Pyeon CH (2015) Experimental benchmarks on thorium-loaded accelerator-driven system (ADS) at Kyoto University Critical Assembly. KURRI-TR(CD)-48
83. Pyeon CH (2017) Experimental benchmarks of neutronics on solid Pb-Bi in accelerator-driven system with 100 MeV protons at Kyoto University Critical Assembly. KURRI-TR-447
84. Pyeon CH, Yamanaka M (2018) Experimental benchmarks of neutron characteristics on uranium-lead zoned core in accelerator-driven system at Kyoto University Critical Assembly. KURNS-EKR-001
85. Pyeon CH (2020) Experimental benchmarks of medium-fast spectrum core (EE1 core) in accelerator-driven system at Kyoto University Critical Assembly. KURNS-EKR-007

Chapter 2
Subcriticality

Kengo Hashimoto

Abstract For a subcritical reactor system driven by a periodically pulsed spallation neutron source in KUCA, the Feynman-α and the Rossi-α neutron correlation analyses are conducted to determine the prompt neutron decay constant and quantitatively to confirm a non-Poisson character of the neutron source. The decay constant determined from the present Feynman-α analysis well agrees with that from a previous analysis for the same subcritical system driven by an inherent source. Considering the effect of a higher mode excited, the disagreement can be successfully resolved. The power spectral analysis on frequency domain is also carried out. Not only the cross-power but also the auto-power spectral density have a considerable correlated component even at a deeply subcritical state, where no correlated component could be previously observed under a 14 MeV neutron source. The indicator of the non-Poisson character of the present spallation source can be obtained from the spectral analysis and is consistent with that from the Rossi-α analysis. An experimental technique based on an accelerator-beam trip or restart operation is proposed to determine the subcritical reactivity of ADS. Applying the least-squares inverse kinetics method to the data analysis, the subcriticality can be inferred from time-sequence neutron count data after these operations.

Keywords Subcriticality · Prompt neutron decay constant · Feynman-α method · Rossi-α method · Power spectral analyses

K. Hashimoto (✉)
Atomic Energy Research Institute, Kindai University, Osaka, Japan
e-mail: kengoh@pp.iij4u.or.jp

© The Author(s) 2021
C. H. Pyeon (ed.), *Accelerator-Driven System at Kyoto University Critical Assembly*,
https://doi.org/10.1007/978-981-16-0344-0_2

13

2.1   Feynman-α and Rossi-α Analyses

2.1.1   *Experimental Settings*

2.1.1.1   Core Configurations

The Feynman-α and the Rossi-α neutron correlation analyses considering a non-Poisson character of the spallation source [1] are carried out in the A-core that has two kinds of fuel assemblies: 1/8"P60EUEU and 1/8"P4EUEU. Figure 2.1 shows a side view of the fuel assembly referred to as 1/8"P60EUEU. Fuel and moderator plates of the assembly were set in a 1.5 mm-thick aluminum sheath and the cross section of these plates within the assembly was the square of 2" (50.8 mm). The fuel assembly consisted of 60 unit cells. The unit cell was composed of two (1/16" thick) uranium plates with Al cladding and one (1/8" thick) polyethylene plate. Each of the fuel assemblies had 120 sheets of the uranium plates. The active height of the core, namely 60 unit cells, was 15" (about 38 cm). Adjacent to both axial sides of the active region of each fuel assembly, about 22" (57 cm) upper and 24" (52 cm) lower polyethylene reflectors were attached, respectively. Another fuel assembly referred to as 1/8" P4EUEU consisted of 4 unit cells with upper and lower polyethylene reflectors. The core configuration employed in this study is shown in Fig. 2.2. The twenty-five 1/8"P60EUEU assemblies and one 1/8"P4EUEU assembly were loaded on a grid plate to constitute a critical reactor core. The core was surrounded with many polyethylene-reflector assemblies.

The subcriticality for the present experiment was adjusted by changing the axial position of the central fuel loading [3], as well as the positions of safety and control rods. These subcritical and critical patterns employed in the present experiments are shown in Table 2.1, where these respective subcriticalities are calculated by the continuous-energy Monte Carlo code MVP version 3 (MVP3) [3, 4] with the nuclear data library JENDL-4.0 [5] are also included. The error of the subcriticality indicates statistical uncertainty $\pm 1\sigma$ of the MVP calculation.

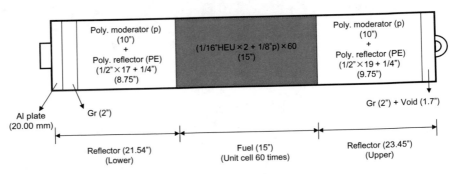

Fig. 2.1   Description of 1/8"P60EUEU fuel assembly (Ref. [2])

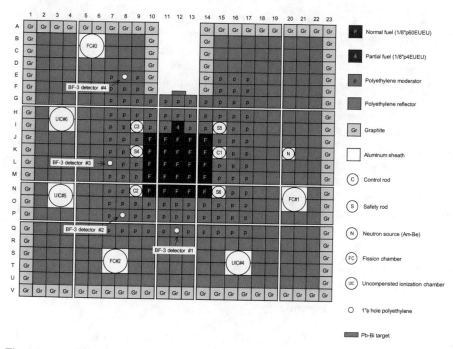

Fig. 2.2 Top view of core configuration and neutron detector location (Ref. [1])

Table 2.1 Experimental patterns of control rods and central fuel loading (Ref. [1])

Subcritical	Axial position [mm]						Central	Subcriticality
Pattern	C1	C2	C3	S4	S5	S6	Loading	[%Δk/k]
Critical	U.L.	678.98	U.L.	U.L.	U.L.	U.L.	C.I.	
A	U.L.	L.L.	U.L.	U.L.	U.L.	U.L.	C.I.	0.307 ± 0.008
B	L.L.	U.L.	U.L.	U.L.	U.L.	U.L.	C.I.	0.725 ± 0.018
C	L.L.	L.L.	L.L.	U.L.	U.L.	U.L.	C.I.	1.483 ± 0.040
D	L.L.	L.L.	L.L.	U.L.	U.L.	L.L.	C.I.	2.056 ± 0.054
E	L.L.	L.L.	L.L.	L.L.	L.L.	L.L.	C.I.	3.189 ± 0.083
F	L.L.	L.L.	L.L.	L.L.	L.L.	L.L.	C.W.	13.604 ± 0.381

L.L.: Lower Limit [0 mm], U.L.: Upper Limit [1200 mm]
C.I.: Completely Inserted, C.W.: Completely Withdrawn

2.1.1.2 Experimental Conditions

Four BF$_3$ proportional neutron counters (LND-202101, 1" dia., 15.47" len.) were used for the present experiment. These BF3 counters on locations (Q, 12), (P, 8), (L, 7), and (E, 8) are referred to as B1, B2, B3, and B4, respectively. The axial center of effective length of these counters was located at the axial center position of active region of the fuel assembly. The present nuclear instrumentation system consisted of conventional detector bias-supply, pre-amplifier, spectroscopy amplifier, and discriminator modules. Finally, signal pulses from these BF_3 neutron counters were fed to a time-sequence data acquisition system, which registered the arriving time of the signal as digital data. The time length of the acquired data for each subcritical pattern was about 30 min.

The pulsed proton beams were supplied by a fixed-field alternating gradient (FFAG) accelerator. The proton beam intensity of the accelerator was set to 30 pA for any subcritical patterns except for pattern A. For only pattern A, we necessarily made the proton beam intensity fall to 12 pA, to reduce counting loss originated from the dead-time effect of neutron counter. Throughout the present accelerator operations, the pulsed repetition frequency and the beam width were set to 30 Hz and 100 ns, respectively.

2.1.2 Formulae for Data Analyses

2.1.2.1 Feynman-α Formula

Rana and Degweker [6] derived the Feynman-α and the Rossi-α formulae for a periodically pulsed non-Poisson source, where delayed neutron contribution was considered and each pulse was assumed to be a delta function. Since the pulse width of 100 ns of our accelerator is much shorter than the time scale of the present correlation analyses, the assumption is acceptable. First, we consider the zero-power transfer function $G(s)$ as follows:

$$\frac{1}{G(s)} = s\left(\Lambda + \sum_{i=1}^{7} \frac{\beta_i}{\lambda_i + s}\right) - \rho, \tag{2.1}$$

where the six group model of delayed neutrons is supposed. When the poles and the residues of the above transfer function are represented by s_i and A_i, respectively, a parameter Y_i of the Feynman-α analysis can be defined as [7].

$$Y_i = 2\frac{\overline{v(v-1)}}{\bar{v}^2} \frac{A_i \, G(\alpha_i)}{\alpha_i}, \tag{2.2}$$

where

$$\alpha_i = -s_i. \tag{2.3}$$

Then, the Feynman-α formula derived by Rana and Degweker [6] can be written as follows:

$$Y(T) = \frac{\bar{v}^2 \lambda_d}{m_1} \sum_{i=1}^{7} Y_i \left(1 - \frac{1 - e^{-\alpha_i T}}{\alpha_i T}\right) \left\{ m_1 \lambda_f + \frac{(m_2 - m_1^2)(-\rho)}{v(v-1)\Lambda} \right\}$$

$$+ \frac{\bar{v}^2 \lambda_d m_1 (-\rho)}{v(v-1)\Lambda} \sum_{i=1}^{7} Y_i \left\{ \frac{e^{-\alpha_i \left(T - \frac{[fT]}{f}\right)} + e^{-\frac{\alpha_i}{f}} e^{\alpha_i \left(T - \frac{[fT]}{f}\right)} - 1 - e^{-\frac{\alpha_i}{f}}}{T \alpha_i \left(1 - e^{-\frac{\alpha_i}{f}}\right)} \right\},$$

$$+ \frac{\lambda_d m_1 \Lambda}{(-\rho)} \left\{ 1 + 2[fT] - \frac{[fT]}{fT}([fT] + 1) - fT \right\} \tag{2.4}$$

where f is pulse repetition frequency and $[f\,T]$ represents largest integer less or equal to $f\,T$. The largest α_7 is a prompt-neutron decay constant to be determined and the other α_i is a decay constant of each delayed-neutron mode. Other notations are conventional except for m_1 and m_2, which are first and second factorial moment of source multiplicity distribution and are defined by the following equations [8], respectively:

$$m_1 = \bar{N}\,\overline{v_{\text{sp}}}, \tag{2.5}$$

$$m_2 = \overline{N(N-1)}\,\overline{v_{\text{sp}}}^2 + \bar{N}\,\overline{v_{\text{sp}}(v_{\text{sp}} - 1)}, \tag{2.6}$$

where N and v_{sp} are number of protons in a pulsed bunch and number of neutrons produced by each spallation event, respectively. These definitions lead to the following expression:

$$m_2 - m_1^2 = \left(\overline{N^2} - \bar{N} - \bar{N}^2\right)\overline{v_{\text{sp}}}^2 + \bar{N}\,\overline{v_{\text{sp}}(v_{\text{sp}} - 1)}. \tag{2.7}$$

The above quantity gives an expression for non-Poisson character of a neutron source and is included in the first term of Eq. (2.4). When the proton number N follows the Poisson distribution, the first term of Eq. (2.7) disappears. The second term is expected to increase with an increase in proton energy but the present energy 100 MeV may lead to a small positive value.

Equation (2.4) is not available for least-squares fitting to the Y data because of a complexity of the delayed neutron terms and many unknown parameters included in the terms. Here, we reduce the rigorous equation to obtain a practical fitting formula. First, the gate-time T of the Feynman-α analysis is restricted within the following range:

$$T \ll \frac{1}{\alpha_i}\, i = 1, 2, \ldots, 6. \tag{2.8}$$

In the above range, the following Maclaurin expansions can be done:

$$e^{-\alpha_i T} \simeq 1 - \alpha_i T + \frac{(-\alpha_i T)^2}{2}, \tag{2.9}$$

$$e^{\pm \alpha_i \left(T - \frac{[fT]}{f}\right)} \simeq 1 \pm \alpha_i \left(T - \frac{[fT]}{f}\right) + \frac{1}{2}\alpha_i^2 \left(T - \frac{[fT]}{f}\right)^2, \qquad (2.10)$$

$$e^{-\frac{\alpha_i}{f}} \simeq 1 - \frac{\alpha_i}{f}. \qquad (2.11)$$

We substitute Eqs. (2.9), (2.10), and (2.11) into Eq. (2.4) to obtain the following final form:

$$
\begin{aligned}
Y(T) =& C_1 \left(1 - \frac{1 - e^{-\alpha T}}{\alpha T}\right) \\
&+ C_2 \left\{ \frac{e^{-\alpha\left(T - \frac{[fT]}{f}\right)} + e^{-\frac{\alpha}{f}} e^{\alpha\left(T - \frac{[fT]}{f}\right)} - 1 - e^{-\frac{\alpha}{f}}/}{T \alpha_p \left(1 - e^{-\frac{\alpha}{f}}\right)} \right\} \\
&+ C_3 \left\{ 1 + 2[fT] - \frac{[fT]}{fT}([fT] + 1) - fT \right\} + C_4 T \qquad (2.12)
\end{aligned}
$$

where

$$C_1 = \bar{v}^2 \lambda_d Y_7 \left\{ \lambda_f + \frac{\left(m_2 - m_1^2\right)(-\rho)}{m_1 v (v - 1) \Lambda} \right\}, \qquad (2.13)$$

$$C_2 = \frac{\bar{v}^2 \lambda_d m_1 (-\rho)}{v (v - 1) \Lambda} Y_7, \qquad (2.14)$$

$$C_3 = \frac{\lambda_d m_1 \Lambda}{(-\rho)} - \frac{\bar{v}^2 \lambda_d m_1 (-\rho)}{v (v - 1) \Lambda} \sum_{i=1}^{6} Y_i, \qquad (2.15)$$

$$C_4 = \bar{v}^2 \lambda_d \sum_{i=1}^{6} \frac{Y_i \alpha_i}{2} \left\{ \lambda_f + \frac{\left(m_2 - m_1^2\right)(-\rho)}{m_1 v (v - 1) \Lambda} \right\}, \qquad (2.16)$$

$$\alpha = \alpha_7. \qquad (2.17)$$

In this Feynman-α analysis, Eq. (2.12) is fitted to the Y data to obtain the prompt-neutron decay constant α and the four coefficients (C_1, C_2, C_3, C_4).

2.1.2.2 Rossi-α Formula

In the same manner as a derivation of the practical Feynman-α formula, the Rossi-α formula proposed by Rana and Degweker [6] can be reduced. Their formula can be written as follows:

$$p\left(\tau\right) = \frac{\lambda_d \bar{v}^2}{2} \sum_{i=1}^{7} \alpha_i Y_i e^{-\alpha_i \tau} \left\{ \lambda_f + \frac{\left(m_2 - m_1^2\right)\left(-\rho\right)}{m_1 v \left(v - 1\right) \Lambda} \right\}$$

$$+ \frac{\lambda_d m_1 \bar{v}^2 \left(-\rho\right)}{2 v \left(v - 1\right) \Lambda} \sum_{i=1}^{7} \alpha_i Y_i \left\{ \frac{e^{-\alpha_i \left(\tau - \frac{[f\tau]}{f}\right)} + e^{-\frac{\alpha_i}{f}} e^{\alpha_i \left(\tau - \frac{[f\tau]}{f}\right)}}{\left(1 - e^{-\frac{\alpha_i}{f}}\right)} \right\}. \quad (2.18)$$

The conditional counting probability $p(\tau)\Delta\tau$ is a probability that, given a neutron count at a time, there is a subsequent count in $\Delta\tau$ around time τ later. First, the time interval τ of the Rossi-α analysis is restricted within the following range:

$$\tau \ll \frac{1}{\alpha_i} \quad i = 1, 2, \ldots, 6. \quad (2.19)$$

Under the above time-interval range, the following Maclaurin expansions can be done:

$$e^{-\alpha_i \tau} \simeq 1 - \alpha_i \tau, \quad (2.20)$$

$$e^{\pm \alpha_i \left(\tau - \frac{[f\tau]}{f}\right)} \simeq 1 \pm \alpha_i \left(\tau - \frac{[f\tau]}{f}\right) + \frac{1}{2} \alpha_i^2 \left(\tau - \frac{[f\tau]}{f}\right)^2. \quad (2.21)$$

We substitute the above equations into Eq. (2.18) to obtain the final form:

$$p(\tau)\Delta\tau = C_5 e^{-\alpha\tau} + C_6 \left\{ \frac{e^{-\alpha\left(\tau - \frac{[f\tau]}{f}\right)} + e^{\alpha\left(\tau - \frac{[f\tau]+1}{f}\right)}}{\left(1 - e^{-\frac{\alpha}{f}}\right)} \right\} + C_7 - C_8 \tau, \quad (2.22)$$

where

$$C_5 = \frac{\lambda_d \bar{v}^2}{2} \alpha_7 Y_7 \left\{ \lambda_f + \frac{\left(m_2 - m_1^2\right)\left(-\rho\right)}{m_1 v \left(v - 1\right) \Lambda} \right\} \Delta\tau, \quad (2.23)$$

$$C_6 = \frac{\lambda_d m_1 \bar{v}^2 \left(-\rho\right)}{2 v \left(v - 1\right) \Lambda} \alpha_7 Y_7 \Delta\tau, \quad (2.24)$$

$$C_7 = \frac{\lambda_d \bar{v}^2}{2} \sum_{i=1}^{6} \alpha_i Y_i \left\{ \lambda_f + \frac{\left(m_2 - m_1^2\right)\left(-\rho\right)}{m_1 v \left(v - 1\right) \Lambda} \right\} \Delta\tau$$

$$+ \frac{f \lambda_d m_1 \bar{v}^2 \left(-\rho\right)}{v \left(v - 1\right) \Lambda} \sum_{i=1}^{6} Y_i \Delta\tau, \quad (2.25)$$

$$C_8 = \frac{\lambda_d \bar{v}^2}{2} \sum_{i=1}^{6} \alpha_i^2 \, Y_i \left\{ \lambda_f + \frac{(m_2 - m_1^2)(-\rho)}{m_1 \, v \, (v - 1) \, \Lambda} \right\} \Delta \tau, \qquad (2.26)$$

$$\alpha = \alpha_7. \qquad (2.27)$$

In this Rossi-α analysis, Eq. (2.22) is fitted to the $p(\tau)\Delta\tau$ data to obtain the prompt-neutron decay constant α and the four coefficients (C_5, C_6, C_7, C_8).

2.1.3   Results and Discussion

2.1.3.1   Feynman-α Analyses

Time-sequence counts data within a time interval (gate time) of 1 ms were generated from the arriving time data registered, and then the count data within longer gate times were synthesized by the moving-bunching technique [9] to calculate a gate-time dependence of the Y defined as variance-to-mean ratio minus 1 of the count's data. Figure 2.3 shows a gate-time T and a subcriticality dependence of the Y obtained by the Feynman-α analysis, where neutron counter is B1. The gate-time dependence of the Y is oscillatory due to the periodicity of the pulsed source. The Y of the slightly subcritical pattern B tends to increase with a lengthening in gate time, while that of the deeply subcritical patterns C, D, and E scarcely have the increasing trend and the difference of the amplitude among these patterns is small. At pattern F, whose Y is not drawn in Fig. 2.3, the Y scarcely has the increasing trend and the amplitude is slightly smaller than that at pattern E. The least-squares fits of Eq. (2.12) to the Y data are included in Fig. 2.3, where the fitted curves are in very good agreement with the Y data. The result obtained from counter B2 and B3 was similar to the above observation obtained from counter B1 but that from counter B4 was entirely different.

Fig. 2.3 Subcriticality
dependence of Y value of
counter B1 (Ref. [1])

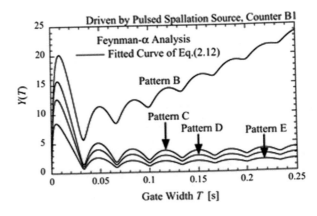

Fig. 2.4 Subcriticality dependence of Y value of counter B4 (Ref. [1])

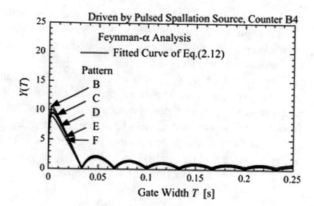

Figure 2.4 shows a gate-time T and a subcriticality dependence of the Y obtained by the Feynman-α analysis, where neutron counter is B4. In a shorter gate-time range than about 0.03 s, a slight difference in the Y among subcritical patterns can be observed. In the longer range, however, the Y data of respective patterns are overlapped. This feature suggests that counter B4 hardly detects fission neutrons which have an information of the subcriticality and most of the neutron counts must be generated from the detection of source neutrons arriving from the target.

In Fig. 2.5, the prompt-neutron decay constant α_p obtained from the present Feynman-α analysis under the pulsed spallation source is compared with the average decay constant done from the previous analysis under a stationary source inherent in nuclear fuels [2], where the previous three decay constants from three neutron counters (B1, B2, B3) are averaged to obtain the average and the standard deviation.

Fig. 2.5 Comparison of respective prompt neutron decay constants obtained from Feynman-α analyses under pulsed spallation and stationary inherent source (Ref. [1])

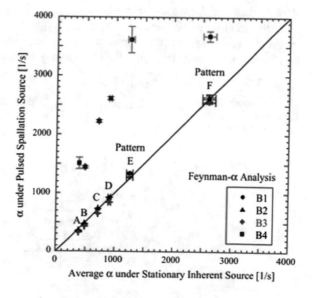

The error of the present decay constant represents statistical uncertainty $\pm 1\sigma$ derived from a nonlinear least-squares method. Since counter B4 was placed far from the core and consequently had very small detection efficiency, the previous analysis for the counter was unsuccessful.

The present decay constants of counters B1, B2, and B3 agree well with the previous average decay constant, while the decay constant of counter B4 is much larger than the previous one. As mentioned in Fig. 2.4, the counter B4 placed far from the fuel region and closely to the target must scarcely detect fission neutrons and consequently have little information about fission chain. Most of neutron counts of the counters B1, B2, and B3 located closely to the fuel region are expected to be generated from the detection of fission neutrons.

2.1.3.2   Rossi-α Analyses

Figure 2.6 shows a time-interval τ and subcritical-pattern dependences of the conditional counting probability $p(\tau)\Delta\tau$ obtained by the Rossi-α analysis, where neutron counter is B1. The least-squares fits of Eq. (2.22) to the counting probability data are included in this figure, where the fitted curves seem to be in good agreement with the data but the fittings are unsuccessful. Figure 2.7 shows an enlarged view near the second peak in Fig. 2.6. A considerable difference between the counting probability data and Eq. (2.22) fitted to the data can be observed. The probability data have a smooth convex top at every integral multiple of pulse period, while the fitted curve of Eq. (2.22) has a sharp cusp arising from a delta-function-like shape of pulsed neutron. Since the pulse width of 100 ns of our accelerator is much shorter than the time scale of the present correlation analyses, the assumption that each pulse has a delta-function-like shape is acceptable. Here, we consider the reason why the Rossi-α data had the smooth convex top.

In the previous pulsed-neutron-source experiments [10–12], a considerable delay in counter response to neutron generation has been observed. The delay has been considered to originate primarily from a slowing down and a thermalization time of

Fig. 2.6 Subcriticality dependence of $P(\tau)\Delta\tau$ of counter B1 (Ref. [1])

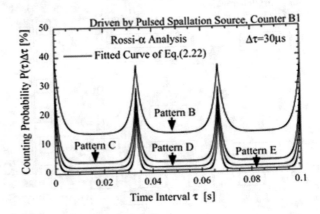

Fig. 2.7 Enlarged view near a peak in Fig. 2.6 (Ref. [1])

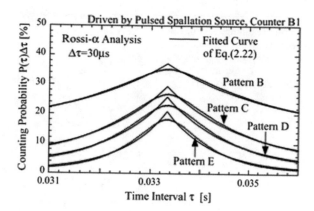

high-energy source neutrons for moderation to thermal energy and from a diffusion time of the thermalized source neutrons for arrival at core. The delay could be also interpreted as a higher harmonics effect [13, 14].

In the previous pulsed-neutron-source experiments mentioned above, the decay data deviated from a single exponential curve were masked to determine the fundamental prompt-neutron decay constant from a least-squares fitting of a conventional formula based on the one-point kinetics model. We tried to apply this masking technique to the present Rossi-α analysis. As shown in Fig. 2.8, the data around each smooth convex top were masked for a least-squares fitting of the present Rossi-α formula. The cusps of these fitted curves appear sharper than those of Fig. 2.7. The correlation coefficient is an indication of a goodness of the least-squares fitting. In this study, we employed the mask width with the maximum coefficient. The optimal mask width for every counter and every subcritical pattern was determined.

In Fig. 2.9, the prompt-neutron decay constant α obtained from the present Rossi-α analysis under the pulsed spallation source is compared with the average decay constant done from the previous analysis under a stationary source inherent in nuclear fuels [2], where the masking technique is not applied to the present analysis. The

Fig. 2.8 Masking data around peaks for least-squares fitting (Ref. [1])

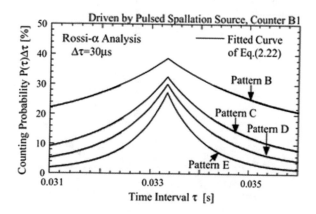

Fig. 2.9 Comparison of
respective prompt-neutron
decay constants obtained
from Rossi-α analyses under
pulsed spallation and
inherent source (Ref. [1])

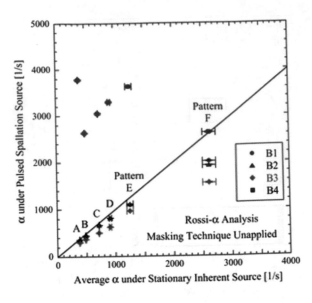

present decay constant is in poor agreement with the previous one. This disagreement could be resolved by applying the masking technique, as shown in Fig. 2.10. Except for counter B4, the decay constant obtained from the present analysis with the masking technique is in good agreement with that done from the previous analysis. The counter B4 placed far from the fuel region and closely to the neutron source derives a large decay constant of source neutrons in reflector. As shown in Fig. 2.5, the decay constant obtained from the present Feynman-α analysis, except for counter

Fig. 2.10 Effect of masking
data around peaks on
prompt-neutron decay
constant obtained from
Rossi-α analysis under
pulsed spallation source
(Ref. [1])

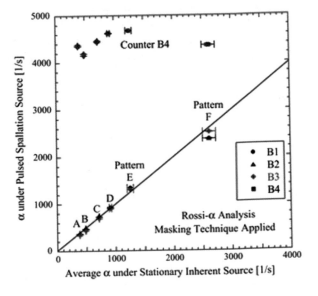

B4, well agreed with the previous decay constant. We can have this fortunate agreement because the respective negative and positive sharp cusps of the uncorrelated second and third terms of Eq. (2.12) barely cancel out, as shown in Fig. 2.5. In contrast, the present Rossi-α formula has no cancelation mechanism and has sharp cups at every integral multiple of pulse period.

2.1.3.3 Comparison of Correlation Amplitude Between Spallation and Poisson Sources

Many authors [6, 8, 15–18] theoretically showed that the non-Poisson spallation source enhanced the correlation amplitudes of various reactor noise analyses. Here, we compare the correlation amplitudes obtained from the Feynman-α and Rossi-α analyses under the present non-Poisson spallation source with those under the previous Poisson inherent source [2]. First, the respective prompt correlation amplitudes C_1 and C_5 of the present Feynman-α and Rossi-α formulae are rewritten in the familiar forms. The prompt quantity Y_7 included in the correlation amplitudes can be described as follows [6]:

$$Y_7 = \frac{\overline{\nu(\nu - 1)}}{\bar{\nu}^2 \, \alpha^2}. \tag{2.28}$$

When the above equation is substituted for Eqs. (2.13) and (2.23), the following equations can be, respectively, obtained:

$$C_1 = \lambda_d \frac{\overline{\nu(\nu - 1)}}{\alpha^2} \left\{ \lambda_f + \frac{\left(m_2 - m_1^2\right)(-\rho)}{m_1 \, \overline{\nu(\nu - 1)} \, \Lambda} \right\}, \tag{2.29}$$

and

$$C_5 = \frac{\lambda_d \, \overline{\nu(\nu - 1)}}{2\,\alpha} \left\{ \lambda_f + \frac{\left(m_2 - m_1^2\right)(-\rho)}{m_1 \, \overline{\nu(\nu - 1)} \, \Lambda} \right\} \Delta\tau. \tag{2.30}$$

The respective correlation amplitudes C_{1P} and C_{5P} of the conventional Feynman-α and Rossi-α formulae for a stationary Poisson source can be described as follows [7]:

$$C_{1P} = \lambda_d \, \lambda_f \, \frac{\overline{\nu(\nu - 1)}}{\alpha^2}, \tag{2.31}$$

and

$$C_{5P} = \frac{\lambda_d \, \lambda_f \, \overline{\nu(\nu - 1)}}{2\,\alpha\,X_P} \Delta\tau, \tag{2.32}$$

where

$$\lambda_d = \varepsilon \, \lambda_f. \tag{2.33}$$

Assuming that a detection efficiency ε has no difference between the non-Poisson spallation and the Poisson sources, the following relationships hold:

$$C_1 = C_{1P} + \frac{\lambda_d \left(m_2 - m_1^2\right)(-\rho)}{m_1 \, \alpha^2 \, \Lambda}, \tag{2.34}$$

and

$$C_5 = C_{5P} + \frac{\lambda_d \left(m_2 - m_1^2\right)(-\rho)}{2 \, m_1 \, \alpha \, \Lambda} \, \Delta\tau. \tag{2.35}$$

The respective second terms of the right-hand sides of the above two equations express the enhancement by the non-Poisson character of the spallation source. The enhancement disappears at a critical state and increases with an increase in subcriticality.

Next, the subcriticality dependence of the enhancement is experimentally confirmed. Figure 2.11 shows a comparison of respective prompt correlation amplitudes obtained from the Feynman-α analyses under the present pulsed spallation and the stationary inherent sources. The latter is a stationary Poisson source since the inherent source neutrons are dominantly produced by (α, n) reaction and spontaneous fissions negligibly contribute to the source strength [2]. This figure indicates

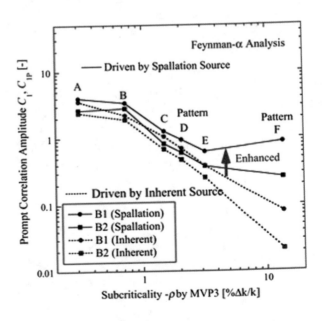

Fig. 2.11 Comparison of respective prompt-neutron correlation amplitudes obtained from Feynman-α analyses under pulsed spallation and stationary inherent sources (Ref. [1])

Fig. 2.12 Comparison of
respective prompt-neutron
correlation amplitudes
obtained from Rossi-α
analyses under pulsed
spallation and stationary
inherent sources (Ref. [1])

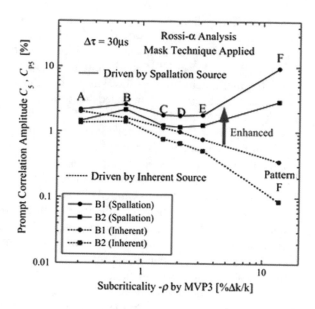

Fig. 2.12 Comparison of respective prompt-neutron correlation amplitudes obtained from Rossi-α analyses under pulsed spallation and stationary inherent sources (Ref. [1])

that the enhancement of the prompt correlation amplitude increases with an increase in subcriticality. The non-Poisson character of the spallation source enhances the amplitude clearly.

Figure 2.12 also shows another comparison of respective prompt correlation amplitudes obtained from the Rossi-α analyses under the present pulsed spallation and the stationary inherent sources. This also indicates that the non-Poisson character of the spallation source significantly enhances the amplitude. The above discussion is confined to the qualitative observation and some quantitative evaluation of the non-Poisson character must be difficult. This is because a detection efficiency ε has a considerable difference between the external spallation and the inherent sources and the efficiency must depend on the subcriticality. A spatially uniform inherent source in fuels hardly excites any higher modes, however, another spatially localized external source significantly excites a higher mode with an increase in subcriticality, as could be observed in a source multiplication measurement [2]. In next Sect. 2.1.3.4, we try to derive a quantitative indicator of the non-Poisson character.

2.1.3.4   Indicator of Non-poisson Character of Spallation Source

Dividing Eq. (2.23) by Eq. (2.24) of the present Rossi-α formula, the following equation can be obtained:

$$\frac{C_5}{C_6} = \frac{\lambda_f \overline{\nu(\nu-1)}\,\Lambda}{m_1(-\rho)} + \frac{m_2 - m_1^2}{m_1^2}. \tag{2.36}$$

K. Hashimoto

The above equation has no detection efficiency and asymptotically approaches the second term with an increase in the subcriticality. The second term quantitatively expresses a non-Poisson character of the spallation source and may be referred to as "the Degweker's factor" of multiplicity distribution of neutrons in a pulse bunch. When the multiplicity follows the Poisson distribution, the factor becomes zero. Degweker et al. theoretically simulated the Rossi-α analysis changing parametrically the above factor to investigate an impact of non-Poisson source [8].

Generally, the ratio of the second factorial moment to the first factorial moment squared of a multiplicity distribution has been employed as an indication characterizing the distribution. As an example, Diven factor [19] is defined as the ratio of the second factorial moment to the first factorial moment squared of a multiplicity distribution of fission neutrons emitted from a fission event and is a useful indication characterizing the multiplicity distribution. Adding 1 to the Degweker's factor, the result is the ratio of the second factorial moment to the first factorial moment squared of the multiplicity distribution of neutrons in a pulse bunch. Consequently, this factor is eligible to employ as the indication characterizing the multiplicity distribution and should be determined experimentally.

Figure 2.13 shows a subcriticality dependence of the ratio C_5/C_6 determined from the present Rossi-α analysis. At the more deeply subcritical system than pattern C, the ratio seems to be an asymptotic value. Seeing the ratio within the subcritical range from pattern C to F, no systematic dependence on the subcriticality can be observed. Averaging the ratios over the subcritical range and over four counters, we obtain the second term, i.e., the Degweker's factor of 0.067 ± 0.011. The non-zero value convinces us that the present spallation source has the non-Poisson character.

Fig. 2.13 Subcriticality dependence of coefficient ratio C_5/C_6 obtained from Rossi-α analysis (Ref. [1])

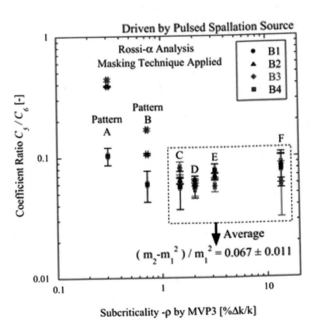

If the accelerator had higher energy proton beam than 100 MeV, the factor would be larger [20].

2.2 Power Spectral Analyses

2.2.1 Experimental Settings

The power spectral analysis on frequency domain is carried out in the same A-core as shown in Fig. 2.2 [21]. Figure 2.14 shows the present signal processing circuit, whose former stage consisted of conventional charge preamplifier (PA), detector bias-supply (HV), spectroscopy amplifier (SA), and single-channel analyzer (SCA) modules. A special count-rate meter (Oken S-1955) input logic pulse train from SCA to output analog signal proportional to instantaneous count rate. The rate meter could be well modeled by a primary delay element which had the time constant of 3.88 ms (break frequency 41.0 Hz). The time constant of the meter is so short that a large portion of reactor noise passes through the filtration. Finally, analog signals from the two count-rate meters were fed to a fast Fourier transform (FFT) analyzer (Ono DS-3200) to obtain auto- and cross-power spectral densities and to record the analog signals as digital data. This FFT analyzer has a highly resolvable analog-to-digital converter whose number of bits and dynamic range are 24 bits and above 110 dB, respectively. An analysis range in frequency from 1.25 to 1000 Hz was specified to obtain 800-point spectral data. Delayed neutrons are expected to contribute hardly to the power spectral density obtained from the FFT analyzer because the above minimum frequency of 1.25 Hz is larger than the 6th decay constant $3.01 \, s^{-1}$ (0.48 Hz) of a delayed neutron data given by Keepin [22].

In each subcritical pattern, time-sequence signal data were acquired for about 10 min. A response function of the above count-rate meter was measured in advance and the auto-and cross-power spectral densities obtained were divided by the auto-power spectral density of the response of the count-rate meter, so as to compensate an influence of the meter. Throughout the present accelerator operations, the pulsed

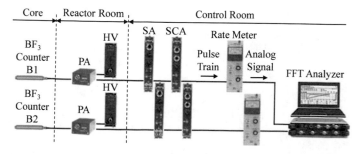

Fig. 2.14 Signal processing circuit for power spectral analysis (Ref. [21])

repetition frequency and the beam width were set to 30 Hz and 100 ns, respectively. The proton beam intensity of an accelerator was set to 30 pA.

2.2.2 Formula for Power Spectral Analyses

2.2.2.1 Degweker's Formula

Degweker and Rana [8] formulated the auto- and cross-power spectral densities for a periodically pulsed non-Poisson source, where delayed neutron contribution was neglected and each pulse was assumed to be a delta function. The neglect of delayed neutron contribution is acceptable as mentioned in the above Sect. 2.2.1. Since the pulse width 100 ns of our accelerator is much shorter than the time scale of the present analysis, the assumption of the delta function is also acceptable. Here, their formulae are rewritten as a function of not angular frequency ω s^{-1} but frequency f Hz. Then, their cross-power spectral density between neutron detector 1 and 2 can be described as follows:

$$\Phi_{12}(f) = \frac{C_1(f)}{(2\pi f)^2 + \alpha^2} + C_2(f) \sum_{n=-\infty}^{\infty} \frac{\delta(f - n f_R)}{(2\pi f)^2 + \alpha^2}, \tag{2.37}$$

where

$$C_1(f) = H(f)\, q^2\, f_R\, \lambda_{d1}\, \lambda_{d2}\, (m_2 - m_1^2 + 2m_1 Y_1), \tag{2.38}$$

$$C_2(f) = H(f)\, q^2\, f_R^2\, \lambda_{d1}\, \lambda_{d2}\, m_1^2, \tag{2.39}$$

$$Y_1 = \lambda_f\, \frac{\overline{\nu(\nu - 1)}}{2\alpha}. \tag{2.40}$$

Their auto-power spectral density of neutron detector 1 can be also described as follows:

$$\Phi_{11}(f) = C_3(f) + \frac{C_4(f)}{(2\pi f)^2 + \alpha^2} + C_5(f) \sum_{n=-\infty}^{\infty} \frac{\delta(f - n f_R)}{(2\pi f)^2 + \alpha^2}, \tag{2.41}$$

where

$$C_3(f) = \frac{H(f)\, q^2\, f_R\, \lambda_{d1}\, m_1}{\alpha}, \tag{2.42}$$

$$C_4(f) = H(f)\, q^2\, f_R\, \lambda_{d1}^2\, (m_2 - m_1^2 + 2m_1 Y_1), \tag{2.43}$$

$$C_5(f) = H(f) q^2 f_R^2 \lambda_{d1}^2 m_1^2, \tag{2.44}$$

where α and f_R represent prompt-neutron decay constant and pulse repetition frequency, respectively. $H(f)$ is a power spectral density of an impulse response of detector and processing-circuit system.

2.2.2.2 Formula Applied to Present Analyses

The auto-power spectral density has a white chamber noise indicated by the first term of Eq. (2.41). The correlated component done by the second term is completely hidden by the chamber noise in a higher frequency range, and this feature suggests a difficulty in estimating the break frequency, i.e., the prompt-neutron decay constant [12, 23]. In previous reactor noise analysis for a stationary source, Nomura [24] proposed the use of two neutron detectors with independent electronic circuits to reduce this spurious white noise and demonstrated the usefulness of his proposal. His original improvement referred to as the two-detector method is identical with the cross-power spectral analysis familiar to signal processing field. Actually, Eq. (2.37) for the cross-power spectral density has no term of the white noise. In the present study, we analyze only the cross-power spectral density free from the white noise.

The two coefficients defined by Eqs. (2.38) and (2.39) include $H(f)$ and depend on frequency f. When the frequency response of the count-rate meter is compensated as mentioned in Sect. 2.2.1, these coefficients are independent of the frequency. For the following expressions, each coefficient is described as a constant. Then, Eq. (2.37) can be rewritten as

$$\Phi_{12}(f) = \frac{C_1}{(2\pi f)^2 + \alpha^2} + C_2 \sum_{n=-\infty}^{\infty} \frac{\delta(f - n f_R)}{(2\pi f)^2 + \alpha^2}. \tag{2.45}$$

The second term of the above equation gives an expression to the uncorrelated delta-function peaks at the multiple of pulse repetition frequency f_R. At frequency of the integral multiple, Eq. (2.45) can be reduced as follows:

$$\Phi_{12}(n f_R) = \frac{C_1}{(2\pi n f_R)^2 + \alpha^2} + \frac{C_2}{(2\pi n f_R)^2 + \alpha^2}, \quad n = 1, 2, 3\dots. \tag{2.46}$$

The second uncorrelated term of the above equation is larger than the first correlated term by over two orders of magnitude [12, 23]. Then, Eq. (2.46) can be simplified as

$$\Phi_{12}(n f_R) \simeq \frac{C_1}{(2\pi n f_R)^2 + \alpha^2}, \quad n = 1, 2, 3\dots. \tag{2.47}$$

From a least-squares fit of Eq. (2.47) to peak data at frequency of the integral multiple of f_R, the prompt-neutron decay constant α and coefficient C_2 can be determined.

Next, a well-known effect of a higher mode excited by the injection of pulsed neutrons [10–12] should be considered. Sakon et al. employed the following equation to consider the effect successfully [12, 23].

$$\Phi_{12}(n f_R) \simeq \frac{C_2}{(2\pi n f_R)^2 + \alpha^2} + \frac{C_{2H}}{(2\pi n f_R)^2 + \alpha_H^2}, \quad n = 1, 2, 3, \ldots. \quad (2.48)$$

When certain a higher prompt mode as well as a fundamental prompt mode are excited, such a higher term as the second term of the above equation can be added to the fundamental term of the power spectral density [25–27]. The prompt-neutron decay constant of the higher mode is represented by α_H. The coefficient C_{2H} includes the eigenfunction and the adjoint eigenfunction of the higher prompt mode. We try to apply Eq. (2.48) as well as Eq. (2.47) to derive the fundamental decay constant α from the uncorrelated peaks.

When the uncorrelated peaks of the cross-power spectral density are masked, Eq. (2.45) can be reduced to the following equation for the remaining data unmasked:

$$\Phi_{12}(f) = \frac{C_1}{(2\pi f)^2 + \alpha^2}. \quad (2.49)$$

From a least-squares fit of Eq. (2.49) to the unmasked data, the prompt neutron decay constant α and coefficient C_1 can be also determined. The above equation gives an expression to the correlated noise component and is identical to the familiar formula for a stationary neutron source.

2.2.3 Results and Discussion

2.2.3.1 Power Spectral Density

Figure 2.15a, b show measured auto-power spectral density of counter B1 and cross-power spectral density between neutron detector B1 and B2, respectively, where subcritical pattern is F. The auto-power spectral density is composed of a continuous correlated component, another constant chamber noise and many delta-function-like peaks at the integral multiple of the repetition frequency, as expected by Eq. (2.41). The correlated component tends to be hidden by the white chamber noise with an increase in frequency and this feature suggests a difficulty in estimating the break frequency, i.e., the prompt-neutron decay constant.

On the other hand, the cross-power spectral density has no white chamber noise, expected by Eq. (2.37). The correlated component is larger than one decade (20 dB). This feature is significantly different from that of the auto-power spectral density as

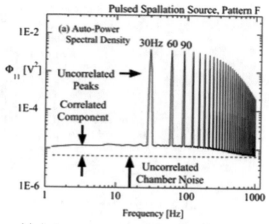

(a) Auto-power spectral density of counter B1

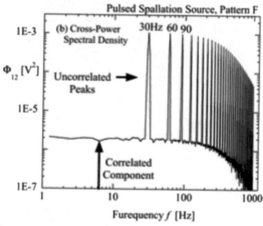

(b) Cross-power spectral density between counters B1 and B2

Fig. 2.15 Power spectral density measured at subcritical pattern F (Ref. [21])

shown above. Obviously, the cross-power spectral density is favorable to estimating the decay constant.

For almost the same subcritical system driven a pulsed 14 MeV Poisson source, previously, a power spectral analysis was carried out [12]. Figure 2.16a, b show the cross-power spectral densities measured previously, where the respective pulse repetition frequencies are 20 and 500 Hz and the subcriticality 13.59 %Δk/k is almost same as that of the present pattern F. In these cross-power spectral densities, no correlated noise component could be observed, and we failed in determining the prompt-neutron decay constant from correlated component. Naturally, the auto-power spectral density also had no correlated component. In contrast to the previous analysis, the present analysis gives a considerable correlated component as shown

K. Hashimoto

Fig. 2.16 Cross-power
spectral density measured
previously at a subcritical
KUCA system driven by a
pulsed DT neutron source
(Ref. [21])

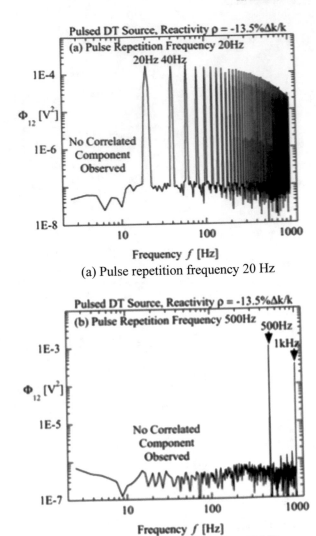

(a) Pulse repetition frequency 20 Hz

(b) Pulse repetition frequency 500 Hz

in Fig. 2.15. The non-Poisson character of the spallation source must enhance the
amplitude of the correlated component over a precision limit of the FFT analyzer.

2.2.3.2 Prompt Neutron Decay Constant

Figure 2.17 shows a least-squares fit of Eq. (2.49) to the correlated noise component
of the cross-power spectral density measured at pattern F. In this fitting, 33 delta-
function-like peaks were masked. No systematical deviation of the fitted curve from
the data can be seen. Figure 2.18 shows a least-squares fit of Eqs. (2.47) and (2.48)

Fig. 2.17 Least-squares fit to correlated component of cross-power spectral density (Ref. [21])

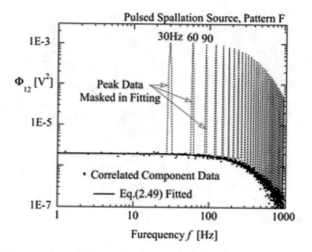

Fig. 2.18 Least-squares fit to uncorrelated peaks of cross-power spectral density (Ref. [21])

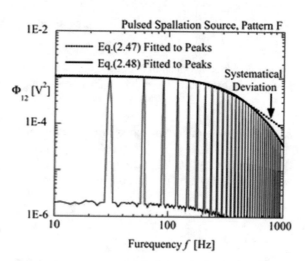

to the peak points of the cross-power spectral density at pattern F. No systematical deviation of fitted Eq. (2.47) from the peak point can be seen in a lower frequency range than roughly 500 Hz, while above this frequency, a systematical deviation of the fitted curve from the peak points is significant. In contrast to Eq. (2.47), no deviation of fitted Eq. (2.48) from the peaks can be observed over the frequency range. At all subcritical patterns, the fittings of Eq. (2.48) were very successful.

In Fig. 2.19, the prompt-neutron decay constant obtained from the present cross-power spectral analysis is compared with the average decay constant done from the previous Rossi-α analysis under the same spallation source [1]. When Eq. (2.47) is fitted to uncorrelated peaks, the present decay constant is in poor agreement with the previous one. Fortunately, the fitting of Eq. (2.48) completely resolves this disagreement. When Eq. (2.49) is fitted to continuous correlated data, the agreement with

Fig. 2.19 Comparison of respective prompt neutron decay constants obtained from present cross-power spectral and previous Rossi-α analyses (Ref. [21])

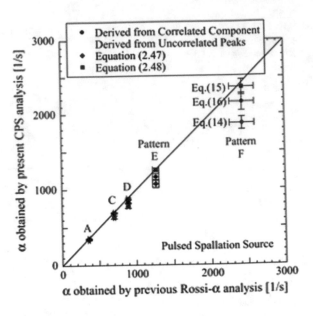

the previous decay constant is not too bad. The analysis for a much longer time than 10 min may be required to enhance the agreement.

2.2.3.3 Indicator of Non-poisson Character of Spallation Source

Dividing Eq. (2.38) by Eq. (2.39) of the formula for cross-power spectral density, the following equation can be obtained:

$$f_R \frac{C_1}{C_2} = \frac{\lambda_f \overline{\nu (\nu - 1)}}{m_1 \alpha} + \frac{m_2 - m_1^2}{m_1^2}. \tag{2.50}$$

The above equation has no detection efficiency and asymptotically approaches to the second term with an increase in the subcriticality, i.e., the prompt-neutron decay constant α. The second term quantitatively expresses a non-Poisson character of the spallation source and may be referred to as "the Degweker's factor" of multiplicity distribution of neutrons in a pulse bunch. When the multiplicity follows the Poisson distribution, the factor becomes zero. Degweker et al. theoretically simulated the Rossi-α analysis changing parametrically the above factor to investigate an impact of non-Poisson source [8]. The Degweker's factor is a useful indication characterizing the multiplicity distribution of the spallation neutrons in a pulsed bunch.

Figure 2.20 shows a subcriticality dependence of the ratio f_R C_1/C_2 determined from the present cross-power spectral analysis. At the more deeply subcritical system than pattern C, the ratio seems to be an asymptotic value. Seeing the ratio within the subcritical range from pattern C to F, no systematic dependence on the subcriticality

Fig. 2.20 Subcriticality dependence of f_R C_1/C_2 (Ref. [21])

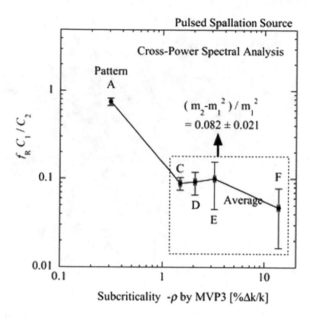

can be observed and these ratios are considered to be the asymptotic value, i.e., the Degweker's factor. Averaging the ratios over the subcritical range from Pattern C to F, we obtained the second term, i.e., the Degweker's factor of 0.082 ± 0.021. This value is consistent with 0.067 ± 0.011 determined from the previous Rossi-α analysis [1].

2.3 Beam Trip and Restart Methods

2.3.1 Experimental Settings

2.3.1.1 Core Configurations

A series of accelerator-beam trip and restart experiments is carried out to determine the subcriticality of a reactor system driven by a pulsed 14 MeV neutron source [28]. This core configuration is shown in Fig. 2.21. The core was composed of 20 regular fuel assemblies and one partial fuel assembly, which were loaded on the grid plate. The fuel and moderator elements of each assembly were set in a 1.5-mm-thick aluminum sheath and the cross section of the elements within the assembly was the square of 2". The regular fuel assembly was composed of 36 unit cells of one (1/16" thick) uranium plate with Al clad and two (1/8" and 1/4" thick) polyethylene plates, while the partial fuel assembly was composed of 12 unit cells of these plates. The partial fuel assembly was employed to adjust the excessive reactivity of the core.

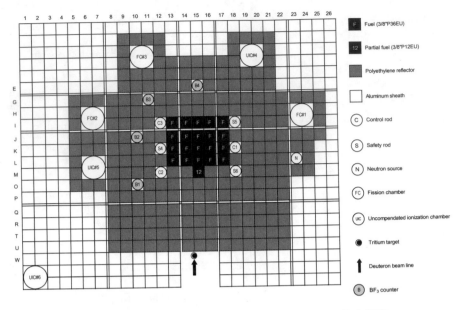

Fig. 2.21 Top view of core configuration and neutron detector location (Ref. [28])

The active height of the core was about 40 cm, with additional about 60 cm upper and lower polyethylene reflectors. The configuration of these fuel assemblies was reported in detail by Pyeon et al. [29].

A pulsed neutron generator was combined with the core, where 14 MeV pulsed DT neutrons were injected into the subcritical system through the polyethylene reflector. The generator consisted of a duoplasmatron-type ion source, a Cockcroft-Walton-type accelerator for deuteron (D$^+$) beam, and tritium (T) target of gas-in-metal type. The pulsed neutrons are generated through the D-T reaction by the pulsed and accelerated D$^+$ beam and T in the target metal. The target was placed outside the polyethylene reflector, as shown in Fig. 2.21. The pulse duration and repetition period of the D$^+$ beam pulse can be remotely controlled by using an arc-pulser installed in the control room of KUCA. The current and acceleration voltage of the D$^+$ beam pulse can also be controlled from the control room. In the present experiment, the major parameters of the accelerator drive were set to 160 keV in beam energy, 0.6–0.8 mA in beam current, 0.8 ms in pulse width, 1 ms in pulse repetition period, and 5.1–9.2 V in arc voltage of the ion source.

2.3.1.2 Experimental Procedures and Conditions

Four BF$_3$ proportional counters (1" dia.) were employed as experimental channels. As shown in Fig. 2.21, these BF$_3$ counters were placed on several positions around the core to measure the reactor response to beam trip and restart operations and to

Table 2.2 Control rod patterns employed in the present experiment (Ref. [28])

Pattern	Rod position			Reactivity
	C1	C2, C3	S4, S5, S6	[%Δk/k]
A	L.L.	U.L.	U.L.	−0.240
B	L.L.	L.L.	U.L.	−0.636
C	L.L.	L.L.	L.L.	−1.577

L.L.: Lower Limit [0 mm], U.L.: Upper Limit [1200 mm]

investigate the spatial dependence. The nuclear instrumentation system consisted of a detector bias supply, a preamplifier, a spectroscopy amplifier, and discriminator modules. Finally, signal pulses from four neutron counters were fed to a multichannel scaler to acquire time-sequence count data. The gate width of the scaler was set to 0.1 s. At a slightly subcritical state, the count rate of each neutron counter was expected to be so high that count losses would be induced by the dead-time effect of the neutron counter. Hence, the acquired count data were corrected for the losses on the basis of the non-paralysable model, where we used the dead time of 4 μs predetermined by an improved Feynman-α analysis [30].

First, an Am–Be neutron source for reactor startup was inserted. Then, the subcriticality for the experiment was adjusted by changing the axial positions of safety rods and control ones. The neutron source was taken out of the core and the injection of pulsed neutrons began. The control rod patterns employed in the experiment are shown in Table 2.2. The reference reactivity, included in this table, was evaluated from the reactivity worth of each rod, whose worth was predetermined by the positive period method and the rod drop one.

In a beam trip experiment, a certain arc voltage of the ion source was suddenly dropped to turn off the D⁺ pulse beam. In the succeeding beam restart experiment, the voltage was rapidly returned to the original value to turn on the beam.

2.3.2 Data Analyses Method

2.3.2.1 Least-Squares Inverse Kinetics Method

First, the theory of the least-squares inverse kinetics method (LSIKM) [31–33] is briefly described. Assuming the zero-power and one-point kinetics model, the time-dependent neutron behavior of a subcritical reactor system driven by an external neutron source can be described as

$$\frac{dN(t)}{dt} = \frac{\rho - \beta}{\Lambda} N(t) + \sum_{k=1}^{6} \lambda_k C_k(t) + S, \tag{2.51}$$

$$\frac{dC_k(t)}{dt} = \frac{\beta_k}{\Lambda} N(t) - \lambda_k C_k(t), \tag{2.52}$$

where $N(t)$ is the neutron density, $C_k(t)$ the concentration of k-th-group precursor, and S the neutron source strength. Other notations are conventional. As the above neutron density, usually, time-sequence count-rate data can be employed to determine the reactivity ρ. The differential term of Eq. (2.51) is usually neglected to simplify the analysis. The assumption is applicable to the KUCA system. Consequently, the discrete form of Eq. (2.51) on time domain is described as

$$N(t_j) = \frac{\Lambda}{\beta - \rho} Q(t_j) + \frac{\Lambda S}{\beta - \rho}, \tag{2.53}$$

$$Q(t_j) = \sum_{k=1}^{6} \lambda_k C_k(t_j), \tag{2.54}$$

where t_j is the jth discrete time. As the time-dependent neutron density $N(t_j)$, the time-sequence count data were employed. The time-dependent precursor density $C_k(t_j)$ of delayed neutrons can be obtained by solving numerically Eq. (2.52). In this study, the implicit time-integration method was employed to obtain the precursor density. When the time-sequence data $N(t_j)$ and $Q(t_j)$ are plotted on the x-y coordinate, two unknown constants, i.e., the reactivity and the source strength, can be determined from the least-squares fitting of Eq. (2.53) to these data. By applying the LSIKM to beam trip data, Eq. (2.53) is reduced to

$$N(t_j) = \frac{\Lambda}{\beta - \rho} Q(t_j), \tag{2.55}$$

where the source strength is set to zero.

The delayed neutron data and prompt-neutron generation time of the present reactor system were generated using the SRAC code system [34], where a three-dimensional, 19-energy-group diffusion calculation was done with JENDL-3.2 nuclear library [35].

2.3.2.2 Integral Count Technique

The integral count technique has been frequently employed to determine the subcriticality from a source jerk experiment and a rod drop one [31, 36]. When a neutron source is rapidly taken out of a subcritical reactor core or a control rod is dropped into a critical core at $t = 0$, the reactivity of the core can be expressed as

$$\rho = -\sum_{k=1}^{6} \frac{\beta_k}{\lambda_k} \frac{N(0)}{\int_0^\infty N(t)\,dt}. \tag{2.56}$$

The above expression is a familiar formula for the integral count technique.

The technique is also applied to the present beam trip experiment. Prior to beam trip operation, the average count rate used as $N(0)$ is measured using a conventional counting scaler. Then, the integral counts after the operation, which can be used as an integral appearing in the denominator of Eq. (2.56), are also measured using the scaler.

2.3.3 Results and Discussion

2.3.3.1 Time-Sequence Data

Figure 2.22 shows the time-sequence $N(t)$ and $Q(t)$ data obtained from counter B4 in a beam trip experiment, where the control rod pattern is A and the beam is turned off at zero time. The time-sequence $N(t)$ data in this figure are indicated as instantaneous count rate at every 0.1 s. After the beam trip, $N(t)$ and $Q(t)$ promptly decrease and then asymptotically tend to zero. The statistical fluctuation of $Q(t)$ defined using Eq. (2.54) is slight, compared with that of $N(t)$. This is because the delayed-neutron emission rate $Q(t)$ is an integral quantity of $N(t)$ over a passing time. The least-squares approximation fits a model function with minimum error on the y-axis assuming no error on the x-axis. Therefore, the time-sequence $Q(t)$ and $N(t)$ data should be assigned to x- and y-axis variables, respectively, for successful least-squares fitting on the x-y coordinate.

Figure 2.23 shows the time-sequence $N(t)$ and $Q(t)$ data obtained from counter B4 in a beam restart experiment, where the control rod pattern is A and the beam is turned on at zero time. After the beam restart, $N(t)$ and $Q(t)$ promptly increase and then asymptotically tend to their individual constants.

2.3.3.2 Least-Squares Fitting

In Fig. 2.24, the above $Q(t)$ and $N(t)$ data are plotted on the x-y coordinate, where the fitted lines are drawn by a straight line. Equations (2.55) and (2.53) were fitted to the data sets of Fig. 2.24a, b, respectively. The fitting is successful and the subcritical reactivity can be determined from the slope of the fitted line.

Table 2.3 summarizes the reactivities obtained from the beam trip and restart experiments. As the errors of these results of the LSIKM, the statistical uncertainties that originated from the least-squares fit were employed, while the uncertainty of delayed-neutron yield β was not taken into account. The errors of the integral count technique were estimated from counting statistics. For comparison, the results obtained by a pulsed neutron experiment [11] are also shown in Table 2.4, where the subcriticality obtained by a conventional analysis technique has a significant counter-position dependence. This dependence is originated from a higher mode excited by

Fig. 2.22 Time-sequence data $N(t)$ and $Q(t)$ in a beam trip experiment (Ref. [28])

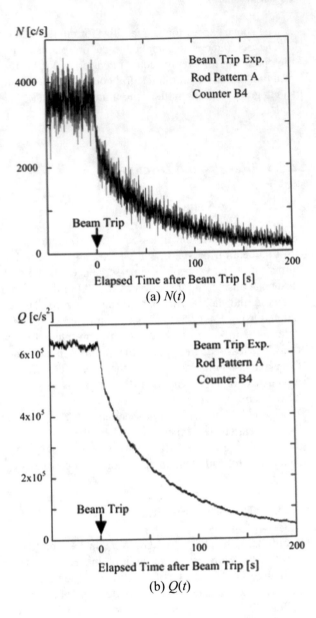

(a) $N(t)$

(b) $Q(t)$

a pulsed neutron source. The mask analysis technique was employed to reduce the dependence.

It is obvious from Table 2.3 that counter B1 significantly overestimates the subcriticality. This feature is originated from a large amount of source neutrons traveling from the DT target. This is because the counter relatively close to the target counts much more the source neutrons during successive pulse injection, and then the source neutrons decay out with a larger decay constant after beam trip and more steeply

Fig. 2.23 Time-sequence data $N(t)$ and $Q(t)$ in a beam restart experiment (Ref. [28])

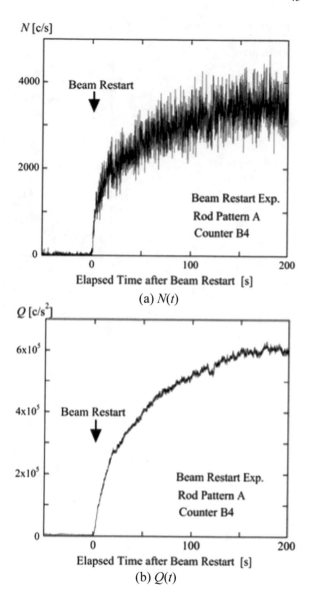

(a) $N(t)$

(b) $Q(t)$

increase after beam restart. The decay of fission neutrons contains information on the reactivity, while that of the source neutrons is free from such information. Generally, the decay constant of the source-neutron mode free from the neutron-production process is larger than that of the prompt mode of the fission neutron [11]. In Fig. 2.25, the decay data of counter B1 are compared with those of B4 in a beam trip experiment, where the control rod pattern is C and the data is normalized in such a way that the average counts before the beam trip become one. This figure shows a larger

Fig. 2.24 Plot of variables
and fitted line on
x-y coordinate (Ref. [28])

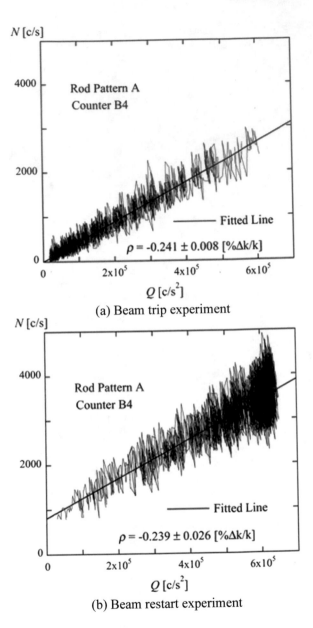

(a) Beam trip experiment

(b) Beam restart experiment

decay of B1 data just after the trip, as noted above. For an actual ADS, a neutron counter should be placed at a position far from a spallation target. Table 2.3 also shows that the results of the LSIKM for beam trip data are consistent with those of the integral count technique except for counter B1.

As seen from Tables 2.3 and 2.4, the present subcriticality obtained using the counters other than B1 has a slight counter-position dependence compared with

Table 2.3 Reactivity obtained by beam trip and restart experiments [%Δk/k] (Ref. [28])

Rod Pattern	Neutron Counter	Beam trip experiment LSIKM	Integral	Beam restart Experiment[a]
A	B1	−0.283 ± 0.010	−0.288 ± 0.020	−0.340 ± 0.034
	B2	−0.244 ± 0.008	−0.232 ± 0.012	−0.242 ± 0.025
	B3	−0.244 ± 0.009	−0.233 ± 0.016	−0.241 ± 0.028
	B4	−0.241 ± 0.008	−0.232 ± 0.012	−0.239 ± 0.026
B	B1	−1.293 ± 0.018	−1.115 ± 0.054	−1.278 ± 0.083
	B2	−0.641 ± 0.008	−0.617 ± 0.026	−0.632 ± 0.035
	B3	−0.631 ± 0.009	−0.612 ± 0.034	−0.625 ± 0.042
	B4	−0.617 ± 0.008	−0.599 ± 0.026	−0.593 ± 0.033
C	B1	−5.082 ± 0.094	−4.065 ± 0.256	−2.698 ± 0.423
	B2	−1.599 ± 0.025	−1.513 ± 0.092	−1.460 ± 0.165
	B3	−1.555 ± 0.033	−1.501 ± 0.120	−1.447 ± 0.201
	B4	−1.504 ± 0.027	−1.431 ± 0.084	−1.355 ± 0.171

[a]LSIKM applied

Table 2.4 Reactivity obtained by pulsed neutron experiments [%Δk/k] (Ref. [28])

Rod Pattern	Neutron Counter	Pulsed neutron experiment Conventional	Mask technique
A	B1	−16.05 ± 0.68	−0.236 ± 0.039
	B2	−0.299 ± 0.042	−0.223 ± 0.038
	B3	−0.153 ± 0.036	−0.227 ± 0.039
	B4	−0.043 ± 0.036	−0.245 ± 0.039
B	B1	−16.25 ± 0.69	−0.644 ± 0.055
	B2	−0.872 ± 0.065	−0.642 ± 0.054
	B3	−0.622 ± 0.053	−0.658 ± 0.055
	B4	−0.359 ± 0.049	−0.668 ± 0.055
C	B1	−18.09 ± 0.73	−1.616 ± 0.092
	B2	−2.342 ± 0.126	−1.554 ± 0.089
	B3	−1.497 ± 0.087	−1.572 ± 0.090
	B4	−0.952 ± 0.075	−1.565 ± 0.089

those obtained in a conventional pulsed neutron experiment and agrees with the result of the pulsed neutron experiment based on the mask analysis within experimental errors. Here, we note a further observation from this table, which is a larger error of the beam restart experiment than that of the beam trip one. From the least-squares fitting of Eq. (2.55) to beam trip data, only the reactivity is determined. From the fitting of Eq. (2.53) to beam restart data, however, not only is the reactivity but also the source strength is simultaneously determined. The additional unknown constant

Fig. 2.25 Difference in decay data between counters B1 and B4 (Ref. [28])

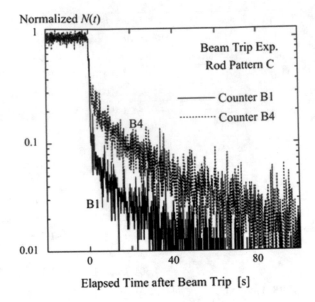

Elapsed Time after Beam Trip [s]

to be inferred is responsible for the larger error of the beam restart experiment. For the smaller error, the trip experiment is advantageous. Nevertheless, the restart experiment is useful for the simultaneous determination of both the reactivity and source strength.

2.4 Conclusion

We derived the Feynman-α and the Rossi-α formulae applicable to the respective correlation data analyses for a pulsed non-Poisson neutron source. These formulae were applied to the Feynman-α and the Rossi-α analyses for a subcritical system driven by a pulsed spallation source in KUCA. The prompt-neutron decay constant determined from the present Feynman-α analysis well agreed with that done from a previous analysis for the same subcritical system driven by an inherent neutron source. However, the decay constant determined from the present Rossi-α analysis was in poor agreement with that done from the above previous analysis. When the data around the convex top were masked for least-squares fitting of the present Rossi- When the data around the convex top of the counting probability distribution were masked for least-squares fitting of the present Rossi-α formula, the disagreement could be successfully resolved. formula, the disagreement could be successfully resolved. When the respective prompt-neutron correlation amplitudes determined from the present Feynman-α and Rossi-α analyses were compared with those done from the previous analyses under the Poisson inherent source, the non-Poisson spallation source definitely enhanced the respective correlation amplitudes.

The Degweker's factor $(m_2 - m_1^2)/m_1^2$ of 0.067 ± 0.011, which is a quantitative indication of the non-Poisson character, could be determined from the present Rossi-α analysis.

The power spectral analysis on frequency domain was conducted in the same A-core as the above Feynman-α and Rossi-α analyses. Not only the cross-power but also the auto-power spectral density had a considerable correlated noise component even at a deeply subcritical state, where no correlated component could be observed under a pulsed DT(14 MeV) neutron source. The non-Poisson character of the spallation source must enhance the correlation amplitude of these power spectral densities. The Degweker's factor of 0.082 ± 0.021 could be determined from the present analysis and was consistent with that obtained by the above Rossi-α analysis.

An experimental technique based on accelerator-beam trip and restart operation was proposed to determine the subcritical reactivity of ADS. A series of these experiments was performed in a subcritical thermal core of KUCA. The results demonstrated the applicability of the proposed technique to the thermal ADS of KUCA. We expect the proposed technique to be applied for an actual ADS in start-up or shut-down operation.

References

1. Nakajima K, Sano T, Hohara S et al (2021) Feynman-α and Rossi-α analyses for a subcritical reactor system driven by a pulsed spallation neutron source in Kyoto University Critical Assembly. J Nucl Sci Technol 58:117
2. Nakajima K, Sano T, Takahashi et al (2020) Source multiplication measurements and neutron correlation analyses for a highly enriched uranium subcritical core driven by an inherent source in Kyoto University Critical Assembly. J Nucl Sci Technol 57:1152
3. Nagaya Y, Okumura K, Sakurai T et al (2016) MVP/GMVP Version3: general purpose Monte Carlo codes for neutron and photon transport calculations based on continuous energy and multigroup methods. JAEA-Data/Code 2016-018
4. Nagaya Y, Okumura K, Mori T (2015) Recent developments of JAEA's Monte Carlo code MVP for reactor physics applications. Ann Nucl Energy 82:85
5. Shibata K, Iwamoto O, Nakagawa T et al (2011) JENDL-4.0: a new library for nuclear science and technology. J Nucl Sci Technol 48:1
6. Rana YS, Degweker SB (2009) Feynman-alpha and Rossi-alpha formulas with delayed neutrons for subcritical reactors driven by pulsed non-poisson sources. Nucl Sci Eng 162:117
7. Williams MMR (1974) Random processes in nuclear reactors. Pergamon Press, Oxford, UK, pp 26–49
8. Degweker SB, Rana YS (2007) Reactor noise in accelerator driven systems-II. Ann Nucl Energy 34:463
9. Okuda R, Sakon A, Hohara S et al (2016) An improved Feynman-α analysis with a moving-bunching technique. J Nucl Sci Technol 53:1647
10. Tonoike K, Miyoshi Y, Kikuchi T et al (2002) Kinetic parameter βeff/ℓ measurement on low enriched uranyl nitrate solution with single unit cores (600φ, 280T, 800φ) of STACY. J Nucl Sci Technol 39:1227
11. Taninaka H, Hashimoto K, Pyeon CH et al (2010) Determination of lambda-mode eigenvalue separation of a thermal accelerator-driven system from pulsed neutron experiment. J Nucl Sci Technol 47:376

12. Sakon A, Hashimoto K, Maarof MA et al (2014) Measurement of large negative reactivity of an accelerator-driven system in the Kyoto University Critical Assembly. J Nucl Sci Technol 51:116
13. Furuhashi A (1962) Characteristic spectra in neutron thermalization. J At Energy Soc Jpn 4:677
14. Takahashi H (1962) Space and time dependent eigenvalue problem in neutron thermalization. EURATOM, EUR-22.e
15. Pazsit I, Yamane Y (1998) Theory of neutron fluctuations in source-driven subcritical systems. Nucl Instrum Methds A 403:431
16. Pazsit I, Yamane Y (1998) The variance-to-mean ratio in subcritical systems driven by a spallation source. Ann Nucl Energy 25:667
17. Behringer K, Wydler P (1999) On the problem of monitoring the neutron parameters of the fast energy amplifier. Ann Nucl Energy 26:1131
18. Kuang ZF, Pazsit I (2000) A quantitative analysis of the Feynman- and Rossi-alpha formulas with multiple emission sources. Nucl Sci Eng 136:305
19. Diven BC, Martin HC, Taschek RF et al (1956) Multiplicities of fission neutrons. Phys Rev 101:1012
20. Letourneau A, Galin J, Goldenbaum F et al (2000) Neutron production in bombardments of thin and thick W, Hg, Pb targets by 0.4, 0.8, 1.2, 1.8 and 2.5 GeV protons. Nucl Instrum Methds Phys Res B 170:299
21. Nakajima K, Sakon A, Sano T et al (2021) Power spectral analysis for a subcritical reactor system driven by a pulsed spallation neutron source in Kyoto University Critical Assembly. J Nucl Sci Technol 58:374
22. Keepin GR, Wimett TF, Zeigler RK (1957) Delayed neutrons from fissionable isotopes of Uranium Plutonium and Thorium. Phys Rev 107:1044
23. Sakon A, Hashimoto K, Sugiyama W et al (2013) Power spectral analysis for a thermal subcritical reactor system driven by a pulsed 14 MeV neutron source. J Nucl Sci Technol 50:481
24. Nomura T (1965) Improvement in S/N ratio of reactor noise spectral density. J Nucl Sci Technol 2:76
25. Akcasu AZ, Osborn RK (1966) Application of Langevin's technique to space- and energy-dependent noise analysis. Nucl Sci Eng 26:13
26. Sheff JR, Albrecht RW (1966) The space dependence of reactor noise I—theory. Nucl Sci Eng 24:246
27. Hashimoto K, Nishina K, Tatematsu A et al (1991) Theoretical analysis of two-detector coherence functions in large fast reactor assemblies. J Nucl Sci Technol 28:1019
28. Taninaka H, Hashimoto K, Pyeon CH et al (2011) Determination of subcritical reactivity of a thermal accelerator-driven system from beam trip and restart experiment. J Nucl Sci Technol 48:873
29. Pyeon CH, Hervault M, Misawa T (2008) Static and kinetic experiments on accelerator-driven system with 14 MeV neutrons in Kyoto University Critical Assembly. J Nucl Sci Technol 45:1171
30. Hashimoto K, Ohya K, Yamane Y (1996) Experimental investigation of dead-time effect on Feynman- α method. Ann Nucl Energy 23:1099
31. Taninaka H, Hashimoto K, Ohsawa T (2010) An extended rod drop method applicable to subcritical reactor system driven by neutron source. J Nucl Sci Technol 47:351
32. Hoogenboom JE, van der Sluijs AR (1988) Neutron source strength determination for on-line reactivity measurements. Ann Nucl Energy 15:553
33. Tamura S (2003) Signal fluctuation and neutron source in inverse kinetics method for reactivity measurement in the sub-critical domain. J Nucl Sci Technol 40:153
34. Okumura K, Kaneko K, Tsuchihashi K (1996) SRAC95; general purpose neutronics code system. JAERI-Data/Code 96-015
35. Nakagawa T, Shibata K, Chiba S et al (1995) Japanese evaluated nuclear data library version 3 revision-2 JENDL-3.2. J Nucl Sci Technol 32:1259
36. Hogan WS (1960) Negative-reactivity measurements. Nucl Sci Eng 84:518

Chapter 3
Reactor Kinetics

Cheol Ho Pyeon

Abstract In static and kinetic experimental analyses, the reactivity effect of introducing a neutron guide has been examined with various materials and adjustments of the beam window. With the objective of improving the KUCA core characteristics, the implementation of the neutron guide is predicted to increase the fast neutrons in directing the fuel region. With regard to the kinetic characteristics, the subcriticality and the prompt neutron decay constant are monitored for several core configurations and detector positions. The KUCA core is equipped to make locally a hard spectrum core region with the combined use of ^{235}U fuel, a polyethylene moderator, and a Pb–Bi reflector for criticality. In this study, the first attempt is made to examine experimentally the characteristics of kinetics parameters in ADS comprised of ^{235}U-fueled and Pb–Bi-zoned core, and spallation neutrons generated by an injection of 100 MeV protons onto the solid Pb–Bi target. Online monitoring of reactivity has been deduced in real time by the inverse kinetic method on the basis of the one-point kinetic equation with measured neutron signals in the core. Here, measurements by the one-point kinetic equation are validated through the subcriticality evaluation with the PNS histogram and the methodology by the inhour equation.

Keywords α-fitting method · Pulsed-neutron source method · Inverse kinetic method · Kalman filter method

3.1 α-Fitting Method

3.1.1 Experimental Settings

3.1.1.1 Core Configurations

For safety reasons particular to KUCA, the tritium target is located not at the center of the core but at a peripheral position in the critical assembly. This location is very

C. H. Pyeon (✉)
Institute for Integrated Radiation and Nuclear Science, Kyoto University, Osaka, Japan
e-mail: pyeon.cheolho.4z@kyoto-u.ac.jp

© The Author(s) 2021
C. H. Pyeon (ed.), *Accelerator-Driven System at Kyoto University Critical Assembly*,
https://doi.org/10.1007/978-981-16-0344-0_3

different from the current design of ADS where a target is located at the center of the core for effective utilization of the generated neutrons. Consequently, the introduction of a neutron guide is requisite for effectively directing the high-energy neutrons generated from the tritium target to the center of the core for experiments on ADS with 14 MeV neutrons. The neutron guide, which is very similar to the neutron shield and the beam duct [1, 2], is composed of several shielding materials, including iron, boron, polyethylene, the beam duct, and a special fuel assembly with a void.

Five major core settings are presented here: a core without the neutron guide (Fig. 3.1a; Reference core: Case 1); a core including only streaming void (SV) in the fuel region (Fig. 3.1b; SV core: Case 2); and three cores with the neutron guide including SV (Fig. 3.1c–f; neutron-guided core with beam window: Cases 3, 4, 5, and 6, respectively). Numerals 12, 20, 22, 26, and 24 correspond to fuel plates in the partial assembly/ies used to reach criticality (Fig. 3.1a–f, respectively).

Each fuel assembly was set in a $2.1'' \times 2.1''$ and 1.5 mm thick aluminum (Al) sheath; the cross section of the elements within the assembly was $2'' \times 2''$. The fuel assemblies constituting the core are reproduced in Fig. 3.2. The standard fuel assembly shown (F; Fig. 3.2a) was composed of 36 unit cells of $1/16''$ thick and a 93 wt% enriched uranium plate with Al clad and two ($1/8''$ and $1/4''$) thick polyethylene plates. The active height of the core was about $16''$, with additional about $23''$ and $21''$ upper and lower polyethylene reflectors, respectively. Case 1 (Fig. 3.1a) was composed of 20 regular fuel assemblies and one partial fuel assembly of 12 fuel unit cells. In Cases 2–6 (Fig. 3.1b–f, respectively), the fuel region consisted of 18 regular fuel assemblies, SV assembly composed of one $5.08 \times 5.08 \times 5.08$ cm center void (Fig. 3.2b), 32 fuel unit cells, and two partial fuel assemblies. Details of the partial fuel assembly of (14, M and 16, M) used for Case 5 are presented in Fig. 3.2e. The active part was centered with the rest of the core using an Al cell identical to the fuel cell but with Al replacing the fuel plates.

The purpose of the neutron guide was to reproduce the conditions of a high-energy neutron beam entering the fuel region from an isotropic source. Thus, the role of the SV assemblies was to direct the highest possible number of the high-energy neutrons generated in the target to the center of the fuel region, in order to improve neutron multiplication. Moreover, to reproduce a high-energy source, it was necessary to reduce the thermal component of the external neutron source, i.e., the neutrons were moderated before they reached the fuel region. This was achieved by shielding unnecessary fast neutrons and by capturing parasite thermal neutrons. For deflecting unnecessary fast neutrons, the close vicinity in front of the target included an ion block around the guide void to shield the fast neutrons by inelastic scattering. For capturing parasite thermal neutrons, polyethylene blocks containing 10 wt% boron around the guide void were included around the Fe shielding near the target and in the two rows next to the assemblies. The rest of the neutron guide consisted of polyethylene assemblies and one void space. The detailed composition of the neutron guide was presented in Ref. [1].

14 MeV neutrons were produced with a yield of about 8×10^8 s^{-1} from the tritium target in pulsed mode. The duty ratio and the duration of irradiation were adjusted

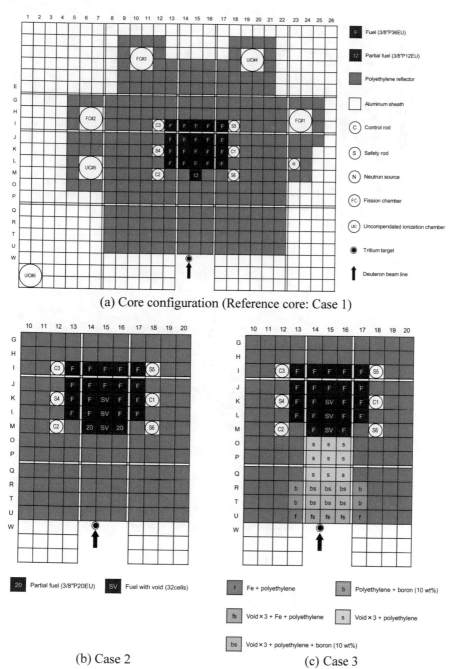

(a) Core configuration (Reference core: Case 1)

(b) Case 2

(c) Case 3

Fig. 3.1 ADS core configurations with 14 MeV neutrons shown in Table 3.1(Ref. [2])

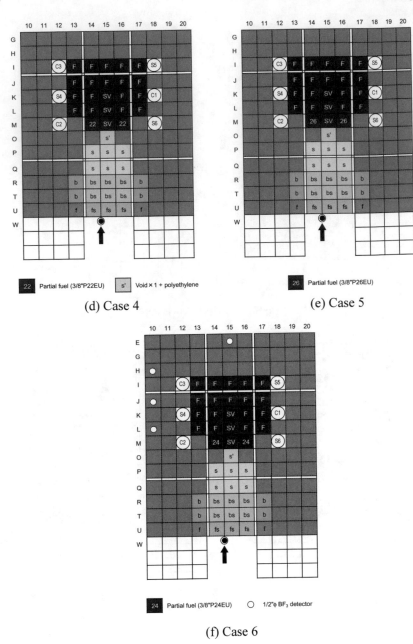

(d) Case 4   (e) Case 5

(f) Case 6

Fig. 3.1 (continued)

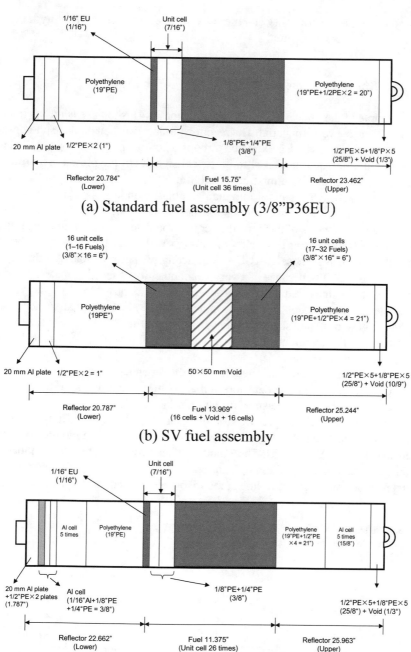

(a) Standard fuel assembly (3/8"P36EU)

(b) SV fuel assembly

(c) Partial fuel assembly (26)

Fig. 3.2 Fuel assemblies (Ref. [2])

depending on the subcriticality and the requisite measurements. The duty ratio was limited to 1% as a design limit.

3.1.1.2 Kinetics Parameters

The principle of the optical fiber neutron detector [3, 4] is to have neutrons interact with a neutron converter material. The reaction product can then produce photons in a scintillating material which is extracted through a plastic optical fiber, multiplied into a photo-multiplier, and converted to electrical signals. In the present experiments, detectors were formed by a mixture of lithium-6 (^{6}Li) enriched LiF and ZnS(Ag) scintillator pasted at the tip of a 1 mm diameter plastic optical fiber. ^{6}Li was selected for its large ^{6}Li$(n, t)^4$He cross sections for thermal neutrons.

The main advantages of the optical fiber neutron detector are not only its relative simplicity and low building cost, but also its very small size, which allows it to be used in small cores such as those in KUCA with negligible perturbation. The drawbacks include low sensibility because of a very small quantity of reacting material and the treatment of the signal required to remove as much as possible of the noise from γ-ray interferences without losing too much of the valuable signal. A schema of the detection settings and size references is shown in Fig. 3.3. Levels of the thermal neutron flux in the measurements were between 10^6 and 10^4 s^{-1} cm^{-2}, for the duration of irradiation in hours.

To provide information on detector position dependency of the measurements, each core was set with three detectors: Fiber #1 in the fuel region, Fiber #2 in the boundary region between the standard and partial fuel assemblies, and Fiber #3 in the neutron guide region (Fig. 3.4). Moreover, another fiber with a ThO$_2$ scintillator being fission reactive to the high-energy neutrons was used as a monitor of source intensity of 14 MeV neutrons. The quantity of neutron converter and scintillating material, in the order of milligram, cannot be made identical for all detectors and, thus, precludes making an absolute comparison between count rates.

Fig. 3.3 Schema of an optical fiber detection system (Ref. [2])

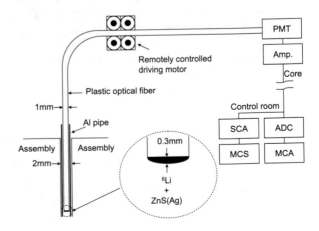

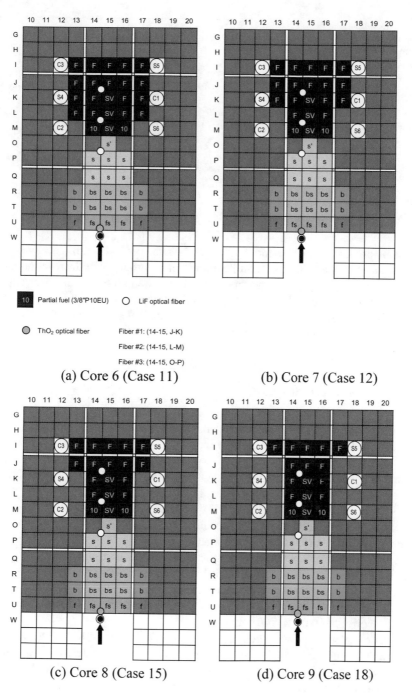

(a) Core 6 (Case 11)   (b) Core 7 (Case 12)

(c) Core 8 (Case 15)   (d) Core 9 (Case 18)

Fig. 3.4 Top view of the configuration of A-core kinetic experiments shown in Table 3.1 (Ref. [2])

Fig. 3.5 Comparison of measured and calculated prompt neutron decay constants (Ref. [2])

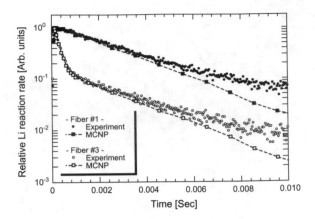

Using prompt neutron performance (Fig. 3.5), the prompt neutron decay constant was deduced by the least-square fitting of the time distribution of the reaction rates to an exponential function over the time optimal duration. Subcriticality was deduced from the prompt neutron decay constant by the extrapolated area ratio method [5]. For normalization and comparison between experimental integral results, the duration of irradiation and the duty ratio of the beam were used.

3.1.2 Numerical Simulations

The numerical calculations were executed with the Monte Carlo multi-particle transport code, MCNP-4C3 [6]. The effect on neutronic parameters resulting from the difference between nuclear data libraries was evaluated by considering JENDL-3.3 [7] and ENDF/B-VI.2 [8] for transport. Dosimetry files JENDL-3.1 [9] and ENDF/B-V were used for reaction rate calculation, regardless of the library used for transport. The source was represented by a 14 MeV neutron punctual isotropic source. Since the effects of their reactivity are not negligible, the irradiation samples and the In wire were included in the simulated geometry and transport calculation; reaction rates were deduced from tallies taken in a similar way as in the static experiments. Although better in the core region, an overall statistical error of 5% was retained in the reaction rate in the presented results. The results of eigenvalue calculations were obtained after 2,000 active cycles of 10,000 histories each. The deduced subcriticalities have statistical errors of 0.02 % $\Delta k/k$.

Kinetics calculations were conducted using MCNP-4C3 with JENDL-3.3. Since such calculations are known to have a tendency to overestimate effective multiplication factors, reactivity adjustment was taken into consideration before conducting source calculations: the density of the fuel was artificially reduced by 5%. This adjustment was estimated in the reflected core such that a critical state calculation gives a k_{eff} equivalent to 1 (in effect 0.99985 ± 0.00025).

For the kinetics calculations, time tallies were taken at the equivalent detector position, without the actual inclusion of the detector in the calculation geometry. The following is an alternative for the preceding highlighted part: Point detectors with a sphere of exclusion 1 mm in diameter, as well as a flux tally in a 5 cm long and 1 mm in diameter cylindrical volume, were centered at the detector position for estimating the validity of the use of the point detector tally and the optimization of calculation time.

3.1.3   Results and Discussion

3.1.3.1   Subcriticality

Positions of the control and safety rods in Cores 1–9 are shown in Table 3.1. The results of measured and calculated subcriticalities are presented in Table 3.2. A comparison of the experiments and the calculations demonstrates the ability of MCNP calculations to reproduce subcriticality levels within 10% C/E. The simultaneous changes in the irradiation samples, the fuel mass, and the core geometry limit precise conclusions on possible influences on the precision of calculations. The apparent better C/E values for ENDF/B-VI.2 are not conclusive when a 5% relative experimental error is taken into account. Namely, in Cases 1 and 3, there was found to be a discrepancy of about 10% error between two libraries, due to no irradiation sample in (15, K; Fig. 3.1c) and a large size window in front of fuel

Table 3.1   Control and safety rod positions in Cores 1–9 (dimension in mm) (Ref. [2])

Core	Case	C1	C2	C3	S4	S5	S6
1	1	0.00	0.00	0.00	1200.00	1200.00	1200.00
2	2	0.00	0.00	0.00	1200.00	1200.00	1200.00
3	3	0.00	0.00	0.00	1200.00	1200.00	1200.00
4	4	0.00	0.00	0.00	1200.00	1200.00	1200.00
5	5	0.00	0.00	0.00	1200.00	1200.00	1200.00
6	10	0.00	0.00	0.00	1200.00	1200.00	1200.00
	11	0.00	0.00	0.00	0.00	0.00	0.00
7	12	1200.00	1200.00	1200.00	1200.00	1200.00	1200.00
	13	0.00	0.00	0.00	1200.00	1200.00	1200.00
	14	0.00	0.00	0.00	0.00	0.00	0.00
8	15	1200.00	1200.00	1200.00	1200.00	1200.00	1200.00
	16	0.00	0.00	0.00	1200.00	1200.00	1200.00
	17	0.00	0.00	0.00	0.00	0.00	0.00
9	18	1200.00	1200.00	1200.00	1200.00	1200.00	1200.00

0.00: Lower limit (Full insertion), 1200.00: Upper limit (Full withdrawal)

Table 3.2 Comparison of measured and calculated subcriticalities (Ref. [2])

Case	Experiment ($\%\Delta k/k$)	JENDL-3.3 ($\%\Delta k/k$)	C/E	ENDF/B-VI.2 ($\%\Delta k/k$)	C/E
1	0.79 ± 0.06	0.75 ± 0.02	0.95	0.82 ± 0.02	1.04
2	0.68 ± 0.05	0.68 ± 0.02	1.01	0.67 ± 0.02	0.99
3	0.89 ± 0.06	0.98 ± 0.02	1.09	0.91 ± 0.02	1.02
4	0.70 ± 0.05	0.76 ± 0.02	1.09	0.73 ± 0.02	1.04
5	1.76 ± 0.12	1.71 ± 0.02	0.97	1.72 ± 0.02	0.97

Table 3.3 Comparison of measured subcriticalities obtained by the source multiplication method and calculated subcriticalities by MCNP (Ref. [2])

Case	Experiment ($\%\Delta k/k$)	JENDL-3.3 ($\%\Delta k/k$)	C/E	ENDF/B-VI.2 ($\%\Delta k/k$)	C/E
11	0.72 ± 0.05	0.69 ± 0.02	0.96	0.69 ± 0.02	0.97
12	2.72 ± 0.19	2.88 ± 0.02	1.06	2.88 ± 0.02	1.06
15	5.96 ± 0.42	6.44 ± 0.02	1.08	6.50 ± 0.02	1.09
18	8.66 ± 0.61	10.46 ± 0.02	1.21	10.54 ± 0.02	1.22

region, respectively. Nonetheless, the results agree with the validation of subcriticality measurement techniques by the rod drop and the positive period methods for such cores down to about 2 $\%\Delta k/k$.

Measurements by the source multiplication method (Table 3.3) show the results of subcriticalities in the experiments and the calculations for the neutron-guided core (Fig. 3.1f) with zero to three extracted SV assemblies. The measured subcriticalities were obtained by the source multiplication method down to about 9 $\%\Delta k/k$. Nonetheless, a reliable relative discrepancy within 10% C/E between experiments and calculations was confirmed only down to 6 $\%\Delta k/k$. Therefore, as a measurement methodology for the KUCA core with the neutron guide, this source multiplication method proved convenient and complementary to the rod drop and the calibration curve methods and reliable on a relatively large subcriticality range of up to 6 $\%\Delta k/k$.

3.1.3.2 Kinetic Parameters

Adjustment of the fuel density in the calculation was first considered for the neutron-guided core, leading to a correcting factor of 5%. Only for cases of very small reactivity did the relative difference appear within the order of calculation precision. For consistency, the following results are those obtained by applying the same correction of 5% to all calculations.

A representative selection of the calculation results of the subcriticalities is shown in Table 3.4 with the C/E values for each detector. Although better for small subcriticality, an overall 10% in relative error was taken into account for a comparison of the subcriticalities. For a comparison between the detectors, remarkably Fiber #1

Table 3.4 Comparison of measured subcriticalities obtained by the area ratio method and calculated subcriticalities by MCNP (JENDL-3.3) (Ref. [2])

Case	MCNP ($\%\Delta k/k$)	Fiber #1 ($\%\Delta k/k$)	C/E	Fiber #2 ($\%\Delta k/k$)	C/E	Fiber #3 ($\%\Delta k/k$)	C/E
10	0.97 ± 0.03	0.99 ± 0.01	0.98	0.96 ± 0.01	1.01	0.99 ± 0.01	1.03
11	1.83 ± 0.03	1.88 ± 0.02	0.97	2.15 ± 0.02	0.85	1.78 ± 0.02	1.03
12	2.55 ± 0.03	2.55 ± 0.03	1.00	3.12 ± 0.03	0.82	2.42 ± 0.02	1.05
13	3.45 ± 0.03	3.40 ± 0.03	1.02	3.25 ± 0.03	1.06	3.63 ± 0.04	0.95
14	4.15 ± 0.03	4.49 ± 0.04	0.93	4.00 ± 0.04	1.04	4.60 ± 0.05	0.90
15	6.24 ± 0.03	5.89 ± 0.06	1.06	6.54 ± 0.07	0.95	6.87 ± 0.07	0.91
16	6.76 ± 0.03	6.59 ± 0.07	1.03	10.01 ± 0.10	0.67	7.56 ± 0.08	0.89
17	7.41 ± 0.03	7.55 ± 0.08	0.98	8.18 ± 0.08	0.91	8.64 ± 0.09	0.86
18	10.38 ± 0.03	10.24 ± 0.10	1.01	12.28 ± 0.12	0.85	11.93 ± 0.12	0.87

appeared little affected by the increase in the subcriticality, with the discrepancy being within 7% even for the largest subcriticality, while Fibers #2 and #3 reached about 30%. This tendency can be clearly seen in Fig. 3.6.

The results allowed some estimation of the relative effects among the various parameters suspected to influence the measurement of the subcriticality by the neutron pulsed technique and the prompt neutron decay constant. Although the same value of the delayed neutron fraction was used for the area ratio method (Table 3.5), the results from Fiber #2 allowed the evaluation of the variation in the delayed neutron fraction to be in the smallest order of the uncertainty over the subcriticality. The measurements from Fiber #2 (Fig. 3.7), although underestimating the subcriticality (Fig. 3.6), tended to give a relatively good evaluation of the prompt neutron decay constant compared with those from Fiber #3. Fiber #1, however, gave a marked underestimation of the prompt neutron decay constant of less than 7%. Finally, Fiber

Fig. 3.6 Measured and calculated results of subcriticalities in KUCA kinetic experiments (Ref. [2])

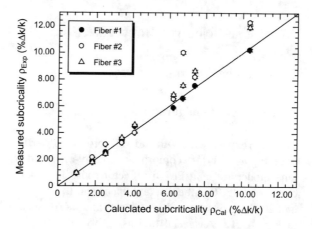

Table 3.5 Comparison of C/E values in measured and calculated neutron decay constants (Ref. [2])

Case	Fiber #1	Fiber #2	Fiber #3
10	1.02 ± 0.02	1.12 ± 0.03	1.07 ± 0.02
11	1.16 ± 0.02	1.11 ± 0.03	1.18 ± 0.03
12	1.05 ± 0.02	1.07 ± 0.03	1.09 ± 0.02
13	1.04 ± 0.03	1.07 ± 0.04	1.12 ± 0.04
14	1.07 ± 0.03	1.04 ± 0.04	1.14 ± 0.04
15	1.16 ± 0.03	1.18 ± 0.05	1.21 ± 0.04
16	1.11 ± 0.04	1.03 ± 0.05	1.06 ± 0.04
17	0.88 ± 0.03	0.96 ± 0.05	0.83 ± 0.03
18	1.13 ± 0.03	1.06 ± 0.03	1.12 ± 0.03

Fig. 3.7 Measured and calculated results of neutron decay constants in KUCA kinetic experiments (Ref. [2])

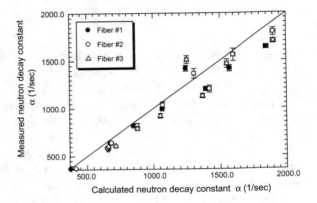

#2 showed a good evaluation of the prompt neutron decay constant, considering the detector position dependency of prompt neutron decay constant measurements.

3.2 Pulsed-Neutron Source Method

3.2.1 Experimental Settings

3.2.1.1 Core Configurations

The ADS experiments with 100 MeV protons (Pb–Bi target) were carried out in the A-core (Fig. A2.11) [10] comprising a highly-enriched uranium (HEU) fuel, a polyethylene moderator, and Pb–Bi reflector rods. The fuel assembly "F" (3/8″p36EU) is composed of 36 unit cells, and upper and lower polyethylene blocks are about 25″ and 20″ long, respectively, in an aluminum (Al) sheath 2.1″ × 2.1″ × 60″, as shown in Fig. A2.12. A special fuel assembly "f" (1/8″ p5EUEU <1/8″Pb-Bi30EUEU>

1/8″p15EUEU) shown in Fig. 2A.13 is composed of a total of 60 unit cells: 30 unit cells with HEU plate 1/8″ thick and Pb–Bi plate 3.426 mm thick, and 30 unit cells with HEU plate 1/8″ thick and a polyethylene plate 1/8″ thick. As shown in Fig. 2A.14, numeral 16″ (3/8″p16EU) corresponds to the number of fuel plates in the partial fuel assembly used for reaching critical mass. The neutron spectra were numerically obtained by the MCNP calculations, when 100 MeV protons are injected onto the Pb–Bi target, as shown in Fig. 3.8a, b.

Subcriticality was attained by full insertion of control and safety rods, and the substitution of fuel assemblies for polyethylene ones, as shown in Table 2A.13 and Fig. 2A.16: an insertion (Cases II-1, II-2, and II-3 ranged between 1160 and 2483 pcm in subcriticality) of control and safety rods, and the substitution (Cases II-4, II-5, and II-6 ranged between 4812 and 11556 pcm) of fuel assemblies for polyethylene moderators. In Cases II-1, II-2, and II-3, the subcriticality was deduced experimentally with the combination of the worth of control (C1, C2, and C3) and safety (S4, S5,

Fig. 3.8 Comparison between neutron spectra at Pb–Bi target, Fibers #1 and #2 (Ref. [10])

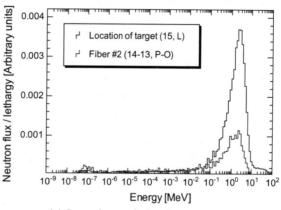

(a) Locations of Pb-Bi target and Fiber #2

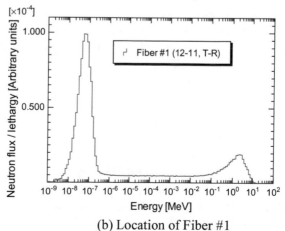

(b) Location of Fiber #1

and S6) rods by the rod drop method and its calibration curve by the positive period method. Furthermore, in Cases II-4, II-5, and II-6, the subcriticality was numerically obtained by the MCNP6.1 [11] code with the JENDL-4.0 [12] library, because the reactivities of control and safety rods were varied by the substitution of fuel assembly rods for polyethylene ones.

The Pb–Bi target was located inside the core at the location (15, L) shown in Fig. 2A.11 on the basis of the characteristics of the location of the target outside the core, as discussed in the previous study [13]. Note that the location of the original target is not easily moved to the center of the core, because control and safety rods are fixed in the core to function as the control driving system at KUCA. The Pb–Bi target was 50 mm in diameter and 18 mm thick. The main characteristics of proton beams were 100 MeV energy, 0.7 nA intensity, 40 mm beam spot, 20 Hz beam repetition, 100 ns beam width, and 1.0×10^7 s^{-1} neutron yield.

3.2.1.2   Measurements

During the injection of 100 MeV protons onto the Pb–Bi target located at (15, L) shown in Fig. A2.11, the time evolution of prompt and delayed neutron behavior was examined by the optical fiber detectors [14] set at three locations: Fiber #1 at (12-11, T-R) in Fig. A2.16 between the polyethylene moderator rods, Fiber #2 at (14-13, P-O) in Fig. A2.16 outside, and Fiber #3 at (15-14, O-M) inside the ^{235}U-fueled and Pb–Bi-zoned core. The optical fiber was shaped with a mixture of lithium-6-enriched LiF and ZnS (Ag) scintillator pasted at its 1 mm diameter tip.

From the results of neutron signals shown in Fig. 3.9, prompt neutron decay constant α was deduced from the exponential function fitting of the PNS measurements in the region of prompt neutron behavior as follows:

$$N = C_{\text{PNS}} \cdot \exp(-\alpha t) + B_{\text{PNS}}, \tag{3.1}$$

where N indicates the counting rate of the neutron signal, and C_{PNS} and B_{PNS} the constant values obtained by the least-squares fitting. Additionally, subcriticality $\rho_\$$ in dollar units was deduced by the PNS method, on the basis of the following theoretical background: in the area ratio method [15], subcriticality $\rho_\$$ in dollar units was determined by the ratio of two prompt and delayed components in the decay of neutron density as follows:

$$\rho_\$ = \frac{\rho}{\beta_{\text{eff}}} = -\frac{A_p}{A_d}, \tag{3.2}$$

where ρ indicates the subcriticality in pcm units, β_{eff} the effective delayed neutron fraction, A_p the area of the decay curve by prompt neutrons, and A_d the area of delayed neutrons. For reducing the spatial higher mode components of neutron flux, the extrapolated area ratio method [5] was introduced into the measurement of subcriticality as follows:

Fig. 3.9 Time evolution by the PNS method (Ref. [10])

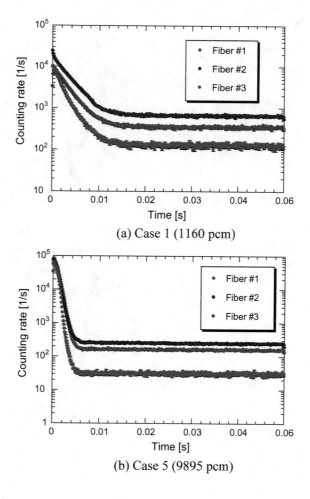

(a) Case 1 (1160 pcm)

(b) Case 5 (9895 pcm)

$$\rho_\$ = \frac{\rho}{\beta_{\text{eff}}} = -\exp(\alpha t_w)\frac{\int_{t_w}^{T} A_p(t)dt}{\int_0^{T} A_d(t)dt}, \tag{3.3}$$

where T indicates the measurement time, and t_w the waiting time for reducing the higher mode components of neutron flux.

Another approach to the α value was attempted by the Feynman-α method [16] with the use of neutron noise data shown in Fig. 3.10. The α value was deduced from the least-squares fitting for the Y value with gate width t_g, on the basis of the theoretical background by taking delayed neutron effects into account [17], as follows:

$$Y = \frac{C_{\text{Noise}}}{\alpha^2}\left(1 - \frac{1 - \exp(-\alpha t_g)}{\alpha t_g}\right)$$

Fig. 3.10 Neutron noise data by the Feynman-α method (Ref. [10])

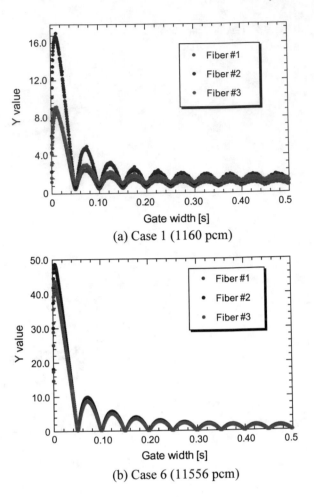

(a) Case 1 (1160 pcm)

(b) Case 6 (11556 pcm)

$$+\frac{B_{\text{Noise}}}{t_g}\sum_{n=1}^{\infty}\left(\frac{1}{n^4\alpha^2T_0^2+4n^6\pi^2}\sin^2\left(\frac{n\pi}{T_0}W\right)\sin^2\left(\frac{n\pi}{T_0}t_g\right)\right), \tag{3.4}$$

where C_{Noise} and B_{Noise} indicate the constant values obtained from the noise data, and W and T_0 the pulsed width as a fitting parameter and the pulsed frequency obtained from proton beam characteristics (20 Hz), respectively. In Eq. (3.4), the maximum value of n of the second term was set as 1000 leaving a margin from the saturation of the fitting results by setting about 300 in the maximum value. Moreover, the α value of the second term in Eq. (3.4) was acquired as the fitting parameter from experimental noise data by the Feynman-α method.

Table 3.6 Comparison between measured and calculated (MCNP6.1 with JENDL-4.0) reactivities [pcm] of excess and control rods (C1, C2, and C3) (Ref. [10])

Reactivity	Calculation	Experiment	C/E[a]
Excess	149 ± 11	149 ± 3	1.00 ± 0.08
C1	567 ± 11	549 ± 3	1.03 ± 0.02
C2	189 ± 11	194 ± 1	0.98 ± 0.06
C3	498 ± 11	483 ± 2	1.03 ± 0.02

[a]Calculation/Experiment

3.2.1.3 Numerical Simulations

Numerical calculations were performed by the Monte Carlo transport code, MCNP6.1 together with JENDL-4.0 for transport and with JENDL/HE-2007 [18] for high-energy protons and spallation neutrons. Here, in MCNP6.1, since the effects of reactivity by neutron detectors (optical fiber, FC, and UIC detectors) and control (safety) rods are not negligible, neutron detectors and control (safety) rods were included in the simulated geometry and transport calculations. The precision of numerical reactivities of excess and control rods (C1, C2, and C3) in pcm units was attained by the eigenvalue calculations within a relative difference of 3% between the experiment and the calculation, as shown in Table 3.6, with a total number of 1×10^8 histories and a statistical error of less than 5 pcm. Note that the relative difference of 3% in C/E values was attributable to numerical reactivity by MCNP with 20 pcm at most in the KUCA core.

3.2.2 Results and Discussion

3.2.2.1 Kinetics Parameters

The prompt neutron decay constant was attained by the PNS and the Feynman-α methods shown in Eqs. (3.1) and (3.2), respectively, with the use of neutron signals obtained from the three optical fibers at the locations in the core shown in Fig. 2A.11. As shown in Table 3.7, well-known findings were observed with a small difference between the results of Fibers #1 and #2, from the viewpoint of three issues: the neutron spectrum (Figs. 4.5a, b) and the subcriticality measurement methods (PNS and Feynman-α methods), on the subcriticality ranging between 1160 and 2483 pcm (Cases II-1 to II-3). A small difference between Fibers #1 and #2 was found in the measurements, because one-point reactor approximation is assumed to be valid in the shallow level of subcriticality. Conversely, on the subcriticality ranging between 4812 and 11556 pcm (Cases II-4 to II-6), a notable difference was observed in the experimental results between two methods, although the detector position dependency and neutron spectrum were small on the prompt neutron decay constant,

Table 3.7 Measured prompt neutron decay constants α [s^{-1}] in Cases II-1 to II-6 (Ref. [10])

	PNS method			Feynman-α method		
Case	Fiber #1	Fiber #2	Fiber #3	Fiber #1	Fiber #2	Fiber #3
II-1	398 ± 3	327 ± 3	495 ± 10	401 ± 9	367 ± 6	554 ± 17
II-2	507 ± 3	453 ± 3	685 ± 11	498 ± 5	464 ± 4	700 ± 7
II-3	673 ± 4	632 ± 2	1020 ± 8	655 ± 2	620 ± 2	877 ± 5
II-4	983 ± 4	971 ± 3	1378 ± 18	815 ± 6	822 ± 5	1029 ± 12
II-5	1665 ± 9	1681 ± 5	1828 ± 23	1365 ± 9	1400 ± 7	1669 ± 17
II-6	1911 ± 7	1931 ± 3	2061 ± 38	1556 ± 13	1636 ± 10	1917 ± 22

as compared with Fibers #1 and #2. For Fiber #3, a strong influence of spallation neutrons at the location of the Pb–Bi target was observed in the results of the prompt neutron decay constant by both PNS and Feynman-α methods, as compared with those of Fibers #1 and #2.

Subcriticality in dollar units was deduced experimentally by the extrapolated area ratio method with the use of prompt and delayed neutron components and by the α-fitting method [19] with the α value (the Feynman-α method). Nonetheless, attention was paid to the conversion coefficient β_{eff} of subcriticality in dollar units to one in pcm units, where β_{eff} was obtained by MCNP6.1 with JENDL-4.0 in each subcritical state of the core shown in Fig. 2A.16.

As shown in Table 3.8, the measured subcriticality by the PNS and the Feynman-α methods showed good agreement with the calculated one, with an error around 10% in the C/E (calculation/experiment) value, in the subcriticality ranging between 1160 and 2483 pcm, at the locations of Fiber #2. Additionally, in a comparison between Fibers #1 and #2, the detector position dependency by both the PNS and the Feynman-α methods was found to be attributable to placing the fiber detectors at different locations. At the locations of Fibers #1 and #2, with subcriticality ranging between 4812 and 11556 pcm, the deeper the subcriticality, the less accurate were the experimental results of measurements by both the PNS and the Feynman-α methods. The reason for this tendency was considered to be the effect of spallation neutrons on the neutron flux becoming greater with the deep subcriticality. As a consequence, the discrepancy between the experiments and the calculations is identified as the limitation of the applicability of the measurement method to a deep subcriticality level of over or about 10000 pcm. For Fiber #3, a large influence on spallation neutrons, as well as on the measurements of prompt neutron decay constant, was observed with the two measurement methods, although the Feynman-α method showed very good agreement in Cases II-5 and II-6. Neutron noise data are assumed to be dominant over the Poisson distribution, demonstrating that the effect of spallation neutrons becomes stronger with the deep subcriticality. In the case of the deep subcriticality, kinetic parameters of ρ and Λ shown in Tables 3.8 and 3.9, respectively, were numerically obtained by the MCNP eigenvalue calculations. Subsequently, values of kinetic parameters should be exactly acquired by the time-dependent MCNP calculations

Table 3.8 Comparison between subcriticality ρ [pcm] of reference, PNS, and Feynman-α methods (Ref. [10])

		PNS method		
Case	ρ [pcm]	Fiber #1	Fiber #2	Fiber #3
II-1	1160 ± 5	1224 ± 10 (0.95 ± 0.01)	1153 ± 8 (1.01 ± 0.01)	2108 ± 59 (0.55 ± 0.02)
II-2	1684 ± 6	1880 ± 15 (0.90 ± 0.01)	1604 ± 11 (1.05 ± 0.01)	4490 ± 109 (0.38 ± 0.01)
II-3	2483 ± 6	2824 ± 22 (0.88 ± 0.01)	2250 ± 15 (1.10 ± 0.01)	10347 ± 116 (0.24 ± 0.01)
II-4	4812 ± 6	5656 ± 62 (0.85 ± 0.01)	4177 ± 31 (1.15 ± 0.01)	22939 ± 1279 (0.21 ± 0.01)
II-5	9895 ± 6	12819 ± 137 (0.77 ± 0.01)	8187 ± 54 (1.21 ± 0.01)	14985 ± 1136 (0.66 ± 0.05)
II-6	11556 ± 6	16312 ± 160 (0.71 ± 0.01)	9738 ± 64 (1.19 ± 0.01)	19109 ± 2563 (0.60 ± 0.08)
		Feynman-α method		
Case	ρ [pcm]	Fiber #1	Fiber #2	Fiber #3
II-1	1160 ± 5	1491 ± 57 (0.87 ± 0.03)	1283 ± 36 (0.90 ± 0.03)	2445 ± 105 (0.47 ± 0.02)
II-2	1684 ± 6	2024 ± 28 (0.83 ± 0.01)	1814 ± 24 (0.93 ± 0.01)	3250 ± 41 (0.52 ± 0.01)
II-3	2483 ± 6	2881 ± 15 (0.86 ± 0.01)	2677 ± 11 (0.93 ± 0.01)	4203 ± 29 (0.59 ± 0.01)
II-4	4812 ± 6	4336 ± 38 (1.11 ± 0.01)	4384 ± 35 (1.10 ± 0.01)	5740 ± 77 (0.84 ± 0.01)
II-5	9895 ± 6	8087 ± 61 (1.22 ± 0.01)	8316 ± 50 (1.19 ± 0.01)	10110 ± 113 (0.98 ± 0.01)
II-6	11556 ± 6	9633 ± 88 (1.20 ± 0.01)	10175 ± 68 (1.14 ± 0.01)	12096 ± 150 (0.96 ± 0.01)

Bracket means the C/E (calculation/experiment) value

Table 3.9 Comparison between $\rho_\$$ [\$], β_{eff} [pcm], and Λ [$\times 10^{-5}$ s] by the PNS method and MCNP calculations (Ref. [10])

	$\rho_\$$ [\$]	β_{eff} [pcm]	Λ [$\times 10^{-5}$ s]	
Case	Fiber #2 (PNS)	MCNP6.1	MCNP6.1	Fiber #2 (PNS)
II-1	1.47 ± 0.01	785 ± 4	4.88 ± 0.01	5.93 ± 0.08
II-2	2.04 ± 0.01	785 ± 4	4.77 ± 0.01	5.28 ± 0.09
II-3	2.87 ± 0.01	783 ± 5	4.64 ± 0.01	4.80 ± 0.05
II-4	5.30 ± 0.02	788 ± 5	5.16 ± 0.01	5.11 ± 0.06
II-5	10.03 ± 0.03	816 ± 5	5.43 ± 0.01	5.53 ± 0.05
II-6	12.08 ± 0.03	806 ± 5	5.50 ± 0.01	5.46 ± 0.05

in the previous study [20], to get a better understanding of the time behavior of the kinetic parameters.

Furthermore, most of the C/E values demonstrated a notably greater tendency of the calculation results to the deep subcriticality attributable to the effect of the spallation neutrons on neutron flux, whereas the value of Fiber #1 by the PNS method showed a decreasing tendency.

3.2.2.2 Evaluation of β_{eff}/Λ

For the α-fitting method, prompt neutron decay constant α is easily deduced by combining subcriticality ρ in pcm units, effective delayed neutron fraction β_{eff}, and neutron generation time Λ as follows:

$$\alpha = \frac{\beta_{eff} - \rho}{\Lambda}. \tag{3.5}$$

Using the relationship between ρ and $\rho_\$$ in pcm and dollar units shown in Eq. (3.2), respectively, $\rho_\$$ can be expressed as follows:

$$\rho_\$ = -\frac{\Lambda}{\beta_{eff}}\left(\alpha - \frac{\beta_{eff}}{\Lambda}\right). \tag{3.6}$$

Assuming that the experimental results of α and $\rho_\$$ validate the accuracy of kinetics parameters, the value of Λ was deduced by Eq. (3.6), as shown in Table 3.9, with the combined use of the measured α and $\rho_\$$, and the calculated β_{eff}. Through the experimental analyses of the ADS with spallation neutrons in KUCA, a previous study [21] has clearly demonstrated that the value of β_{eff} has a small effect on the evaluation of subcriticality in pcm units converted from that in dollar units, as compared with the results of numerical subcriticality in pcm units. Here, considering the characteristics of β_{eff}, attention was directed to another kinetic parameter, Λ, obtained with the combined use of α and $\rho_\$$ by varying the subcriticality. The deduction of Λ in Fiber #2 was based on Eq. (3.6), by using the value of β_{eff} obtained from the MCNP6.1 calculations, since the experimental results of Fiber #2 by the PNS method showed a relatively good agreement with the numerical simulations, as described in Sect. 3.2.2.1. As shown in Table 3.9, a comparison between the experimental and the numerical values of Λ demonstrated the same tendency as the acceptable relative difference between MCNP6.1 and Fiber #2 by varying the subcriticality. Furthermore, using the experimental results of α and $\rho_\$$ of Fibers #1 and #2 by the PNS method, the fitting lines (Fiber #1: $\rho_\$ = -12.27E\text{-}03\alpha + 4.10$; Fiber #2: $\rho_\$ = -6.82E\text{-}03\alpha + 1.27$) by Eq. (3.5) were obtained with the values of the gradient (Fiber #1: $-12.27E\text{-}03$; Fiber #2: $-6.82E\text{-}03$) of Λ/β_{eff} as shown in Fig. 3.11, as compared with the results of MCNP6.1, Fibers #1 and #2 shown in Table 3.10. The experimental fitting of Fiber #2 demonstrated good agreement with the MCNP6.1 calculations by varying the subcriticality, although the relative difference between the experiments

Fig. 3.11 Linearity between prompt neutron decay constant α [s^{-1}] and subcriticality $\rho_\$$ [\$] (Ref. [10])

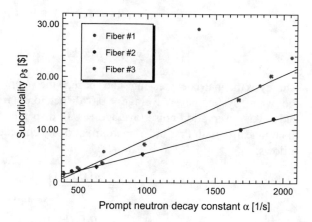

Table 3.10 Comparison between Λ/β_{eff} [$\times 10^{-3}$] values of MCNP6.1, Fibers #1 and #2 (Fitting in Fig. 3.11) by the PNS method (Ref. [10])

Case	MCNP6.1	Fiber #1	Fiber #2
II-1	6.22 ± 0.01	12.27 ± 0.08	6.82 ± 0.05
II-2	6.08 ± 0.01		
II-3	5.93 ± 0.01		
II-4	6.55 ± 0.01		
II-5	6.65 ± 0.01		
II-6	6.82 ± 0.01		

and the calculations was 15% at most. From the results in Fig. 3.11, and Tables 3.9 and 3.10, the kinetic parameters were easily deduced by the combination of experiments and calculations, and verified by the PNS and the α-fitting methods.

From the results in Tables 3.6, 3.7, 3.8, 3.9 and 3.10, these experimental benchmarks are expected to play an important role in the study of weight functions related to the physical interpretation and correction factors of the experimental results, from the theoretical and the numerical aspects, respectively, as well as of detector position dependency, neutron spectrum, and subcriticality measurement methods on kinetic parameters.

3.3   Inverse Kinetic Method

3.3.1   Theoretical Background

In the extended Kalman filter (EKF), the state space x (k) in time step k (1, 2, …) and the observation equation y (k) are expressed as follows:

$$\mathbf{x}(k+1) = \mathbf{f}(\mathbf{x}(k)) + \mathbf{b}v(k), \tag{3.7}$$

$$y(k) = h(\mathbf{x}(k)) + w(k), \tag{3.8}$$

where f is the matrix expressing state space, $\mathbf{b}$ the vector distributing system noise, v the (white) system noise (average $= 0.0$, dispersion $= \sigma_v^2$), h the observable matrix, and w the observation noise (average $= 0.0$, dispersion $= \sigma_w^2$). EKF is applicable to nonlinearity with first-order approximation by considering derivative A and c^T as follows:

$$\mathbf{A}(k) = \left. \frac{\partial \mathbf{f}(\mathbf{x}(k))}{\partial \mathbf{x}} \right|_{\mathbf{x} = \hat{\mathbf{x}}^-(k)} \tag{3.9}$$

$$\mathbf{c}^T(k) = \left. \frac{\partial h(\mathbf{x}(k))}{\partial \mathbf{x}} \right|_{\mathbf{x} = \hat{\mathbf{x}}^-(k)}, \tag{3.10}$$

where $\hat{\mathbf{x}}^-$ is the priori estimate (prediction estimation of x in time step k based on collected experience until time $k - 1$). The procedure of the general Kalman filter technique is divided into a prediction step and a filtering step. For the prediction step, the priori estimate is evaluated with the use of state estimate $\hat{\mathbf{x}}$ in a previous time step as follows:

$$\hat{\mathbf{x}}^-(k) = \mathbf{f}(\hat{\mathbf{x}}(k-1)). \tag{3.11}$$

Next, the priori error covariance matrix P^- is evaluated as follows:

$$\mathbf{P}^-(k) = \mathbf{A}(k)\mathbf{P}(k-1)\mathbf{A}^T(k) + \sigma_v^2 \mathbf{b}\mathbf{b}^T, \tag{3.12}$$

where P is a posteriori error covariance matrix. Here, in the first step, initial values of $\hat{\mathbf{x}}^-$ and P^- are requisite to perform EKF. For the filtering step, the Kalman gain $\mathbf{g}$ is determined as follows:

$$\mathbf{g}^-(k) = \frac{\mathbf{P}^-(k)\mathbf{C}(k)}{\mathbf{C}^T(k)\mathbf{P}^-(k)\mathbf{C}(k) + \sigma_w^2}. \tag{3.13}$$

The state estimate is evaluated with the use of observation results and the priori state estimate by the most likelihood parameter $\mathbf{g}$ as follows:

$$\hat{\mathbf{x}}(k) = \hat{\mathbf{x}}^-(k) + \mathbf{g}(k)\left(y(k) - h(\hat{\mathbf{x}}^-(k))\right). \tag{3.14}$$

Finally, in the next step, the posteriori error covariance matrix is prepared as follows:

$$\mathbf{P}(k) = \left(\mathbf{I} - \mathbf{g}(k) - \mathbf{C}^T(k)\right)\mathbf{P}^-(k). \tag{3.15}$$

In this study, initial values of Eqs. (3.11) and (3.12) were prepared by the inverse kinetic method based on one-point kinetic equations as follows:

$$\frac{dn(t)}{dt} = \frac{\rho(t) - \beta_{\text{eff},i}}{\Lambda} n(t) + \sum_{i=1}^{6} \lambda_i C_i(t),$$ (3.16)

$$\frac{dC_i(t)}{dt} = \frac{\beta_{\text{eff}}}{\Lambda} n(t) + \lambda_i C_i(t),$$ (3.17)

where n is the neutron density, ρ the reactivity, $\beta_{\text{eff},i}$ the effective delayed neutron fraction of the i-th group, λ_i the delayed neutron decay constant of the i-th group, and C_i the density of delayed neutron precursor. For obtaining reactivity in Eq. (3.16) at every time step, the time variation of C_i in Eq. (3.17) is expressed with a backward difference as follows:

$$\frac{C_i(k) - C_i(k-1)}{\Delta k} = \frac{\beta_{\text{eff},i}}{\Lambda} n(k) + \lambda_i C_i(k) \quad (k = 2, 3, \cdots),$$ (3.18)

ρ is obtained by substituting $C_i(k)$ in Eq. (3.18) for Eq. (3.16) as follows:

$$\rho(t) = \frac{dn(t)}{dt}\bigg|_{t=1} \frac{\Lambda}{n(k)} + \beta_{\text{eff}}$$
$$- \frac{\Lambda}{n(k)} \sum_{i=1}^{6} \frac{\lambda_i}{1 + \lambda_i \Delta k} \left(\frac{\beta_{\text{eff},i}}{\Lambda} n(k) \Delta k C_i(k-1) \right).$$ (3.19)

Here, when an experiment is assumed to have started at a critical state, the initial value (at $k = 1$) of C_i is estimated as follows:

$$C_i(1) = \frac{\beta_{\text{eff},i}}{\lambda_i \Lambda} n(1).$$ (3.20)

When monitoring the subcriticality by the EKF technique in the critical core, the result of Eq. (3.20) was used as the initial values.

In ADS experiments with a stable external neutron source, the one-point kinetic equation on neutron derivative is changed as follows:

$$\frac{dn(t)}{dt} = \frac{\rho(t) - \beta_{\text{eff}}}{\Lambda} n(t) + \sum_{i=1}^{6} \lambda_i C_i(t) + S_{\text{eff}},$$ (3.21)

where S_{eff} is the effective strength of the stable neutron source. By substituting C_i in Eq. (3.18) for Eq. (3.21), as the same procedure in Eq. (3.19), ρ is expressed in ADS experiments as follows:

$$\rho(t) = \frac{dn(t)}{dt}\bigg|_{t=1} \frac{\Lambda}{n(k)} + \beta_{eff} - \frac{\Lambda}{n(k)} \sum_{i=1}^{6} \frac{\lambda_i}{1 + \lambda_i \Delta k} \left(\frac{\beta_{eff,i}}{\Lambda} n(k) \Delta k C_i(k-1) \right)$$
$$- \frac{\Lambda}{n(k)} S_{eff}. \tag{3.22}$$

Here, assuming that the external neutron source is stable, S_{eff} was determined as follows:

$$S_{eff} = -\frac{\rho(1)}{\Lambda} n(1). \tag{3.23}$$

When monitoring subcriticality by the EKF technique with the external neutron source, $\rho(1)$, C_i in Eq. (3.20) and S_{eff} in Eq. (3.23) were used as the initial values.

3.3.2 Experimental Settings

3.3.2.1 Critical Core

Transient experiments [22] were carried out in the uranium-polyethylene (EE1) core at KUCA shown in Fig. A4.1. The core was constituted by fuel rods (1/8″p60EUEU), made of an HEU (2″ × 2″ × 1/16″) and a polyethylene moderator (p; 2″ × 2″ × 1/8″) in an aluminum sheath 2.1″ × 2.1″ × 60″, as shown in Fig. A4.2a. The core spectrum was hard in the polyethylene-moderated core at KUCA (an H/U: hydrogen/uranium ratio of approximately 50 in the thermal reactor).

Time evolution of neutron signals was obtained by an optical fiber detector (Eu:LiCaAlF6 scintillator) [23] set at the core center for monitoring reactivity based on one-point kinetic approximation (to prevent measuring higher mode components in neutron flux and variation of detector efficiency).

The transient experiments were conducted after attaining a critical state with the C2 control rod (with all the other control and safety rods withdrawn); C1 control rod (Fig. 4.1) was then dropped from the fully withdrawal position to the fully inserted position. The neutron counts were obtained every 1 s.

3.3.2.2 Subcritical Core

The ADS transient experiment was carried out with the subcritical core at $k_{eff} = 0.97$ (target range of the subcriticality monitoring in ADS) shown in Fig. 3.12. The 14 MeV neutrons were generated by the injection of deuteron beam (intensity 0.15 mA, pulsed width 90 μs, and pulsed frequency 100 Hz) onto a tritium target located at (14–15, Y; Fig. A4.1).

Fig. 3.12 Description of the subcritical core with 14 MeV neutrons at KUCA (Ref. [22])

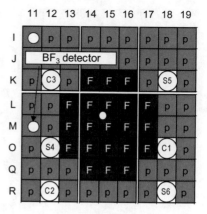

In the preparation of the transient experiment, all control and safety rods were withdrawn, and 14 MeV neutrons were then injected into the subcritical core. After 500 s, the C1 control rod was (slowly) inserted by an actuator-driven mechanism from the fully withdrawal position to the fully inserted position. The BF$_3$ detector used in this experiment was placed at (11, M; Fig. 3.12). Time evolution of the neutron count was obtained every 1 s.

3.3.3 Transient Analyses

3.3.3.1 Calibration in Critical Core

To perform the EKF technique [24, 25] in a time-variable system, setting the variance values of system noise and observation noise is requisite to be set. Furthermore, an initial priori error covariance matrix is needed for the calibration of the filter. The validity of the initial conditions in EKF parameters was evaluated by comparing the results by the rod drop method with those of the EKF technique, and those of the inverse kinetic method in the transient experiment (C1 rod drop) with the critical core.

Numerical analyses of kinetics parameters were conducted with the use of MCNP6.1 together with ENDF/B-VII.1 [26] (total histories were 5E + 08 (5E + 05 history per cycle and 1E + 03 active cycle)). The variance value of system noise was set only for the reactivity of 8E-08 and the observation noise was set as for the neutron count obtained at the time step for the variance value. Furthermore, the error covariance matrix was set zero except for the reactivity (1E-07) as the EKF parameters.

Here, the variable on the state-space model was set as follows:

$$\mathbf{x}(k) = {}^t\big(n(k)\ C_1(k)\ C_2(k)\ C_3(k)\ C_4(k)\ C_5(k)\ C_6(k)\ \rho(k)\big), \qquad (3.24)$$

where t indicates transposition of the matrix. Furthermore, function $f(x(k))$ in Eq. (3.7) is described as follows:

$$\mathbf{f}(\mathbf{x}(k)) = \begin{pmatrix} n(k) + T\left(\frac{\rho(k) - \beta_{eff}}{\Lambda} n(k) + \sum_{i=1}^{6} \lambda_i C_i(k)\right) \\ C_1(k) + T\left(\frac{\beta_{eff,1}}{\Lambda} n(k) - \lambda_1 C_1(k)\right) \\ C_2(k) + T\left(\frac{\beta_{eff,2}}{\Lambda} n(k) - \lambda_2 C_2(k)\right) \\ C_3(k) + T\left(\frac{\beta_{eff,3}}{\Lambda} n(k) - \lambda_3 C_3(k)\right) \\ C_4(k) + T\left(\frac{\beta_{eff,4}}{\Lambda} n(k) - \lambda_4 C_4(k)\right) \\ C_5(k) + T\left(\frac{\beta_{eff,5}}{\Lambda} n(k) - \lambda_5 C_5(k)\right) \\ C_6(k) + T\left(\frac{\beta_{eff,6}}{\Lambda} n(k) - \lambda_6 C_6(k)\right) \\ \rho(k) \end{pmatrix}, \qquad (3.25)$$

where T is the time resolution according to the forward difference: 1.0 in this study. Finally, function $A(k)$ is expressed as follows:

$$\mathbf{A}(k) = \begin{pmatrix} 1 + T\left(\frac{\rho(k) - \beta_{eff}}{\Lambda}\right) & T\lambda_1 & T\lambda_2 & T\lambda_3 & T\lambda_4 & T\lambda_5 & T\lambda_6 & \frac{T h(\mathbf{x}(1))}{\Lambda} \\ T\frac{\beta_{eff,1}}{\Lambda} & 1 - T\lambda_1 & 0 & 0 & 0 & 0 & 0 & 0 \\ T\frac{\beta_{eff,2}}{\Lambda} & 0 & 1 - T\lambda_2 & 0 & 0 & 0 & 0 & 0 \\ T\frac{\beta_{eff,3}}{\Lambda} & 0 & 0 & 1 - T\lambda_3 & 0 & 0 & 0 & 0 \\ T\frac{\beta_{eff,4}}{\Lambda} & 0 & 0 & 0 & 1 - T\lambda_4 & 0 & 0 & 0 \\ T\frac{\beta_{eff,5}}{\Lambda} & 0 & 0 & 0 & 0 & 1 - T\lambda_5 & 0 & 0 \\ T\frac{\beta_{eff,6}}{\Lambda} & 0 & 0 & 0 & 0 & 0 & 1 - T\lambda_6 & 0 \\ 0 & 0 & 0 & 0 & 0 & 0 & 0 & 1 \end{pmatrix}. \qquad (3.26)$$

Note that element $(1, 8)$ in Eq. (3.26) was approximately $\frac{T h(x(1))}{\Lambda}$ ($h(x(1))$ indicates the initial neutron count in the experiment) instead of $\frac{T n(k)}{\Lambda}$. This value was introduced to take into account the strong nonlinearity in the model. In the estimate with the use of $\frac{T n(k)}{\Lambda}$, the results were unreliable and divergent.

The transient experiment was started at the critical state; C1 control rod was then dropped into the core, inducing a rapid decrease in the neutron counts shown in Fig. 3.13. Importantly, the EKF technique reproduced measured count distribution (Fig. 3.13). The results of subcriticality monitoring revealed fluctuation in the result by the inverse kinetic method shown in Fig. 3.14, demonstrating that slight variation in the neutron count in the region of the low count rate greatly affected the estimate of subcriticality. Conversely, the result of the EKF technique notably decreased the fluctuation notably even after the count rate reached almost zero and the estimated value asymptotically approached the reference value, although the overshoot was found when the variation in subcriticality stopped rapidly.

Fig. 3.13 Neutron count rate distribution during C1 drop transient (Ref. [22])

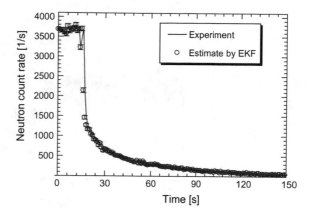

Fig. 3.14 Comparison of subcriticality monitoring between the inverse method and the EKF technique in C1 rod (worth: −888 pcm) drop experiment (Ref. [22])

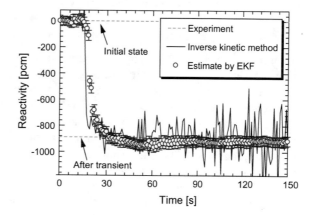

By the transient experiment in the critical core, the validity of the transient analysis was confirmed with the superiority of the filtering technique in reducing the fluctuation of monitoring values.

3.3.3.2 Performance Evaluation

The objective transient experiment was conducted at the subcritical state in the presence of an external neutron source for evaluating the performance of subcriticality monitoring by the EKF technique, at the same initial condition described in Sect. 3.3.3.1. So as to consider the external neutron source in EKF, $T \times S_{\text{eff}}$ was appended to element (1) in Eq. (3.25)

The neutron count distribution was varied gradually compared to the rod drop experiment because of the insertion by actuator-drive, and kept at a constant value at the end of transient behavior in view of the presence of the external neutron source, as shown in Fig. 3.15. The estimate by the EKF technique of the neutron counts showed

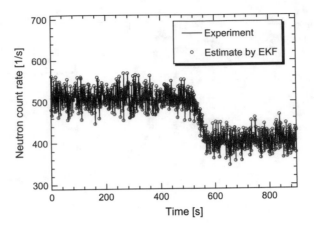

Fig. 3.15 Neutron count rate distribution during C1 drop transient with external neutron source
(Ref. [22])

almost the same distribution as of the experiment. In subcriticality monitoring by the
inverse kinetic method shown in Fig. 3.16, the fluctuation in the monitoring values
increased remarkably after the transient as the neutron count decreased. In EKF, the
monitoring values importantly followed the subcriticality in the transient behavior
without fluctuation. Moreover, the estimated subcriticality by the EKF technique
was found to be significantly more accurate than that by the inverse kinetic method,
demonstrating the validity of the estimate by EKF and the reliability of subcriticality
monitoring.

In the transient experiment with the external neutron source, the S_{eff} values should
be considered variable in addition to the detection efficiency since these values can
be varied by the insertion of control rods. Thus, to improve monitoring accuracy, in

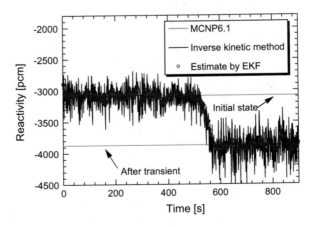

Fig. 3.16 Comparison of subcriticality monitoring between the inverse method and the EKF
technique in ADS transient experiment (Ref. [22])

future studies, the modeling of the space-state equation has to be modified on the basis of S_{eff} and detection efficiency.

3.4   Conclusion

Kinetic experiments on ADS with 14 MeV neutrons were conducted at KUCA. The kinetic experimental and numerical analyses revealed the following: Measurement and calculation methods proven reliable for the evaluation of subcriticality effects down to 6 %$\Delta k/k$. Anticipating an ADS subcriticality level around 3 %$\Delta k/k$, the measurement methodology and the calculation precision were considered convenient for the study of ADS at KUCA. Moreover, optical fiber detectors were considered promising in the evaluation of the subcriticality and the prompt neutron decay constant evaluation at KUCA, although detector position dependency was observed in kinetic measurements by using the optical fiber detection system.

In the ADS experiments with spallation neutrons, the measurement of kinetics parameters was conducted by both the PNS and Feynman-α methods with the use of optical fiber detectors, under subcriticality ranging between 1160 and 11556 pcm. The results confirmed the validity of the prompt neutron decay constant and the subcriticality in dollar units through the deduction of kinetics parameters. The detector position dependency, neutron spectrum, and subcriticality measurement methods still remained, however, in these ADS experiments.

Transient experiments were carried out to evaluate the performance of subcriticality monitoring by the EKF technique in the presence of an external neutron source. To ensure the conditions in EKF, a basic transient experiment was conducted by dropping the control rod at the critical state. In a comparison of the subcriticality by the inverse kinetic method and the EKF technique, the advantage of EKF was indicated over the robustness of the estimate with good accuracy even at low counts. Additionally, the initial condition used in the filtering technique was confirmed as valid through comparison with measured subcriticality by the rod drop method in the basic transient experiment. In the transient experiments in the presence of an external neutron source, the EKF technique was applied to subcriticality monitoring (time resolution: 1 s), with the use of the initial condition in the basic experiments. The EKF technique significantly indicated a significant advantage in reducing fluctuation and estimating more accurately compared with the inverse method, demonstrating the applicability to the subcriticality monitor.

References

1. Pyeon CH, Hirano Y, Misawa T et al (2007) Preliminary experiments on accelerator-driven subcritical reactor with pulsed neutron generator in Kyoto University Critical Assembly. J Nucl Sci Technol 44:1368

2. Pyeon CH, Hervault M, Misawa T et al (2008) Static and kinetic experiments on accelerator-driven system with 14 MeV neutrons in Kyoto University Critical Assembly. J Nucl Sci Technol 45:1171
3. Mori C, Osada T, Yanagida K et al (1994) Simple and quick measurement of neutron flux distribution by using an optical fiber with scintillator. J Nucl Sci Technol 31:248
4. Yamane Y, Uritani A, Misawa T et al (1999) Measurement of the thermal and fast neutron flux in a research reactor with a Li and Th loaded optical fibre detector. Nucl Instrum Methods A 432:403
5. Gozani T (1962) A modified procedure for the evaluation of pulsed source experiments in subcritical reactors. Nukleonik 4:348
6. Briesmeister JF Editor (2000) MCNP—a general Monte Carlo N-particle transport code, version 4C. LANL Report LA-13709-M
7. Shibata K, Kawano T, Nakagawa T et al (2002) Japanese evaluated nuclear data library version 3 revision-3: JENDL-3.3. J Nucl Sci Technol 39:1125
8. Rose PF (1991) ENDF-201, ENDF/B-VI summary documentation. BNL-NCS-17541, 4th Edition
9. Nakajima Y (1991) JNDC WG on activation cross section data: JENDL activation cross section file. JAERI-M 91–032:43
10. Pyeon CH, Yamanaka M, Endo T et al (2017) Experimental benchmarks on kinetic parameters in accelerator-driven system with 100 MeV protons at Kyoto University Critical Assembly. Ann Nucl Energy 105:346
11. Goorley JT, James MR, Booth TE et al (2013) Initial MCNP6 release overview—MCNP6 version 1.0. LANL, LA-UR-13-22934
12. Shibata K, Iwamoto O, Nakagawa T et al (2011) JENDL-4.0: a new library for nuclear science and technology. J Nucl Sci Technol 48:1
13. Lim JY, Pyeon CH, Yagi T et al (2012) Subcritical multiplication parameters of the accelerator-driven system with 100 MeV protons at the Kyoto University Critical Assembly. Sci Technol Nucl Install 2012:395878
14. Yagi T, Pyeon CH, Misawa T (2013) Application of wavelength shifting fiber to subcriticality measurements. Appl Radiat Isot 72:11
15. Sjöstrand N (1956) Measurement on a subcritical reactor using a pulsed neutron source. Arkiv för Fysik 13:233
16. Kitamura Y, Misawa T, Unesaki H et al (2000) General formulae for the Feynman-α method with the bunching technique. Ann Nucl Energy 27:1199
17. Kitamura Y, Pázsit I, Wright J et al (2005) Calculations of the pulsed Feynman- and Rossi-alpha formulae with delayed neutrons. Ann Nucl Energy 32:671
18. Takada H, Kosako K, Fukahori T (2009) Validation of JENDL high-energy file through analyses of spallation experiments at incident proton energies from 0.5 to 2.83 GeV. J Nucl Sci Technol 46:589
19. Simmons BE, King JS (1958) A pulsed technique for reactivity determination. Nucl Sci Eng 3:595
20. Cullen DE, Clouse CJ, Procassini R et al (2003) Static and dynamic criticality: are they different? UCRL-TR-201506
21. Yamanaka M, Pyeon CH, Misawa T (2016) Monte Carlo approach of effective delayed neutron fraction by k-ratio method with external neutron source. Nucl Sci Eng 184:551
22. Yamanaka M, Watanabe K, Pyeon CH (2020) Subcriticality estimation by extended Kalman filter technique in transient experiment with external neutron source at Kyoto University Critical Assembly. Eur Phys J Plus 135:256
23. Watanabe K, Kawabata Y, Yamazaki A et al (2015) Development of an optical fiber type detector using a Eu:liCaAlF6 scintillator for neutron monitoring in boron neutron capture therapy. Nucl Instrum Methods A 802:1
24. Venerus JC, Bullock TE (1970) Estimation of the dynamic reactivity using digital Kalman filtering. Nucl Sci Eng 40:199

25. Shimazu Y, van Rooijen WFG (2014) Quantitative performance comparison of reactivity esti-
    mation between the extended Kalman filter technique and the inverse point kinetic method.
    Ann Nucl Energy 66:161
26. Chadwick MB, Herman M, Oblozinsky P et al (2011) ENDF/B-VII.1 nuclear data for science
    and technology: cross sections, covariances, fission product yields and decay data. Nucl Data
    Sheets 112:2887

Chapter 4
Effective Delayed Neutron Fraction

Masao Yamanaka

Abstract In kinetic analyses on ADS, although adjoint flux distribution is defined under the existence of an external neutron source, an issue of the proper determination of the weighting function still remains in the definition to obtain the kinetics parameters in the fixed-source calculations. Here, an alternative methodology is proposed with the combined use of the k-ratio method and the reaction rates obtained by the fixed-source calculations, when the subcriticality level or the spectrum of the external neutron source is varied. In ADS experiments, the measurement of β_{eff} is expected to provide complementary verification of the calculation and reliability of nuclear data. Then, the formulation of the Rossi-α method in the pulsed-neutron source has been already available for application to the subcriticality measurement in the pulsed-neutron source (PNS) experiments. Accordingly, the methodology is applied uniquely to deduce the β_{eff} value with the pulsed-neutron source (spallation neutrons), with the combined use of the results of experiments and calculations. Using parameters α and $\rho_\$$, the values of $\beta_{\text{eff}}/\Lambda$ are deduced at near-critical configurations through experimental analyses. To estimate the numerical precision of Λ, the value of $\beta_{\text{eff}}/\Lambda$ is used as an index of Λ evaluation that is defined by a ratio of Λ values in the super-critical and subcritical states.

Keywords Effective delayed neutron fraction · k-ratio method · Rossi-α method · Pulsed-neutron source · Neutron generation time

M. Yamanaka (✉)
Institute for Integrated Radiation and Nuclear Science, Kyoto University, Osaka, Japan
e-mail: yamanaka.masao.6z@kyoto-u.ac.jp

© The Author(s) 2021
C. H. Pyeon (ed.), *Accelerator-Driven System at Kyoto University Critical Assembly*,
https://doi.org/10.1007/978-981-16-0344-0_4

83

4.1 Dependency of External Neutron Source

4.1.1 Experimental Settings

4.1.1.1 Uranium-Fueled ADS Core

The critical EE1 core (reference core) was assembled at the KUCA-A core and was made up of 25 fuel rods surrounded by polyethylene reflectors as shown in Fig. A2.1. Each fuel rod (1/8″p60EUEU) was composed of highly enriched uranium (HEU; 2″ × 2″ and 1/16″ thick) and a polyethylene moderator (PE; 2″ × 2″ and 1/4″ thick) as shown in Fig. A2.2. The core was selected for considering the variation of β_{eff} along the subcritilicality level. For the measurement of subcriticality, protons accelerated to 100 MeV were injected onto a disk-type tungsten (W) target in order to generate spallation neutrons. The accelerator was operated in pulsed mode and the repetition rate of the pulse was 20 Hz. The time width of the pulsed proton beam was 100 ns. The averaged proton current was 50 pA. The target was located at (15, H; Fig. A2.1) grid. The diameter and the thickness of the target were 50 mm and 12 mm, respectively. The subcriticality was measured by the extrapolated area ratio method [1] without considering the spatial effects. The neutron signals were obtained with the use of a BF_3 detector inserted diagonally at (10, U; Fig. A2.1) to the core for the measurements of the subcriticality. For the reference core in the critical experiment, excess reactivity and control rod worth (C1, C2, and C3) were measured by the positive period method and the rod drop method, respectively. Experimental analyses [2] were available to examine the precision of eigenvalue calculations by the Monte Carlo method and the accuracy of measured subcriticality by the extrapolated area ratio method. To achieve deep subcriticality, some of the fuel rods "F" (Fig. A2.1) were substituted for polyethylene reflectors and configured as shown in Fig. A2.3d, f, and the subcriticality level then ranged between about 1300 and 7500 pcm.

4.1.1.2 Thorium-Fueled ADS Core

In this experiment, different external neutron sources (spallation neutrons by the injection of 100 MeV protons onto the W target, and 14 MeV neutrons by the injection of deuteron beams onto the tritium target) were used for considering the variation of β_{eff} caused by the spectrum of external neutron source in the subcritical estimation. The subcritical core at $k_{eff} \simeq 0.85$ (Th-HEU-5PE core shown in Figs. A5.3 and A5.4) was composed of the polyethylene reflectors, fuel rods of thorium (Th; 2″ × 2″ and 1/8″ thick), HEU, and PE moderators, as shown in Fig. A5.7. 14 MeV neutrons were produced by 0.4 mA deuteron beam, 10 Hz pulsed frequency, and 10 μs pulsed width. 100 MeV proton beams were injected onto the W target at 50 mm spot size, 10 pA intensity, 20 Hz pulsed frequency, and 100 ns pulsed width. The subcriticality was measured by the same method as that in uranium-fueled ADS core. Here, the measured subcriticality could be affected by spatial effects especially in such a deep subcritical core. Then, the neutron signals were obtained with the use of an optical fiber detector [3] at the core center (Figs. A5.3 and A5.4) in Th-HEU-5PE core. The optical fiber (1 mm diam. and 200 mm long) was coated with a powdered mixture of ^{6}LiF (95% enrichment) for detection of thermal neutrons based on ^{6}Li$(n, t)^4$He reactions and ZnS(Ag) for scintillation.

4.1.2   Numerical Simulations

4.1.2.1   Eigenvalue Calculations

To ensure the accuracy of eigenvalue calculations with the use of MCNPX-2.5.0 [4] and ENDF/B-VII.0 [5], experimental measurements of the excess reactivity and control rod worth of C1, C2, and C3 were carried out for comparing measured and calculated reactivities in the reference core. In the experiments, the critical state was attained by partial insertion of control rod C2 (full withdrawal of C1, C3, S4, S5, and S6), and excess reactivity was deduced from the coupling with control rod worth (C2 rod) and its integral calibration curve obtained by the positive period method. Control rod worth of S4, S5, and S6 was regarded as the same as those of C1, C2, and C3 obtained by the rod drop method, respectively, because of the symmetrical configuration of the rods, as shown in Fig. A2.1.

The MCNPX eigenvalue calculations were performed in a total of 1E + 08 histories (1E + 03 active cycles of 1E + 05 each); the statistical errors were less than 9 pcm. As shown in Table 4.1, the calculated excess reactivity and control rod worth of C1, C2, and C3 reproduced the measured ones within a relative difference of 5% in the C/E (calculation/experiment) value. Subcriticality was experimentally deduced from the combination of the excess reactivity and worth of inserted control rods as shown in Table 4.2. Furthermore, the accuracy of experimental analyses was verified within 5% through the comparison between the results of MCNPX and those of the experiments. Thus, the subcriticality obtained by MCNPX will be regarded as the reference subcriticality where it was not able to be measured with the excess reactivity and the control rod worth. In such deep subcriticality, the subcriticality by the area ratio method could compare with the reference one.

Table 4.1 Comparison between calculated and measured excess reactivities and control rod worth (Ref. [2])

Reactivity	MCNPX-2.5.0 [pcm]	Experiment [pcm]	C/E*
Excess	258 ± 10	257 ± 13	1.00 ± 0.13
C1 rod	839 ± 15	812 ± 24	1.03 ± 0.02
C2 rod	508 ± 15	506 ± 15	1.00 ± 0.03
C3 rod	135 ± 15	139 ± 4	0.97 ± 0.11

C/E*: calculation/experiment (β_{eff} = 807 ± 11 [pcm] and Λ = 30.5 ± 0.1 [μs] by MCNP6.1 with ENDF/B-VII.0)

Table 4.2 Subcriticality deduced by excess reactivity and control rod worth in the experiments and by MCNPX-2.5.0 in Fig. A2.1 (Ref. [2])

Case	# of fuel rods	Rod insertion pattern	Subcriticality [pcm]		C/E
			MCNPX-2.5.0	Experiment	
I-1	25	C1, C2, C3	1258 ± 11	1200 ± 36	1.05 ± 0.02
I-2	25	C1, C2, C3, S4	2099 ± 11	2011 ± 60	1.04 ± 0.01
I-3	25	C1, C2, C3, S4, S5, S6	2774 ± 11	2657 ± 80	1.04 ± 0.01
I-4	21	–	2737 ± 11	–	–
I-5	21	C1, C2, S4, S6	4858 ± 11	–	–
I-6	19	–	5299 ± 11	–	–
I-7	19	C1, C2, S4, S6	7457 ± 12	–	–

4.1.2.2 Reaction Rates

The delayed neutron fraction is defined as the ratio of the delayed neutron generation to the total neutron generation as follows:

$$\beta = \frac{\langle F^d \phi \rangle}{\langle F \phi \rangle}, \tag{4.1}$$

where ϕ is forward angular flux, $< >$ the integration over angle, space, and energy, respectively. F, F^d, and F^p are production operators of total, delayed, and prompt neutron generations by fission reactions, respectively, as follows:

$$F = \int dE' \chi (E' \rightarrow E) \nu (E') \Sigma_f (E'), \tag{4.2}$$

$$F^d = \int dE' \chi^d (E' \rightarrow E) \nu^d (E') \Sigma_f (E'), \tag{4.3}$$

$$F^p = F - F^d, \tag{4.4}$$

where χ and χ^d are the energy spectra of total and delayed neutrons, respectively; ν and ν^d the neutron yields for total and delayed neutrons, respectively; E the energy of scattered neutrons; E' induced neutron energy; and Σ_f the macroscopic fission cross sections. And β_{eff} is defined as the ratio of the contribution of delayed neutrons to total neutrons for reactivity with the use of adjoint angular flux ϕ^+ for weighting function as follows:

$$\beta_{\text{eff}} = \frac{\langle \phi^+, F^d \phi \rangle}{\langle \phi^+, F \phi \rangle}. \tag{4.5}$$

The effective multiplication factor k_{eff} and prompt multiplication factor k_p are then expressed by the neutron fluxes as follows:

$$k_{\mathrm{eff}} = \frac{\langle \phi^+, \, F\phi \rangle}{\langle \phi^+, \, L\phi \rangle}, \tag{4.6}$$

$$k_p = \frac{\langle \phi_p^+, \, F^p\phi_p \rangle}{\langle \phi_p^+, \, L\phi_p \rangle}, \tag{4.7}$$

where L is loss operator account for leakage, absorption, and scattering, and ϕ_p and ϕ_p^+ are the forward and adjoint fluxes taking into account prompt neutrons, respectively. In the k-ratio method [6], the multiplication factors k_{eff} and k_p are used to obtain an approximate value of β_{eff} as follows:

$$\beta_{eff,\, eigen} = \frac{\langle \phi^+, \, (F-F^p)\phi \rangle}{\langle \phi^+, \, F\phi \rangle} = 1 - \frac{\langle \phi^+, \, F^p\phi \rangle}{\langle \phi^+, \, F\phi \rangle} \approx 1 - \frac{k_p}{k_{eff}}. \tag{4.8}$$

Here, two multiplication factors k_{RR} and $k_{\mathrm{RR},\, p}$ by total and prompt neutrons are newly defined with the use of reaction rates, respectively, as follows:

$$k_{\mathrm{RR}} = \frac{\langle F\phi \rangle}{\langle L'\phi \rangle}, \tag{4.9}$$

$$k_{\mathrm{RR},p} = \frac{\langle F^p\varphi_p \rangle}{\langle L'\varphi_p \rangle}. \tag{4.10}$$

Since scattered neutrons are eventually absorbed in the core and the reflector or leaked out from the core, denominator in Eqs. (4.9) and (4.10) can be expressed in terms of leak out and the absorption reactions only. When numerator and denominator are interpreted as integrated reaction rates of the destruction and fission operators, respectively, the loss operator of L' is defined, taking into account leakage and absorption, as follows:

$$L' = \vec{\Omega} \cdot \nabla + \Sigma_a(E), \tag{4.11}$$

where $\vec{\Omega}$ is the direction of the neutron flight, and Σ_a the macroscopic absorption cross sections. Then, $\beta_{\mathrm{eigen}}^{\mathrm{RR}}$ deduced by the reaction rates was expressed approximately by substitution of the multiplication factors by Eqs. (4.9) and (4.10) in Eq. (4.8) as follows:

$$\beta_{\mathrm{eigen}}^{\mathrm{RR}} \approx 1 - \frac{k_{\mathrm{RR},\, p}}{k_{\mathrm{RR}}}. \tag{4.12}$$

Table 4.3 Comparison between effective multiplication factors k_{eff} by MCNPX-2.5.0 and k_{RR} in Eq. (4.9) with reaction rate calculations (Ref. [2])

Core	Case	k_{eff} (MCNPX-2.5.0)	k_{RR} (MCNPX-2.5.0)
EE1 core* (Spallation neutrons)	I-1	0.99093 ± 0.00009	0.99096 ± 0.00013
	I-2	0.98274 ± 0.00009	0.98269 ± 0.00013
	I-3	0.97627 ± 0.00009	0.97619 ± 0.00013
	I-4	0.97662 ± 0.00009	0.97655 ± 0.00012
	I-5	0.95680 ± 0.00009	0.95667 ± 0.00012
	I-6	0.95278 ± 0.00009	0.95281 ± 0.00012
	I-7	0.93358 ± 0.00009	0.93353 ± 0.00012
Th-HEU-5PE (Spallation neutrons)	II-1	0.86397 ± 0.00008	0.86387 ± 0.00012
Th-HEU-5PE (14 MeV neutrons)	II-2	0.84924 ± 0.00008	0.84923 ± 0.00012

*$k_{eff} = 1.00344 \pm 0.00009$ (at critical state by MCNP2.5.0 with ENDF/V-VII.0)

The multiplication factor k_{RR} thus deduced should be compared with the results in k_{eff} shown in Eq. (4.6) by the Monte Carlo calculations to examine the validity of k_{RR} obtained by Eq. (4.9). Reaction rate calculations were performed by MCNPX-2.5.0. Comparing the results of k_{eff} and k_{RR} in Eqs. (4.6) and (4.9), respectively, the results by the reaction rates in Eq. (4.9) revealed fairly good agreement with a difference of 10 pcm with those in Eq. (4.6) by MCNPX, as shown in Table 4.3.

The results revealed that the proposed methodology of k_{RR} with the use of reaction rates in Eq. (4.9) is appropriate through the comparison with the effective multiplication factor by the eigenvalue calculations. Note that $(n, 2n)$ and $(n, 3n)$ reactions were not considered for the estimation of the neutron productions, because there was no relevant difference between considering them or not in the uranium- and the thorium-loaded cores, respectively, in the eigenvalue calculations shown in Eq. (4.12).

4.1.3 k-Ratio Method

4.1.3.1 Reaction Rates by Eigenvalue Calculations

On the basis of the theoretical preparation discussed in Sect. 4.1.2, $\beta_{eff, eigen}$ and β_{eigen}^{RR}, as defined in Eqs. (4.8) and (4.12), can be obtained from the eigenvalue calculations, compared with the reference one obtained by MCNP6.1 [7]. Through the comparison with β_{eff} by MCNP6.1 shown in Table 4.4, the estimation of $\beta_{eff,eigen}$ in Eq. (4.8) demonstrated that the methodology is valid in the subcriticality level of interest. Also, the value of β_{eigen}^{RR} in Eq. (4.12) with the use of reaction rates showed good agreement with those by MCNP6.1, as shown in Table 4.3, indicating a high precision of the reaction rates and an applicability of the k-ratio method with reaction rates. From the

Table 4.4 Comparison between β_{eff} by MCNP6.1, $\beta_{eff,eigen}$ by the k-ratio method in Eq. (4.8), β_{eigen}^{RR} by the k-ratio method with reaction rate calculations (eigenvalue calculation) in Eq. (4.12), β_{source}^{RR} by the k-ratio method with reaction rate calculations (fixed-source calculation) in Eq. (4.20), and β (eigenvalue and fixed-source calculations) in Eq. (4.1) (Ref. [2])

Case	Effective delayed neutron fraction (and delayed neutron fraction) [pcm]			
	β_{eff} (MCNP6.1)	$\beta_{eff,eigen}$ (MCNPX-2.5.0)	β_{eigen}^{RR} (MCNPX-2.5.0)	β_{source}^{RR} (MCNPX-2.5.0)
I-1	817 ± 11	797 ± 9	791 ± 13 (649 ± 22)	809 ± 43 (637 ± 32)
I-2	794 ± 11	801 ± 9	795 ± 13 (649 ± 22)	873 ± 45 (636 ± 33)
I-3	814 ± 11	811 ± 9	807 ± 13 (649 ± 22)	880 ± 46 (636 ± 33)
I-4	806 ± 11	810 ± 9	809 ± 13 (649 ± 22)	826 ± 55 (637 ± 36)
I-5	822 ± 11	825 ± 9	838 ± 13 (649 ± 22)	819 ± 57 (637 ± 37)
I-6	808 ± 11	824 ± 9	806 ± 13 (649 ± 22)	865 ± 81 (636 ± 44)
I-7	829 ± 11	801 ± 9	783 ± 13 (649 ± 22)	792 ± 86 (635 ± 45)
II-1	808 ± 11	795 ± 9	803 ± 9 (652 ± 24)	910 ± 11 (645 ± 14)
II-2	798 ± 11	802 ± 9	802 ± 9 (653 ± 24)	881 ± 7 (635 ± 9)

results in Table 4.4, the proposed methodology by the k-ratio method was confirmed valid on the basis of the reaction rates by the eigenvalue calculations.

4.1.3.2 Reaction Rates by Fixed-Source Calculations

From the definition of β_{eigen}^{RR} shown in Eq. (4.12), the methodology using eigenvalue calculations is extended into a methodology using the fixed-source calculations. In the fixed-source problem, the transport equation is expressed by introducing source term s to acquire neutron flux formed in the subcritical core ϕ_{sub} as follows:

$$L\,\phi_{sub} = F\,\phi_{sub} + s, \qquad (4.13)$$

where L and F indicate the loss and fission operators, respectively. To distinguish source and fission neutrons, ϕ_{sub} was assumed to be expressed in terms of ϕ_{source} and ϕ_{core} as follows:

$$\phi_{\text{sub}} = \phi_{\text{source}} + \phi_{\text{core}}, \tag{4.14}$$

where ϕ_{source} means the neutron flux corresponding to the source neutrons, and ϕ_{core} the neutron flux corresponding to the fission chain reactions. When the source problem is solved by applying $\nu = 0$ to Eq. (4.13), the multiplied angular flux is not formed, and the source term is expressed as follows:

$$s = L \, \phi_{\text{source}}. \tag{4.15}$$

Substituting Eq. (4.15) for Eq. (4.13), the source term is replaced by the product of L and ϕ_{source}, and the neutron balance equation is expressed as follows:

$$L \, \phi_{\text{sub}} = F \, \phi_{\text{sub}} + L \, \phi_{\text{source}}. \tag{4.16}$$

Here, on the basis of the manner in which effective eigenvalues are calculated with the neutron flux corresponding to fission neutrons, a pseudo multiplication factor k^{pseudo} is defined in the pseudo eigenvalue calculations with the use of neutron flux under the existence of an external neutron source as follows:

$$L \, \phi_{\text{core}} = \frac{1}{k^{\text{pseudo}}} F \, \phi_{\text{core}}. \tag{4.17}$$

Then, pseudo multiplication factors $k_{\text{RR}}^{\text{pseudo}}$ and $k_{\text{RR}}^{\text{pseudo}}$ are obtained by substituting Eqs. (4.14) and (4.16) in Eq. (4.17), with the use of reaction rates by the fixed-source calculations, in the same manner as in Eqs. (4.9) and (4.10), respectively, as follows:

$$k_{\text{RR}}^{\text{pseudo}} = \frac{\langle F \, \phi_{\text{core}} \rangle}{\langle L \, \phi_{\text{core}} \rangle} = \frac{\langle F \, \phi_{\text{sub}} \rangle - \langle F \, \phi_{\text{source}} \rangle}{\langle L' \, \phi_{\text{sub}} \rangle - \langle L' \, \phi_{\text{source}} \rangle}, \tag{4.18}$$

$$k_{\text{RR}, p}^{\text{pseudo}} = \frac{\langle F^P \, \phi_{\text{core}, p} \rangle}{\langle L \, \phi_{\text{core}, p} \rangle} = \frac{\langle F^P \, \phi_{\text{sub}, p} \rangle - \langle F^P \, \phi_{\text{source}, p} \rangle}{\langle L' \, \phi_{\text{sub}, p} \rangle - \langle L' \, \phi_{\text{source}, p} \rangle}. \tag{4.19}$$

The $k_{\text{RR}}^{\text{pseudo}} (k_{\text{RR}, p}^{\text{pseudo}})$ can be calculated with two runs considering total (prompt) neutrons: for the calculation of ϕ_{sub} ($\phi_{\text{sub}, p}$) with standard fixed-source calculation and for that of ϕ_{source} ($\phi_{\text{source}, p}$) with the fixed-source calculation under the option of $\nu = 0$. Finally, $\beta_{\text{source}}^{\text{RR}}$ can be obtained with the reaction rates by the fixed-source calculations by substituting Eqs. (4.18) and (4.19) for Eq. (4.8) as follows:

$$\beta_{\text{source}}^{\text{RR}} \approx 1 - \frac{k_{\text{RR}, p}^{\text{pseudo}}}{k_{\text{RR}}^{\text{pseudo}}}. \tag{4.20}$$

4.1.3.3   Results and Discussion

Fixed-source calculations were performed by MCNPX-2.5.0 with 1E + 09 total histories with nuclear data libraries of ENDF/B-VII.0 and JENDL/HE-2007 [8] as shown in Table 4.5. For the case of the spallation neutrons, since neutrons over 20 MeV neutrons could be produced by the 100 MeV proton injections into the W target, JENDL/HE-2007 has a full advantage in the accuracy of the particle transport for neutrons over the energy of 20 MeV. However, because the yields for total and prompt neutrons are not provided in JENDL/HE-2007, ENDF/B-VII.0 was used for fissile and fissionable nuclei. β values calculated with the fluxes by eigenvalue and fixed-source calculations showed independency on the subcriticality, as shown in Table 4.4. However, the β values were decreased by the flux in fixed-source calculation. Then, β_{source}^{RR} in Eq. (4.20) deduced by the fixed-source calculations indicated different values from β (fixed-source calculation), and these values were varied larger than those of β_{eff} by the eigenvalue calculations (MCNP6.1). In Cases I-1 to I-7, the increase of buckling in the core can be the reason to increase β_{eff} and β_{source}^{RR} by the fuel rod replacement (Cases I-1, I-4, and I-6) and by the control rod insertion (Cases I-3, I-5, and I-7) because of especially increasing the leakage of prompt neutrons having higher energy compared to delayed neutrons. The neutron flux distribution was distorted by the injection of spallation neutrons having dominantly lower energy with the comparison of 14 MeV neutrons in Case II-1. This distorted flux distribution was considered to be induced by the leakage of prompt neutrons, resulting in the increase of β_{source}^{RR}.

The target results of the measured subcriticality (pcm units) in the uranium-fueled core were obtained from measured subcriticality in dollar unit multiplied by β_{eff} (MCNP6.1), β_{eigen}^{RR}, and β_{source}^{RR} in Eqs. (4.11) and (4.19) (Table 4.4), respectively, as shown in Table 4.6. The measured subcriticality with the use of β_{source}^{RR} in Eq. (4.20) for conversion from dollar units into pcm units by the fixed-source calculations showed good agreement with the reference subcriticality within a relative difference of 10% in the variation of the subcriticality level. And, β_{source}^{RR} worked well for the results of measured subcriticality, comparing those of reference subcriticality. In Cases I-4 to I-7 (Table 4.6), there was a slight difference between β_{eff} and β_{source}^{RR}.

In deep subcritical cores, the measured subcriticality in the area ratio method is generally considered inaccurately obtained in pcm units to compare with reference one because an assumption is imposed on the measurements: all source neutrons induce the fission reactions and neuron signals originate from correlated neutrons to

Table 4.5 List of nuclear data libraries for calculation of β_{source}^{RR} in particle transport simulations of the ADS experiments (Ref. [2])

	Neutrons	Protons
Spallation neutrons	JENDL/HE-2007 and ENDF/B-VII.0 (for U and Th only)	JENDL/HE-2007
14 MeV neutrons	ENDF/B-VII.0	–

Table 4.6 Results of subcriticality on the basis of effective delayed neutron fractions estimated by MCNP6.1, $\beta_{\mathrm{eigen}}^{\mathrm{RR}}$, and $\beta_{\mathrm{source}}^{\mathrm{RR}}$ (Ref. [2])

Case	Reference Subcriticality [pcm]	Measured subcriticality [pcm]		
		β_{eff} (MCNP6.1)	$\beta_{\mathrm{eigen}}^{\mathrm{RR}}$ (MCNPX-2.5.0)	$\beta_{\mathrm{source}}^{\mathrm{RR}}$ (MCNPX-2.5.0)
I-1	1200 ± 36	1147 ± 17 (1.05 ± 0.02)	1111 ± 20 (1.08 ± 0.02)	1136 ± 61 (1.06 ± 0.06)
I-2	2012 ± 60	1847 ± 27 (1.09 ± 0.02)	1850 ± 30 (1.09 ± 0.02)	2032 ± 105 (0.99 ± 0.05)
I-3	2657 ± 80	2419 ± 35 (1.09 ± 0.02)	2400 ± 40 (1.11 ± 0.02)	2615 ± 136 (1.02 ± 0.05)
I-4	2722 ± 32	2679 ± 39 (1.02 ± 0.02)	2688 ± 44 (1.02 ± 0.02)	2745 ± 183 (1.00 ± 0.07)
I-5	4891 ± 50	5036 ± 84 (0.96 ± 0.02)	5134 ± 88 (0.95 ± 0.02)	5015 ± 354 (0.97 ± 0.07)
I-6	5291 ± 54	5489 ± 85 (0.97 ± 0.02)	5475 ± 95 (0.97 ± 0.02)	5877 ± 554 (0.90 ± 0.09)
I-7	7474 ± 75	8421 ± 144 (0.89 ± 0.02)	7952 ± 149 (0.94 ± 0.02)	8050 ± 874 (0.93 ± 0.10)

(): Calculation/experiment

the fission multiplication. And, in the case of different external neutron sources for the deep subcritical core, as shown in Table 4.7, subcriticalities were compared in terms of the k_{eff}. The difference of the reference k_{eff} between spallation and 14 MeV neutrons is considered to be caused by the slight difference in the core configuration of the air gap shown in Figs A5.3 and A5.4. Here, the subcriticality with $\beta_{\mathrm{source}}^{\mathrm{RR}}$ was observed to reveal a comparative tendency through the comparison of the subcriticality with β_{eff} within the C/E value of 2%. The applicability of the proposed methodology was also confirmed with the variation of the external neutron sources.

Table 4.7 Comparison of k_{eff} and their C/E values estimated by MCNP6.1, $\beta_{\mathrm{eigen}}^{\mathrm{RR}}$, and $\beta_{\mathrm{source}}^{\mathrm{RR}}$ in different external neutron sources (Ref. [2])

Case	Reference k_{eff}	Measured k_{eff}		
		β_{eff} (MCNP6.1)	$\beta_{\mathrm{eigen}}^{\mathrm{RR}}$ (MCNPX-2.5.0)	$\beta_{\mathrm{source}}^{\mathrm{RR}}$ (MCNPX-2.5.0)
II-1	0.86397 ± 0.00008	0.89707 ± 0.00387 (0.96 ± 0.01)	0.89945 ± 0.00392 (0.96 ± 0.01)	0.88559 ± 0.00418 (0.98 ± 0.01)
II-2	0.84924 ± 0.00008	0.84775 ± 0.02543 (1.00 ± 0.03)	0.84516 ± 0.02581 (1.00 ± 0.03)	0.83243 ± 0.02743 (1.02 ± 0.03)

(): Calculation/experiment

Finally, these results demonstrated the fact that proper values for subcriticality determination with the area ratio technique were successfully obtained for the subcriticality estimation with the use of the proposed methodology by the fixed-source calculations.

4.2 Measurement

4.2.1 Nelson Number Method

4.2.1.1 Formulation of Rossi-α Method

In the measurement methodology for β_{eff}, the Nelson number method based on the Rossi-α method [9] provides the advantage of reducing the parameters that are considered difficult to obtain experimentally, including detection efficiency, fission rate at the core center, and the number of neutrons. The methodology assumes, however, that β_{eff} is measured near the critical state and by locating the external neutron source at the core center. It is also applicable to the PNS experiments by modifying the formulation of the Rossi-α method. In the PNS experiments, the formulation by the Rossi-α method is already available for processing neutron signals by the methodology used in the analysis with the pulsed-neutron source [10, 11].

In the Rossi-α method, the joint probability $P(t_1,t_2)$ between two neutron signals detected at times t_1 and t_2 is evaluated by categorizing the same fission chain reactions into correlated probability P_C and the different fission chain reactions and the neutron sources into uncorrelated probability P_U, as follows:

$$P(t_1, t_2)\, dt_1\, dt_2 = P_C(t_1, t_2)\, dt_1\, dt_2 + P_U(t_1, t_2)\, dt_1\, dt_2. \qquad (4.21)$$

P_C in the existence of the pulsed-neutron source can then be expressed as follows:

$$P_C(t_1, t_2) = g\left(\frac{\lambda_d\, \lambda_f\, \langle \nu_p\, (\nu_p - 1)\rangle}{2\,\alpha}\right) e^{-\alpha\,(t_2 - t_1)}\, dt_1\, dt_2, \qquad (4.22)$$

where α is the prompt neutron decay constant, $\langle \nu_p\, (\nu_p - 1)\rangle$ the second moment of the prompt neutron multiplicity distribution for induced prompt fission neutrons, λ_d the detection efficiency for neutron, λ_f the detection efficiency of a fission reaction, and g the correction factor taking into account the variation in the probability of detecting correlated counts originating from neutrons of different worth [9]. For uncorrelated probability P_U, the formulation of its signal is sensitive to the pulsed shape of the external neutron source [10]. In the present study, the shape was regarded as the Gaussian function, and the uncorrelated probability is represented by constant term $P_{U,\text{const}}$ and trigonometric term $P_{U,\text{trig}}$ as follows:

$$P_U(t_1, t_2)\, dt_1\, dt_2 = P_{U,\,\text{const}}(t_1, t_2)\, dt_1\, dt_2 + P_{U,\,\text{trig}}(t_1, t_2)\, dt_1\, dt_2$$

$$= \frac{\lambda_d\, g * S \Lambda \sqrt{2\pi}\sigma}{(-\rho)\, T_0}\, dt_1\, dt_2$$

$$+ \frac{\lambda_d\, g * S(-\rho)\, T_0}{2\Lambda\sqrt{2\pi}\sigma} \sum_{n=1}^{\infty} \frac{1}{\alpha^2 + \left(\frac{2n\pi}{T_0}\right)^2}\, e^{-\left(\frac{2n\pi}{T_0}\right)^2 \sigma^2} \cos\left\{ \left(\frac{2n\pi}{T_0}\right)(t_2 - t_1) \right\} dt_1\, dt_2,$$

(4.23)

where S is the intensity of source neutrons, T_0 the pulsed period, ρ the reactivity, Λ the generation time, σ the pulsed width, and $g*$ the correction factor for spatial and energy distribution of the source neutrons [9]. With the result of the Rossi-α method in the PNS experiments, the intensity of P_C and P_U is obtained from fitting by Eq. (4.21). Since P_C is predicted to decay rapidly; however, the fitting is considered difficult for obtaining both intensities together. Accordingly, the uncorrelated terms were deduced by first fitting in the region, where P_C is sufficiently decayed in Eq. (4.23), with fitting parameters B, D, and σ as follows:

$$P_U(t_1, t_2)dt_1 dt_2 = B dt_1 dt_2$$

$$+ D \sum_{n=1}^{1000} \frac{1}{\alpha^2 + \left(\frac{2n\pi}{T_0}\right)^2}\, e^{-\left(\frac{2n\pi}{T_0}\right)^2 \sigma^2} \cos\left\{ \left(\frac{2n\pi}{T_0}\right)(t_2 - t_1) \right\} dt_1 dt_2,$$

(4.24)

where B is the fitting parameter for the constant value shown in Eq. (4.23) as follows:

$$B = \frac{\lambda_d g * S \Lambda \sqrt{2\pi}\sigma}{(-\rho)T_0},$$

(4.25)

and the upper value of summation was set as 1000 leaving a margin from the saturation of the fitting results by setting about 300 in the upper value. With the fitting results of B, D, and σ, P_C is deduced by subtracting uncorrelated terms from the result of the Rossi-α method in the PNS experiments shown in Eq. (4.21) as follows:

$$P_C(t_1, t_2)dt_1 dt_2 = P(t_1, t_2)dt_1 dt_2$$

$$- \left[B dt_1 dt_2 + D \sum_{n=1}^{1000} \frac{1}{\alpha^2 + \left(\frac{2n\pi}{T_0}\right)^2}\, e^{-\left(\frac{2n\pi}{T_0}\right)^2 \sigma^2} \cos\left\{ \left(\frac{2n\pi}{T_0}\right)(t_2 - t_1) \right\} dt_1 dt_2 \right],$$

(4.26)

Here, the intensity of correlated probability C is obtained by fitting Eq. (4.26) as follows:

$$P_C(t_1, t_2)dt_1 dt_2 = C e^{-\alpha(t_2 - t_1)} dt_1 dt_2,$$

(4.27)

where C is expressed as follows:

$$C = g\left(\frac{\lambda_d \lambda_f \langle v_p(v_p - 1)\rangle}{2\alpha}\right). \tag{4.28}$$

4.2.1.2 Estimation of β_{eff}

For the estimation of β_{eff}, the parameters of B and C obtained by fitting with Eqs. (4.24) and (4.28), respectively, are used with the value of α, which is defined with the use of prompt multiplication factor k_p and neutron lifetime l, as follows:

$$\alpha = \frac{\beta_{\text{eff}} - \rho}{\Lambda} = \frac{1 - k_p}{l} = \frac{1 - (1 - \beta_{\text{eff}})k_{\text{eff}}}{l} = \frac{\beta_{\text{eff}}}{l}\frac{1 - \rho_\$}{1 - \rho_\$\beta_{\text{eff}}}, \tag{4.29}$$

where k_{eff} is the effective multiplication factor and $\rho_\$$ the reactivity in dollar units. Further, λ_f and Λ can be rewritten with the use of β_{eff} and k_{eff} as follows:

$$\lambda_f = \frac{1}{l_f} = \frac{k_p}{l\langle v_p\rangle} = \frac{(1 - \beta_{\text{eff}})k_{\text{eff}}}{l\langle v_p\rangle}, \tag{4.30}$$

$$\Lambda = \frac{l}{k_{\text{eff}}}, \tag{4.31}$$

where l_f is the mean time between fission iterations and $\langle v_p\rangle$ is the average number of prompt neutrons released per fission. With the use of Eqs. (4.29), (4.30), and (4.31), the intensity of correlated probability C in Eq. (4.27) can be expressed as follows:

$$\begin{aligned}
C &= \frac{g\lambda_d\langle v_p(v_p - 1)\rangle k_p}{2\alpha l\langle v_p\rangle} = \frac{g\lambda_d\langle v_p(v_p - 1)\rangle k_p}{2\alpha\Lambda\langle v_p\rangle k_{\text{eff}}} \\
&= \frac{g\lambda_d\langle v_p(v_p - 1)\rangle(1 - \beta_{\text{eff}})k_{\text{eff}}}{2(1 - \rho_\$)\beta_{\text{eff}}\langle v_p\rangle k_{\text{eff}}} \\
&= \frac{g\lambda_d\langle v_p(v_p - 1)\rangle(1 - \beta_{\text{eff}})}{2(1 - \rho_\$)\beta_{\text{eff}}\langle v_p\rangle}.
\end{aligned} \tag{4.32}$$

Also, B shown in Eq. (4.25) can be rewritten with the use of $\rho_\$$ and α, as follows:

$$B = \frac{\lambda_d\, g * S\,\Lambda\,\sqrt{2\pi}\,\sigma}{(-\rho_\$)\,T_0} = \frac{\lambda_d\, g * S\,(1 - \rho_\$)\sqrt{2\pi}\,\sigma}{(-\rho_\$)\,T_0\,\alpha}. \tag{4.33}$$

Here, the Nelson number is defined with the combination of α, B in Eq. (4.33) and C in Eq. (4.32) as follows:

$$N = \left(\frac{2\sqrt{2\pi}\,\sigma g * S\langle v_p \rangle}{g\langle v_p(v_p-1)\rangle T_0} \right)\left(\frac{C}{\alpha B} \right) = \frac{(1-\beta_{\text{eff}})(-\rho_\$)}{\beta_{\text{eff}}(1-\rho_\$)^2}, \tag{4.34}$$

and correction factors g and $g*$ were obtained with the following calculations [6]:

$$g = \frac{\int \mathbf{P(r)}\,d\mathbf{r} \int \mathbf{P(r)}\,\mathbf{I}^2(\mathbf{r})\,d\mathbf{r}}{\left(\int \mathbf{P(r)}\,\mathbf{I(r)}\,d\mathbf{r} \right)^2}, \tag{4.35}$$

$$g* = \frac{\int \mathbf{S(r)}\,\mathbf{I_q(r)}\,d\mathbf{r} \int \mathbf{P(r)}\,d\mathbf{r}}{\int \mathbf{P(r)}\,\mathbf{I(r)}\,d\mathbf{r} \int \mathbf{S(r)}\,d\mathbf{r}}, \tag{4.36}$$

$$\mathbf{P(r)} = \int \Sigma_f(\mathbf{r},\,E)\,\phi(\mathbf{r},\,E)\,dE, \tag{4.37}$$

$$\mathbf{I(r)} = \int \chi(\mathbf{r},\,E)\,\phi^+(\mathbf{r},\,E)\,dE, \tag{4.38}$$

$$\mathbf{I_q(r)} = \int \chi_q(\mathbf{r},\,E)\,\phi^+(\mathbf{r},\,E)\,dE, \tag{4.39}$$

where ϕ and ϕ^+ are the forward and adjoint fluxes, respectively, Σ_f the macroscopic fission cross section, and χ and χ_q the fission spectrum and the spectrum of the external neutron source, respectively. From the relation between N and β_{eff} shown in Eq. (4.24), β_{eff} is obtained experimentally as follows:

$$\beta_{\text{eff}} = \frac{-\rho_\$}{N(1-\rho_\$)^2 - \rho_\$}. \tag{4.40}$$

The procedure for the deduction of β_{eff} in the PNS experiments is shown in Fig. 4.1.

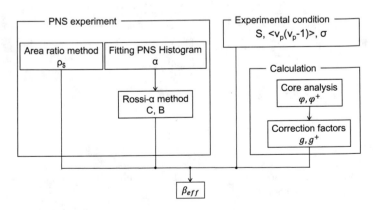

Fig. 4.1 Procedure for deduction of β_{eff} in PNS experiments (Ref. [12])

4.2.2   Experimental Settings

4.2.2.1   ADS Experiments

The A-core was used for the measurement of β_{eff} in the ADS experiments with 100 MeV protons [12]. The core comprised 25 fuel rods surrounded by polyethylene reflectors as shown in Fig. A2.1. Each fuel rod (1/8″p60EUEU) was made up of an HEU (2″ × 2″ and 1/16″ thick) and a polyethylene moderator (PE; 2″ × 2″ and 1/8″ thick) in an Al sheath 54 × 54 × 1524 mm as shown in Fig. A2.2. The core spectrum was a relatively hard one with an H/U (hydrogen/uranium) ratio of approximately 50 in a thermal reactor.

A proton accelerator (FFAG accelerator) was operated to inject 100 MeV protons (beam spot 50 mm, intensity 50 pA (Cases II-1 to II-3 in Fig. A2.3a–c) and 75 pA (Cases II-4 to II-7 in Figs. A2.3d, g) and pulsed frequency 20 Hz) onto a tungsten target (W) (50 mm diam. and 12 mm thick) located at position (15, H; Fig. A2.1) to generate spallation neutrons.

Time evolution according to the injection of spallation neutrons was obtained from the signals of a BF$_3$ detector installed diagonally to the target at position (10, U; Fig. A2.1a) and an optical fiber detector [3] (coated with a powdered mixture of ^{6}LiF (95% enrichment) for detection based on ^{6}Li(n, t)^{4}He reactions and ZnS(Ag) for scintillation in the core) installed at position (15–16, M-N; Fig. A2.1). The PNS experiments were conducted for 10 min in each case to acquire neutron signals for data analyses by the area ratio method and the Rossi-α method. The pulsed width of the neutron source was deduced by fitting for each detector with Eq. (4.23), as shown in Table 4.8. Here, the reason being the pulsed width of the BF$_3$ detector larger than that of the optical fiber is considered as follows: the pulsed width of the source neutrons is large during the transport into detectors; the transported source neutrons are detected, having the width depending on the distance from the neutron source.

4.2.2.2   Subcriticality

Subcriticality was obtained by full insertion of control and safety rods, and by the substitution of the fuel assembly for polyethylene rods, as shown in Table 4.8. The excess reactivity and control rod worth (C1, C2, and C3) were measured by the positive period method and the rod drop method, respectively. In Cases II-1 to II-3, the subcriticality was experimentally deduced with the combined use of control rod worth and its calibration curve obtained by the positive period method. Moreover, in Cases II-4 to II-7, some of the fuel rods "F" (Fig. A2.1) were replaced by polyethylene reflectors and configured as shown in Figs. A2.3d–f. The subcriticality in dollar units was acquired experimentally by the extrapolate area ratio method [1]. The subcriticality level then ranged between 1300 and 7500 pcm.

Table 4.8 Deduced pulsed width and measured subcriticality in dollar units by fuel rod substitution and control rod insertion (Ref. [12])

Case	# of fuel rods	Rod Insertion pattern	Pulsed width in optical fiber and (BF$_3$) detectors [s]	Subcriticality [$] (methodology)
II-1	25	C1, C2, C3	(2.10 ± 0.05)E-04 ((4.30 ± 0.20)E-04)	1.20 ± 0.03 (control rod worth)
II-2	25	C1, C2, C3, S4	(1.25 ± 0.05)E-04 ((3.10 ± 0.20)E-04)	2.01 ± 0.06 (control rod worth)
II-3	25	C1, C2, C3, S4, S5, S6	(1.00 ± 0.05)E-04 ((2.90 ± 0.20)E-04)	2.66 ± 0.07 (control rod worth)
II-4	21	–	(1.00 ± 0.05)E-04 ((2.70 ± 0.10)E-04)	3.32 ± 0.02 (area ratio method)
II-5	21	C1, C2, S4, S6	(5.12 ± 0.25)E-05 ((2.20 ± 0.05)E-04)	6.13 ± 0.05 (area ratio method)
II-6	19	–	(5.00 ± 0.42)E-05 ((2.75 ± 0.02)E-04)	6.80 ± 0.05 (area ratio method)
II-7	19	C1, C2, S4, S6	(5.00 ± 0.42)E-05 ((2.60 ± 0.02)E-04)	10.16 ± 0.09 (area ratio method)

4.2.3   Results and Discussion

4.2.3.1   Parameters in Rossi-α Method

The PNS experiments were carried out for the neutron noise analyses by the Rossi-α method. Moreover, the PNS histogram was obtained to acquire α values to supplement neutron noise analyses, as shown in Fig. 4.2. To obtain the intensity of the second term in Eq. (4.23), fitting based on Eq. (4.24) is required in the region where

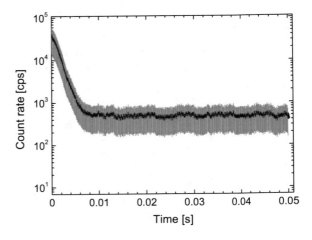

Fig. 4.2 PNS histogram of optical fiber detector in Case II-1 (Ref. [12])

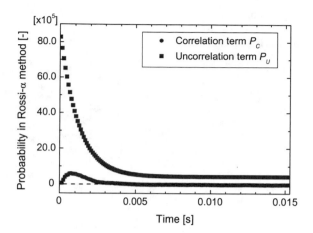

Fig. 4.3 Correlated and uncorrelated probabilities of fiber detector in Case II-1 by Rossi-α method in PNS experiment (Ref. [12])

the correlation probability is negligible. From the PNS histogram shown in Fig. 4.2, the decay of the correlation probability was confirmed at 0.025 s, and intensity B was obtained by fitting for the joint probability by the Rossi-α method in the PNS experiments, on the basis of Eq. (4.25) in the region between 0.025 and 0.125 s. The intensity of the correlation term in Eq. (4.22) was then obtained by subtracting the uncorrelated term from the joint probability in Eq. (4.26) and fitting by Eq. (4.27).

The uncorrelated probability decreased rapidly, compared with the correlated probability as shown in Fig. 4.3. Instead of decreasing exponentially, the correlated probability increased in the vicinity of zero, the reason being the overestimation of the uncorrelated probability in the region arising by varying the shape of the external neutron source from the Gaussian to the pulsed, because the formulation of uncorrelated probability has the sensitivity to the shape of the external neutron source.

In the fitted parameters, α and B values differed from each other when the subcriticality was varied, as shown in Figs. 4.4a, b, respectively, because of the increase of the subcriticality in the α value and the decrease of neutrons in B. For the intensity of correlated probability C, the optical fiber indicated an asymptotic tendency of monotonic decrease, by the change in the subcriticality attributed to control rod insertion and fuel rod replacement in Cases II-1 to II-7, as shown in Fig. 4.4c; the intensity of fission reactions decreased by the subcriticality. The BF$_3$ detector, however, showed the increasing tendency in Cases II-4, II-5, and II-6. Since the low correlated probability of the fission neutrons could be predicted in BF$_3$ installed outside the core, the accuracy of the reconstruction of correlated probability is considered low.

4.2.3.2 Estimation of β_{eff}

To obtain correction factors g and g^* by Eqs. (4.30) and (4.31), respectively, the calculations of reaction rates and adjoint flux were performed by MCNP6.1 together

Fig. 4.4 Parameters obtained by the Rossi-α method in PNS experiments by varying subcriticality (Ref. [12])

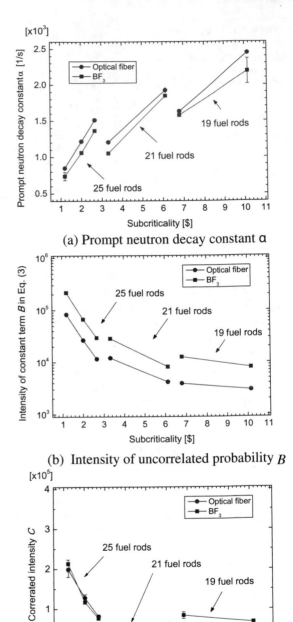

(a) Prompt neutron decay constant α

(b) Intensity of uncorrelated probability B

(c) Intensity of correlated probability C

Table 4.9 Calculated correction factors g and g^* by Eqs. (4.30) and (4.31) with JENDL-4.0 and JENDL/HE-2007 (or ENDF/B-VII.0 and JENDL/HE-2007), respectively (Ref. [12])

Case	g	g^*
1	1.03 ± 0.01 (1.03 ± 0.01)	$(4.29 \pm 0.01)\text{E-}03$ $((4.32 \pm 0.01)\text{E-}03)$
2	1.03 ± 0.01 (1.03 ± 0.01)	$(4.28 \pm 0.01)\text{E-}03$ $((4.31 \pm 0.01)\text{E-}03)$
3	1.04 ± 0.01 (1.04 ± 0.01)	$(4.18 \pm 0.01)\text{E-}03$ $((4.19 \pm 0.01)\text{E-}03)$
4	1.03 ± 0.01 (1.03 ± 0.01)	$(3.84 \pm 0.01)\text{E-}03$ $((3.83 \pm 0.01)\text{E-}03)$
5	1.03 ± 0.01 (1.03 ± 0.01)	$(3.74 \pm 0.01)\text{E-}03$ $((3.78 \pm 0.01)\text{E-}03)$
6	1.03 ± 0.01 (1.03 ± 0.01)	$(2.84 \pm 0.02)\text{E-}03$ $((2.84 \pm 0.01)\text{E-}03)$
7	1.03 ± 0.01 (1.03 ± 0.01)	$(2.75 \pm 0.01)\text{E-}03$ $((2.76 \pm 0.01)\text{E-}03)$

with JENDL-4.0 [13] or ENDF/B-VII.0 (uranium and boron) and JENDL/HE-2007 (all nuclides except uranium and boron); total histories for adjoint flux and reaction rates were 1E + 07 and 1E + 06, respectively; the statistical error of the calculations was less than 1%. The adjoint flux was manually obtained for three-dimensional calculations as follows: an external neutron source in the Watt spectrum of ^{235}U was set inside an HEU plate; the reaction rates for the response of $\nu\Sigma_f$, which is discussed as more appropriate than Σ_f [14] to estimate the adjoint flux, were tallied over the core with NONU option to avoid the neutron multiplication; these reaction rates approximately corresponded to the adjoint flux at the position of the HEU plate; this fixed-source problem was repeated by changing the position of the external neutron source in HEU plate. The correction factor g remained constant on subcriticality, and conversely, the decreasing tendency was indicated on g^* values by varying the subcriticality, as shown in Table 4.9. Here, a slight difference was observed in the correction factors g and g^* between the selection of cross sections, and g and g^* estimated with JENDL-4.0 and JENDL/HE-2007 were used for the β_{eff} measurements.

With the results shown in Table 4.8 and fitted parameters α, B and C, β_{eff} values were deduced by Eq. (4.40) in the ADS experiments. The result of measured β_{eff} is compared with that of calculated β_{eff} by MCNP6.1 as shown in Table 4.10, revealing that the result of the optical fiber at the core center indicates acceptable accuracy of the results by MCNP6.1 within a relative difference of about 13% in the subcritical range of the ADS operation (Cases II-1 to II-7) around $k_{\text{eff}} = 0.93$. With the BF$_3$ detector, the difference between the measured and calculated β_{eff} was very large. The resulting low accuracy is attributable to the variation in the shape of the pulsed-neutron source and the extraction of correlated probability with pulsed width of neutron source based on Eq. (4.24). Thus, the emphasis was placed where the optical fiber detector located at the core center showed good accuracy because the effect of the higher mode in flux distribution is not effective and the correlated neutrons

Table 4.10 Comparison between measured and calculated β_{eff} values (Ref. [12])

Case	k_{eff}^*	Effective delayed neutron fraction β_{eff} [pcm]			
	MCNP6.1	MCNP6.1	Area ratio method	Optical fiber	BF$_3$
II-1	0.99093	817 ± 11	763 ± 21 (6.6 ± 0.2)	870 ± 81 (-6.5 ± 0.6)	911 ± 29 (-11.5 ± 0.4)
II-2	0.98274	794 ± 11	874 ± 26 (-10.0 ± 0.3)	876 ± 58 (-10.3 ± 0.7)	845 ± 33 (-6.5 ± 0.3)
II-3	0.97627	814 ± 11	919 ± 24 (-12.3 ± 0.3)	855 ± 45 (-5.0 ± 0.3)	725 ± 36 (11.0 ± 0.6)
II-4	0.97662	801 ± 11	721 ± 5 (10.0 ± 0.2)	903 ± 109 (-12.7 ± 1.5)	704 ± 28 (12.1 ± 0.5)
II-5	0.95680	826 ± 12	737 ± 6 (10.8 ± 0.2)	915 ± 75 (-10.8 ± 0.9)	303 ± 7 (63.3 ± 1.7)
II-6	0.95278	817 ± 11	729 ± 6 (10.8 ± 0.2)	921 ± 84 (-12.7 ± 1.2)	189 ± 19 (76.9 ± 7.8)
II-7	0.93358	815 ± 11	729 ± 6 (14.3 ± 0.3)	811 ± 72 (-0.4 ± 0.1)	162 ± 14 (90.1 ± 7.1)

(): Relative difference (%); *Standard deviation of 9 pcm

to the same family in fission chain is largely detected. Inversely, when the detector is located outside of the core, the measurement accuracy is considered deteriorated by the difficulty in the extraction of the correlated term related to correlated neutron detections with Eq. (4.26). To estimate the measurement accuracy of β_{eff} by proposed methodology, β_{eff} was compared with the simple estimation by the area ratio method and the relation between reactivity in dollar and pcm units. With the use of the subcriticality in Table 4.8 and k_{eff} in Table 4.10, β_{eff} can be estimated as follows:

$$\beta_{\text{eff}}^{\text{Area ratio method}} = \frac{\rho}{\rho_\$} = \frac{1}{\rho_\$}\left(1 - \frac{1}{k_{\text{eff}}}\right). \tag{4.41}$$

Here, measured β_{eff} by proposed methodology indicated comparable or more accurate results in the comparison with β_{eff} by Eq. (4.41) (named Area ratio method) in Cases II-1 to II-7 for the optical fiber detector, as shown in Table 4.10. Thus, the applicability of the measurement methodology for β_{eff} was demonstrated in the subcritical range of the ADS operation around $k_{\text{eff}} = 0.93$. Furthermore, to obtain good accuracy of β_{eff} values with the proposed methodology, improvement of correlated probability is attainable with the use of a specific shape (Gaussian distribution) of the external neutron source.

4.3   Evaluation of $\beta_{\text{eff}}/\Lambda$

4.3.1   Experimental Settings

4.3.1.1   Core Configurations

ADS experiments were carried out in a uranium-lead (U-Pb) core with 14 MeV neutrons (Fig. A3.1) and spallation neutrons (Fig. A3.2) [15]. The core comprised normal fuel rods (1/8″p60EUEU) of highly enriched uranium (HEU: 2″ × 2″ × 1/16″) and a polyethylene moderator (p: 2″ × 2″ × 1/8″) in an aluminum sheath 2.1″ × 2.1″ × 60″, as shown in Fig. A3.3a, and U-Pb fuel rods composed of an HEU plate and a Pb plate (Pb: 2″ × 2″ × 1/8″) in the center, as shown in Fig. A3.3b. The core spectrum was approximately hard in the central region for the spectrum of the actual ADS, and the driver (normal fuel) region has an H/U (hydrogen/uranium) ratio of approximately 50 in the thermal reactor.

14 MeV neutrons were generated by the injection of a deuteron beam (intensity 0.3 mA, pulsed width 10 μs, and pulsed frequency 20 or 50 Hz) onto a tritium target located at (14–15, Y; Fig. A3.1). A proton accelerator (FFAG accelerator) was operated to inject 100 MeV protons (beam spot 50 mm, intensity 10 pA, and pulsed frequency 20 Hz) onto a lead-bismuth (Pb–Bi) target (50 mm diam. and 18 mm thick) located at (15, D; Fig. A3.2) to generate spallation neutrons.

Time evolution according to the injection of external neutrons was obtained from the signals of four BF$_3$ detectors set around the core. Furthermore, in the case of spallation neutrons, an optical fiber type detector containing a Eu:LiCaAlF6 scintillator [16] was additionally installed at location (10, U; Fig. A3.2) symmetrical to BF$_3$#2 so as to examine the influence of the detector on measured subcriticality.

Subcriticality was obtained by full insertion of control and safety rods, and by the substitution of the fuel assembly for polyethylene rods, as shown in Table 4.11a. In Cases I-1 to I-5, the subcriticality was experimentally deduced with the combined use of control rod worth and its calibration curve obtained by the positive period method. Moreover, in Cases II to VII (Table 4.11b), some of the fuel rods "F" (Figs. A3.1 and A3.2) were substituted for polyethylene reflectors and configured as shown in Figs. A3.6 and A3.7. The subcriticality in dollar units was acquired experimentally by the extrapolated area ratio method [1]. Also, α was obtained by the α-fitting method in PNS experiments. The subcriticality level ranged between 500 and 7500 pcm.

4.3.1.2   Numerical Analyses

Numerical analyses were conducted by using MCNP6.1 together with ENDF/B-VII.1 (total histories 5E + 08: 5E + 05 histories per cycle and 1E + 03 active cycles) and by PARTISN [17] (with mesh size less than 10 × 10 × 10 mm; transport cross section instead of P_L scattering treatment; EO$_{16}$ quadrature for S$_N$ [18]; Fig. 4.5). In PARTISN analyses, seven effective cross Sects. (7-energy group) were generated as

Table 4.11 Subcriticality variation and pulsed-neutron frequency of 14 MeV neutrons and spallation neutrons (Ref. [15])

(a) Cases I-1 to I-5

Case	# of fuel plates	Rod insertion pattern	Pulsed frequency (Hz)	
			14 MeV neutrons	Spallation neutrons
I-1	4560	C1	20	20
I-2	4560	C1, C3	20	20
I-3	4560	C1, C2, C3	20	20
I-4	4560	C1, C3, S5	–	20
I-5	4560	C1, C3, S4, S5	–	20

(b) Cases II to VII

Case	# of fuel plates	Rod insertion*	Pulsed frequency [Hz]	
			14 MeV neutrons	Spallation neutrons
II	4440	–	20	20
III	4200	–	20	20
IV	4080	–	50	20
V	3960	–	20	20
VI	3840	–	50	20
VII	3600	–	20	20

*Full withdrawal of all control (C1, C2, and C3) and safety (S4, S5, and S6) rods

described in Ref. [19] with the SCALE6.2 code system [20], as shown in Fig. 4.6. The validation of the numerical analyses was confirmed through comparison between measured reactivity and calculated reactivity. Here, excess reactivity and control rod worth (C1, C2, and C3) were measured by the positive period method and the rod drop method, respectively, and calculated as follows:

$$\rho_{\text{excess}}^{\text{cal}} = \left(1 - \frac{1}{k_{\text{eff}}^{\text{clean}}}\right) - \left(1 - \frac{1}{k_{\text{eff}}^{\text{critical}}}\right) = \frac{1}{k_{\text{eff}}^{\text{critical}}} - \frac{1}{k_{\text{eff}}^{\text{clean}}}, \qquad (4.42)$$

$$\rho_{\text{rod}}^{\text{cal}} = \left(1 - \frac{1}{k_{\text{eff}}^{\text{critical}}}\right) - \left(1 - \frac{1}{k_{\text{eff}}^{\text{rod}}}\right) - = \frac{1}{k_{\text{eff}}^{\text{rod}}} - \frac{1}{k_{\text{eff}}^{\text{critical}}}, \qquad (4.43)$$

where $\rho_{\text{excess}}^{\text{cal}}$ and $\rho_{\text{rod}}^{\text{cal}}$ are the calculated excess reactivity and control rod worth, respectively, $k_{\text{eff}}^{\text{critical}}$ the value of the effective multiplication factor at the critical state so as to reduce the calculation bias induced by the nuclear data, $k_{\text{eff}}^{\text{clean}}$ the value of effective multiplication factor at the withdrawal of all control rods, and $k_{\text{eff}}^{\text{rod}}$ the value of effective multiplication factor at the insertion of control rod C1, C2, or C3 at the critical state. The difference was confirmed at less than 5% in control rod worth by MCNP6.1, as shown in Table 4.12, although large uncertainty was observed in excess reactivity attributed to estimating small reactivity, considering valid for subsequent analyses. For PARTISN calculations, while the C/E (calculation/experiment) ratio

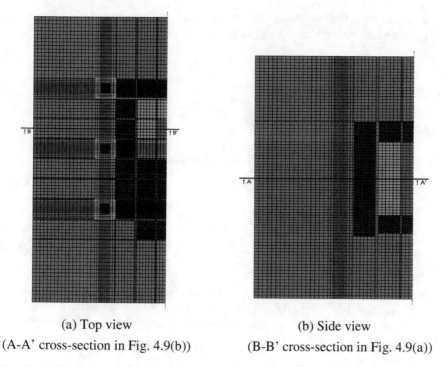

(a) Top view

(A-A' cross-section in Fig. 4.9(b))

(b) Side view

(B-B' cross-section in Fig. 4.9(a))

Fig. 4.5 Calculation geometry with grid mesh in PARTISN (Ref. [15])

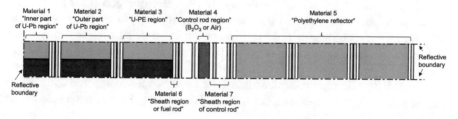

Fig. 4.6 Calculation geometry (2-D) for homogenization with SCALE/NEWT (Ref. [15])

Table 4.12 Comparison between measured and calculated excess reactivity and control rod worth (Ref. [15])

Reactivity	Experiment [pcm]	MCNP6.1 [pcm]	PARTISN [pcm]
Excess	47 ± 5	37 ± 5 (0.77 ± 0.12)	–
C1 rod	840 ± 2	872 ± 6 (1.04 ± 0.01)	983 (1.17 ± 0.01)
C2 rod	646 ± 1	665 ± 6 (1.03 ± 0.01)	806 (1.25 ± 0.01)
C3 rod	226 ± 13	228 ± 6 (1.01 ± 0.01)	229 (1.01 ± 0.06)

(): C/E (calculation/experiment) value

was 1.25 at most, the reproducibility was considered acceptable by comparison with experiments involving variations of subcriticality shown in Table 4.13.

4.3.2 Kinetics Parameters

4.3.2.1 Subcriticality in Dollar Units

The measured $\rho_\$$ was compared with the calculated $\rho_\$^{MCNP}$ by using MCNP6.1 and deduced as follows:

$$
\begin{aligned}
\rho_\$^{MCNP} &= \frac{1}{\beta_{eff}^{MCNP}} \left\{ \left(1 - \frac{1}{k_{eff}^{critical}} \right) - \left(1 - \frac{1}{k_{eff}^{case}} \right) \right\} \\
&= \frac{1}{\beta_{eff}^{MCNP}} \left(\frac{1}{k_{eff}^{case}} - \frac{1}{k_{eff}^{critical}} \right),
\end{aligned}
\tag{4.44}
$$

where k_{eff}^{case} is the effective neutron multiplication factor in each of the cases shown in Table 4.13, and β_{eff}^{MCNP} the effective delayed neutron fraction obtained by MCNP6.1 corresponding to each case. In the comparison between $\rho_\$$ and $\rho_\$^{MCNP}$ by varying the neutron source shown in Figs. 4.7 and 4.8, $\rho_\$^{MCNP}$ values were comparable until deep subcriticality.

Also, notable is that no spatial effect of detector location was observed except for that of BF3#4 detector. The results by BF3#4 detector placed near the neutron source were not reliable since $\rho_\$$ values were significantly overestimated in deep subcriticality (over 6\$) in both 14 MeV and spallation external neutron sources. The large error in deep subcriticality of 9\$ in BF3#1 and BF3#2 was caused by the low count rate of delayed neutrons. As an examination of detector type dependency on $\rho_\$$, LiFCAF fiber detector indicated almost the same $\rho_\$$ value compared with that by BF3#2, validating the measurement results and capability of the λ-mode calculation for $\rho_\$$.

4.3.2.2 Prompt Neutron Decay Constant

Measured α was compared with calculated one (α^{MCNP}) by using MCNP6.1 and deduced as follows:

$$
\alpha^{MCNP} = \frac{1}{\Lambda^{MCNP}} \left\{ \left(\frac{1}{k_{eff}^{case}} - \frac{1}{k_{eff}^{critical}} \right) - \beta_{eff}^{MCNP} \right\},
\tag{4.45}
$$

where Λ^{MCNP} is generation time obtained by MCNP6.1. In addition to α^{MCNP}, the prompt neutron decay constant by the ω-mode calculation with PARTISN ($\alpha^{PARTISN}$) was added to compare the difference between λ-mode and ω-mode calculations, as

Table 4.13 Comparison between subcriticality of experiments and calculations (MCNP6.1 and PARTISN) (Ref. [15])

(a) Cases I-1 to I-5

Case	Experiment	MCNP6.1	Difference* [pcm]	PARTISN	Difference* [pcm]
I-1	0.99292 ± 0.00004	0.99172 ± 0.00008	−43	0.98957	−259
I-2	0.99070 ± 0.00013	0.98948 ± 0.00010	−45	0.98723	−272
I-3	0.98440 ± 0.00013	0.98301 ± 0.00011	−63	0.97858	−514
I-4	0.98848 ± 0.00018	0.98725 ± 0.00011	−46	0.98532	−242
I-5	0.98034 ± 0.00018	0.97882 ± 0.00012	−78	0.97545	−422

(b) Cases II to VII

Case	MCNP6.1	PARTISN	Difference** [pcm]
II	0.99553 ± 0.00004	0.99343	211
III	0.98344 ± 0.00004	0.98169	178
IV	0.96918 ± 0.00004	0.96784	138
V	0.96096 ± 0.00004	0.96048	49
VI	0.95661 ± 0.00004	0.95585	79
VII	0.93272 ± 0.00004	0.93334	−66

*Difference between experiment and calculation $\left(\left(1 - \frac{k_{eff}^{cal}}{k_{eff}} \right) \times 10^5 \text{ [pcm]} \right)$

**Difference of k_{eff} between MCNP6.1 and PARTISN $\left(\left(1 - \frac{k_{eff}^{PARTISN}}{k_{eff}^{MCNP6.1}} \right) \times 10^5 \text{ [pcm]} \right)$

Fig. 4.7 Comparison between measured and calculated (MCNP6.1 with ENDF/B-VII.1) subcriticalities in dollar units with 14 MeV neutrons (Ref. [15])

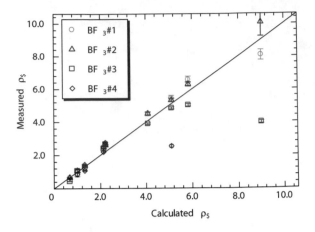

Fig. 4.8 Comparison between measured and calculated (MCNP6.1 with ENDF/B-VII.1) subcriticalities in dollar units with spallation neutrons (Ref. [15])

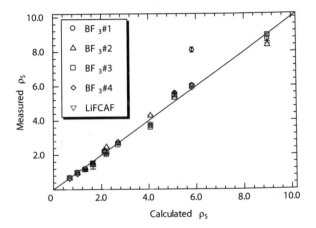

shown in Figs. 4.9 and 4.10. Here, the error of the measurement was evaluated by the fitting error. The large error was obtained in the result by BF_3#4 with spallation neutrons in Fig. 4.10 since the PNS histogram was largely influenced by the decay of the neutron source. As in $\rho_\$$, no dependence on any external neutron source was observed in the measurements, indicating that the decay of neutron flux in the fundamental mode was measured correctly. Also, the results from all detectors were equivalent, except from BF_3#4. Furthermore, the difference in measurement methodology was around 5% between the α-fitting method and the Feynman-α method as described in Ref. [20], demonstrating that the measurement was valid. In comparing α^{MCNP} with measured α, α^{MCNP} was overestimated by 1100 1/s ($k_{eff} = 0.97$). Conversely, $\alpha^{PARTISN}$ agreed with the measured ones, demonstrating that the λ-mode calculation [21] has the possibility to be incapable of evaluating the α value even for target subcriticality in ADS operations through the comprehensive comparisons.

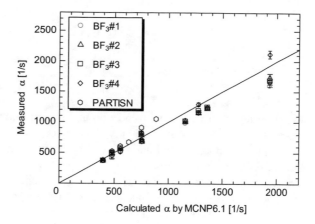

Fig. 4.9 Comparison between measured and calculated (MCNP6.1 with ENDF/B-VII.1 and PARTISN) prompt neutron decay constants in 14 MeV neutrons (Ref. [15])

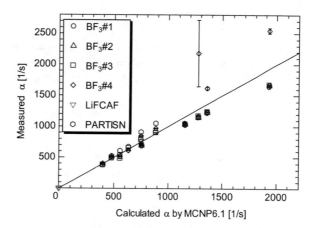

Fig. 4.10 Comparison between measured and calculated (MCNP6.1 with ENDF/B-VII.1 and PARTISN) prompt neutron decay constants in spallation neutrons (Ref. [15])

4.3.3 Results and Discussion

4.3.3.1 Evaluation of β_{eff}/Λ

β_{eff}/Λ as representative of the cores were experimentally deduced by combining α and $\rho_{\$}$ at BF$_3$#2 detector as follows:

$$\left(\frac{\beta_{eff}}{\Lambda}\right)_{EXP} = \left(\frac{\alpha^{EXP}}{1 + \rho_{\$}^{EXP}}\right). \tag{4.46}$$

Also, calculated β_{eff}/Λ values were obtained by combining adjoint-weighted effective delayed neutron fraction $\beta_{eff}^{PARTISN}$ by PARTISN, and the adjoint-weighted generation time $\Lambda^{PARTISN}$ was defined as follows:

$$\beta_{\text{eff}}^{\text{PARTISN}} = \frac{\langle \phi^+(g') \, \chi_d(g') \, \nu_d(g) \, \Sigma_f(g) \, \phi(g) \rangle}{\langle \phi^+(g') \, \chi(g') \, \nu(g) \, \Sigma_f(g) \, \phi(g) \rangle}, \tag{4.47}$$

$$\Lambda^{\text{PARTISN}} = \frac{\langle \frac{1}{v(g)} \phi^+(g) \, \phi(g) \rangle}{\langle \phi^+(g') \, \chi(g') \, \nu(g) \, \Sigma_f(g) \, \phi(g) \rangle}, \tag{4.48}$$

where ϕ and ϕ^+ are the forward and adjoint fluxes in λ-mode or ω-mode calculations, respectively, χ and χ_d the total and delayed neutron fission spectra, respectively, ν and ν_d the number of total and delayed neutrons by fission reaction, respectively, Σ_f the effective fission cross section, v the neutron velocity, < > the integration over energy group and phase space, and g and g' the energy groups. Here, $\chi_d(g')$ was obtained by the extraction of nuclear data of ^{235}U from ENDF/B-VII.1 by NJOY99 [22]. Also, $\nu_d(g)$ was defined as follows:

$$\nu_d(g) = \nu(g) - \nu_p(g), \tag{4.49}$$

where $\nu_p(g)$, which was obtained as in the procedure for $\chi_d(g')$, is the number of prompt neutrons by the fission reaction. Furthermore, $\beta_{\text{eff}}/\Lambda$ values by using MCNP6.1 were obtained by combining kinetics parameters (evaluated with KOPT option in MCNP6.1).

By varying subcriticality, the value of $\beta_{\text{eff}}/\Lambda$ tended to decrease with deep subcriticality as shown in Fig. 4.11. Interestingly, $\beta_{\text{eff}}/\Lambda$ differed even when subcriticality was the same, although the core size was different in Cases I-2 and III. Also, the nonsignificant difference of $\beta_{\text{eff}}/\Lambda$ values was observed indicated between 14 MeV and spallation neutron sources in measurements. Thus, $\beta_{\text{eff}}/\Lambda$ was considered independent of any external neutron source, indicating that the measurement was conducted correctly to extract fundamental-mode components in the time evolution of neutron flux. In the comparison between calculations, the λ-mode calculation by PARTISN showed a bias compared with that by MCNP6.1, and the results were

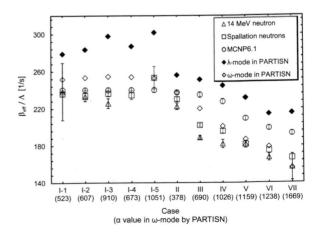

Fig. 4.11 $\beta_{\text{eff}}/\Lambda$ value for α value in ω-mode for subcriticality variation (Ref. [15])

almost the same as the measured ones at shallow subcritical states (Cases I-1 to I-5). At deep subcriticality, however, the λ-mode calculations showed a large difference as compared with the experiments. Conversely, ω-mode calculations correctly evaluated the experiments in the whole range except for Case III, indicating that the kinetics parameters were not correctly evaluated by the λ-mode calculation even for target subcriticality in ADS for Case IV ($k_{\text{eff}} = 0.97$).

Here, $\rho_{\$}^{\text{MCNP}}$ by λ-mode calculation showed agreement with measured $\rho_{\$}$ even in deep subcritical states, as shown in Figs. 4.7 and 4.8. Conversely, as shown in Figs. 4.9 and 4.10, from the comparison of α values, it was observed that the large discrepancy between experiments and calculations as subcriticality becomes deeper. The discrepancy in the comparison between the experiment and the calculation was considered attributable to Λ value for estimating calculated α value by Eq. (4.45). The difference of Λ value between λ-mode and ω-mode calculations could be also explained by comparing between $\rho_{\$}$ and $\beta_{\text{eff}}/\Lambda$ since $\rho_{\$}$ values agreed between the experiment and the λ-mode calculation more accurately than the results of $\beta_{\text{eff}}/\Lambda$ (Fig. 4.11).

4.3.3.2   Discussion

The discussion is focused on the variation of each kinetics parameter by the calculations toward the subcriticality. β_{eff} and Λ were compared in terms of the λ-mode or the ω-mode calculation as shown in Fig. 4.12. The values of β_{eff} were independent of subcriticality in ω-mode calculations; however, λ-mode calculations showed a slightly increased tendency toward subcriticality. Notably, Λ values revealed a different tendency compared with β_{eff} values under λ-mode and ω-mode calculations. Furthermore, β_{eff} values were equivalent in Cases I-5 and IV; by contrast, however, Λ values were different even at almost the same subcritical states, indicating that Λ is sensitive to the size of the core and to the insertion of control rods.

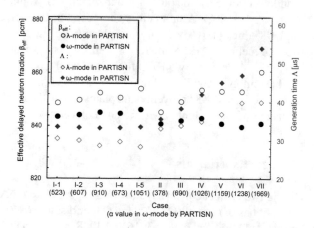

Fig. 4.12 Relationship between calculated β_{eff} and Λ for subcriticality variation (Ref. [15])

In qualitative calculation property, the ω-mode calculation involves $-\alpha/v$ value in absorption term compared to the λ-mode calculation, indicating softer neutron spectrum by decreasing the total absorption cross section of $-\alpha/v$ (since lower the energy, the more increase $-\alpha/v$ value). Furthermore, the adjoint neutron spectrum is estimated softer by the λ-mode calculation to induce fission reactions, resulting in the overestimation of β_{eff} value in the λ-mode calculation since the importance of delayed neutrons is emphasized by the adjoint neutron flux. Furthermore, the different estimation of adjoint neutron spectra is also considered to influence the value of Λ since observation showed an overestimation of β_{eff} and underestimation of Λ in the λ-mode calculation. Accordingly, calculated k_{eff} by the λ-mode calculation is implied to be different from actual neutron multiplication factor (in the subcritical system) by the combined use of α, β_{eff}, and Λ since the spectrum calculation was inadequate in the λ-mode calculations for super-critical and subcritical cores.

Investigation of the cause in the variation of Λ is considered limited since the number of energy groups is insufficient for characterizing the neutron spectra of ϕ and ϕ^+. Thus, further investigation with a higher number of energy groups is needed.

4.4   Neutron Generation Time

4.4.1   *Experimental Settings*

4.4.1.1   Core Configuration

Critical cores set in the A-core (Figs. A4.1 and A4.4) have polyethylene moderator and reflector rods, and three different fuel assemblies: normal "F" (Fig. A4.2a), partial "8", and "4" (Figs. A4.2b and A4.5) corresponding to Figs. A4.1 and A4.4, respectively. The normal fuel assembly "F" is composed of 60 unit cells, and upper and lower polyethylene blocks about $23''$ and $21''$ long, respectively, in an Al sheath ($2.1 \times 2.1 \times 60''$). A unit cell is composed of two HEU fuel plates $2 \times 2''$ square and $1/8''$ thick ($1/16'' \times 2$), polyethylene (p) plate $2 \times 2''$ square and $1/8''$ thick, for normal fuel plate "F." Numeral "8" represents a partial fuel assembly composed of eight unit cells, with two HEU fuel plates and a polyethylene plate as in the normal fuel assembly, providing 52 unit cells of two Al plates $2 \times 2''$ square and $1/8''$ thick ($1/16'' \times 2$), and $1/8''$ polyethylene plates. Also, the numeral "4" corresponds to four unit cells of fuel assembly with 56 unit cells composed of Al and polyethylene plates.

4.4.1.2   Experiments

In the two critical cores [23] shown in Figs. A4.1 and A4.4, criticality was reached at positions of control (C1, C2, and C3) and safety (S4, S5, and S6) rods for the total number of HEU fuel plates: 3016 and 3008, as shown in Tables 4.14 and 4.15, respec-

Table 4.14 Positions of control and safety rods at critical state [mm] (# of HEU plates: 3016; critical core in Fig. A4.1) (Ref. [23])

Control rod			Safety rod		
C1	C2	C3	S4	S5	S6
1200.00	1200.00	630.01	1200.00	1200.00	1200.00

(1200 [mm]: Position of upper limit)

Table 4.15 Positions of control and safety rods at critical state [mm] (# of HEU plates: 3008; critical core in Fig. A4.4) (Ref. [23])

Control rod			Safety rod		
C1	C2	C3	S4	S5	S6
1200.00	723.31	1200.00	1200.00	1200.00	1200.00

(1200 [mm]: Position of upper limit)

tively. Excess reactivity and control rod worth in pcm units were then experimentally obtained by the positive period method and the rod drop method, respectively, with the use of two kinetics parameters β_{eff} and ℓ estimated by numerical calculations. Here, excess reactivity obtained experimentally was about 250 and 100 pcm in Figs. A4.1 and A4.2, respectively, and control rod worth of C1, C2 (Fig. A4.1 or C3 in Fig. A4.2), and C3 (Fig. A4.2 or C2 in Fig. A4.1) was about 880, 150, and 520 pcm, respectively.

In the ADS core (3000 HEU fuel plates) at the subcritical state [23], deuteron beams were injected onto a tritium target set at (14–15, X) in Fig. A4.3a. In the same subcritical core, another external neutron source of 100 MeV proton beams was injected onto a lead-bismuth (Pb–Bi) target set at (15, H) in Fig. A4.6a. Two different external neutron sources (14 MeV neutrons; 100 MeV protons: spallation neutrons) were separately injected into the subcritical core: deuteron beams, 0.1 mA current, 20 Hz repetition rate, 97 μs width, and 1×10^5 s^{-1} neutron yield; 100 MeV protons, 0.1 nA current, 20 Hz repetition rate, 100 ns width, and 1×10^7 s^{-1} neutron yield. The Pb–Bi target was 50 mm in diameter and 18 mm thick. In a series of ADS experiments and during the injection of an external neutron source, α and $\rho_\$$ were measured by the least-square fitting method and the extended area ratio method [1], respectively.

Moreover, to examine the effects of detector position dependence [24] and external neutron source spectrum on measurement results, three optical fiber detectors [16] were set at (14-15, L-M); (13-14, K-L); (12-13, J-K) in Fig. A4.3a and (15-16, O-Q); (16-17, Q-R); (18-19, S-T) in Fig. A4.6a; also, three BF-3 detectors were at (15, H); (11, I); (11, M) in Fig. A4.3a and (15, U); (11, T); (10, O) in Fig. A4.6a.

4.4.1.3 Kinetics Parameters

For the α-fitting method [25], α is easily deduced by combining subcriticality $(-\rho_{pcm})$ in pcm units, β_{eff} and Λ as follows:

$$\alpha = \frac{\beta_{eff} - \rho_{pcm}}{\Lambda}. \tag{4.50}$$

The relationship between ρ_{pcm} and $\rho_\$$ is expressed with the use of β_{eff} as follows:

$$\rho_{pcm} = \rho_\$ \beta_{eff}. \tag{4.51}$$

Substituting Eq. (4.51) for Eq. (4.50), β_{eff}/Λ can be expressed as follows:

$$\frac{\beta_{eff}}{\Lambda} = \frac{\alpha}{1 - \rho_\$}. \tag{4.52}$$

The value of β_{eff}/Λ is easily deduced from the experimental results of α and $\rho_\$$.

4.4.2 Results and Discussion

4.4.2.1 Eigenvalue Calculations

For transport, numerical calculations were performed by the Monte Carlo transport code, MCNP6.1 together with nuclear data libraries, JENDL-4.0 and ENDF/B-VII.1. Here, in MCNP6.1, since the effects of reactivity by neutron detectors (optical fiber detectors, BF-3 detectors, fission chambers, and uncompensated ionization chambers) and control (safety) rods are not negligible, neutron detectors and control (safety) rods were modeled precisely in the simulated geometry and transport calculations. The precision of numerical criticality in pcm units was attained by the eigenvalue calculations with a total of 5×10^8 histories (1×10^5 histories per cycle, 5×10^3 active cycles, and 1×10^2 skip cycles) and a statistical error of about 4 pcm. Also, main kinetics parameters, β_{eff}, Λ and Rossi-α (termed β_{eff}/Λ in MCNP6.1) values were obtained by the eigenvalue calculations, when obtaining the effective multiplication factor by the k-code option.

To confirm the numerical precision of eigenvalue calculations by MCNP6.1, excess reactivity and control rod worth (C1, C2, and C3) were compared with those obtained from the experiments by the positive period method and the rod drop method, respectively. The MCNP eigenvalue calculations in the critical state are not always unit, differing from the experimental results in critical state obtained by nuclear data accuracy and uniformed number density approximation in core materials, although a critical core configuration is more closely simulated by MCNP. Thus, excess reactivity ρ_{excess}^{cal} in pcm units was numerically deduced by the difference between two effective multiplication factors $k_{critical}$ and $k_{super-critical}$, in critical and super-critical (clean) cores, respectively, as follows:

$$\rho_{excess}^{cal} = \frac{1}{k_{critical}} - \frac{1}{k_{super-critical}}. \tag{4.53}$$

Also, control rod worth was numerically deduced by the difference between the critical and the subcritical (rod insertion) cores, as in Eq. (4.53).

Using the kinetics parameters obtained by MCNP6.1, experimental reactivity ρ^{exp} was deduced and compared with numerical reactivity ρ^{cal} obtained by JENDL-4.0 and ENDF/B-VII.1, as shown in Tables 4.16 and 4.17, with the consideration of statistical errors of neutron counts obtained from the neutron detectors and processing of error propagation caused by experimental analyses. For the critical cores (3016 and 3008 HEU plates in Tables 4.14 and 4.15) shown in Figs. A4.1 and A4.4, respectively, numerical reactivity obtained by JENDL-4.0 revealed a fairly good agreement with the experimental reactivity within a relative difference of around 5% through the C/E (calculation/experiment) value of excess reactivity and control rod worth, as shown in Tables 4.16 and 4.18. In terms of ENDF/B-VII.1, the difference between numerical and experimental reactivity was relatively large over 10%, as shown in Tables 4.17 and 4.19. The difference between β_{eff} values by JENDL-4.0 and ENDF/B-VII.1 was attributable to numerical results by the MCNP calculations in fraction β_i ($i = 1$–6: precursor group of delayed neutrons) and decay constant λ_i of delayed neutrons, as shown in Figs. 4.13 and 4.14, respectively.

Table 4.16 Comparison between measured and calculated reactivity [pcm] with JENDL-4.0 (# of HEU plates: 3016; critical core in Fig. A4.1 and Table 4.14) (Ref. [23])

	ρ_{J40}^{cal}	ρ_{J40}^{exp}	C/E ($\rho_{J40}^{cal}/\rho_{J40}^{exp}$)
Excess reactivity	259.9 ± 5.7	271.7 ± 0.1	0.96 ± 0.02
C1 rod worth	876.1 ± 5.7	867.4 ± 3.9	1.01 ± 0.01
C2 rod worth	154.8 ± 5.7	142.8 ± 3.5	1.08 ± 0.05
C3 rod worth	529.2 ± 5.7	506.4 ± 4.1	1.05 ± 0.01

MCNP6.1 with JENDL-4.0: $\beta_{eff} = 813 \pm 10$ [pcm], $\Lambda = 31.67 \pm 0.07$ [µs]

Table 4.17 Comparison between measured and calculated reactivity [pcm] with ENDF/B-VII.1 (# of HEU plates: 3016; critical core in Fig. A4.1 and Table 4.14) (Ref. [23])

	ρ_{E71}^{cal}	ρ_{E71}^{exp}	C/E ($\rho_{E71}^{cal}/\rho_{E71}^{exp}$)
Excess reactivity	253.9 ± 5.6	229.8 ± 0.3	1.10 ± 0.02
C1 rod worth	862.5 ± 5.7	718.8 ± 3.2	1.20 ± 0.01
C2 rod worth	138.2 ± 5.6	118.3 ± 2.9	1.17 ± 0.05
C3 rod worth	513.1 ± 5.7	419.6 ± 4.5	1.22 ± 0.01

MCNP6.1 with ENDF/B-VII.1: $\beta_{eff} = 801 \pm 10$ [pcm], $\Lambda = 31.26 \pm 0.07$ [µs]

Table 4.18 Comparison between measured and calculated reactivity [pcm] with JENDL-4.0 (# of HEU plates: 3008; critical core in Fig. A4.4 and Table 4.15) (Ref. [23])

	$\rho_{\text{J40}}^{\text{cal}}$	$\rho_{\text{J40}}^{\text{exp}}$	C/E ($\rho_{\text{J40}}^{\text{cal}}/\rho_{\text{J40}}^{\text{exp}}$)
Excess reactivity	106.1 ± 5.7	100.2 ± 1.7	1.06 ± 0.06

MCNP6.1 with JENDL-4.0: $\beta_{\text{eff}} = 812 \pm 10$ [pcm], $\Lambda = 31.57 \pm 0.07$ [μs]

Table 4.19 Comparison between measured and calculated reactivity [pcm] with ENDF/B-VII.1 (# of HEU plates: 3008; critical core in Fig. A4.4 and Table 4.15) (Ref. [23])

	$\rho_{\text{E71}}^{\text{cal}}$	$\rho_{\text{E71}}^{\text{exp}}$	C/E ($\rho_{\text{E71}}^{\text{cal}}/\rho_{\text{E71}}^{\text{exp}}$)
Excess reactivity	97.6 ± 5.6	84.7 ± 1.5	1.15 ± 0.07

MCNP6.1 with ENDF/B-VII.1: $\beta_{\text{eff}} = 802 \pm 10$ [pcm], $\Lambda = 31.30 \pm 0.07$ [μs]

Fig. 4.13 Comparison between fraction β_i ($i = 1$–6) by JENDL-4.0 and ENDF/B-VII.1 (# of HEU plates: 3016; critical state in Fig. A4.1 and Table 4.14) (Ref. [23])

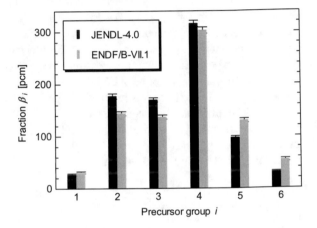

Fig. 4.14 Comparison between decay constant λ_i ($i = 1$–6) by JENDL-4.0 and ENDF/B-VII.1 (# of HEU plates: 3016; critical state in Fig. A4.1 and Table 4.14) (Ref. [23])

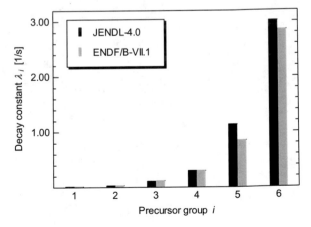

As shown in Tables 4.16, 4.17, 4.18 and 4.19, numerical results (eigenvalue calculations) by MCNP6.1 with JENDL-4.0 demonstrated good agreement with the experimental data of reactivity in the critical cores. JENDL-4.0 was taken as a reference library by comparing it with ENDF/B-VII.1.

4.4.2.2 Experimental Analyses of β_{eff}/Λ

Kinetics parameters β_{eff}, Λ and Rossi-α, and k_{eff} values were obtained by the MCNP eigenvalue calculations and compared with those of JENDL-4.0 and ENDF/B-VII.1, for critical and near-critical states, as shown in Tables 4.20, 4.21, and 4.22. For the two critical cores shown in Figs. A4.1 and A4.4, JENDL-4.0 demonstrated a small difference between the super-critical (Table 4.20) and the critical (Table 4.21) states,

Table 4.20 Comparison between kinetic parameters by MCNP6.1 with JENDL-4.0 and ENDF/B-VII.1 (# of HEU plates: 3016 in Fig. A4.1) (Ref. [23])

	Parameter	JENDL-4.0	ENDF/B-VII.1
Super-critical core (clean core)	β_{eff} [pcm]	813 ± 10	801 ± 10
	Λ [μs]	31.67 ± 0.07	31.26 ± 0.07
	Rossi-α [s^{-1}]	256.89 ± 3.11	256.18 ± 3.11
Critical core (partial insertion of C3 rod)	β_{eff} [pcm]	818 ± 10	800 ± 10
	Λ [μs]	30.91 ± 0.07	30.69 ± 0.07
	Rossi-α [s^{-1}]	264.55 ± 3.23	260.66 ± 3.17

Table 4.21 Comparison between kinetic parameters by MCNP6.1 with JENDL-4.0 and ENDF/B-VII.1 (# of HEU plates: 3008 in Fig. A4.4) (Ref. [23])

	Parameter	JENDL-4.0	ENDF/B-VII.1
Super-critical core (clean core)	β_{eff} [pcm]	812 ± 10	802 ± 10
	Λ [μs]	31.57 ± 0.07	31.30 ± 0.07
	Rossi-α [s^{-1}]	257.22 ± 3.11	256.35 ± 3.11
Critical core (partial insertion of C2 rod)	β_{eff} [pcm]	821 ± 10	801 ± 10
	Λ [μs]	31.33 ± 0.07	30.03 ± 0.07
	Rossi-α [s^{-1}]	262.22 ± 3.17	258.09 ± 3.15

Table 4.22 Comparison between kinetic parameters by MCNP6.1 with JENDL-4.0 and ENDF/B-VII.1 (# of HEU plates: 3000; subcritical core in Figs. A4.3a and A4.6a) (Ref. [23])

	Parameter	JENDL-4.0	ENDF/B-VII.1
Subcritical core (clean core)	β_{eff} [pcm]	806 ± 10	796 ± 10
	Λ [μs]	31.71 ± 0.07	31.42 ± 0.07
	Rossi-α [s^{-1}]	254.05 ± 3.07	253.35 ± 3.09

because of the slight difference in the number of fuel plates: 3016 and 3008. Also, in terms of ENDF/B-VII.1, the three kinetics parameters were almost the same in the two states.

At the subcritical state with 3000 fuel plates, three kinetics parameters showed a meaningful change from the critical state, although the subcriticality was very small and around the criticality (Table 4.22). Kinetic parameters α and $\rho_\$$ were then obtained by the extended area ratio method and the least-square fitting method, respectively, by varying the external neutron source: 14 MeV neutrons (Table 4.23) and spallation neutrons (Table 4.24). Also, as shown in Tables 4.23 and 4.24, detector position dependency was revealed interestingly: the results of Fiber #1 and BF-3 #3 were rather good, and on the contrary, those of Fiber #2, BF-3 #1, and BF-3 #3 looked problematic.

Table 4.23 Measured prompt neutron decay constants α [s^{-1}] deduced by least-square fitting method, subcriticality $\rho_\$$ [$] (dollar units) by extended area ratio method, and $\beta_{\mathrm{eff}}/\Lambda$ [s^{-1}] by α-fitting method (# of HEU plates: 3000; subcritical core with 14 MeV neutrons in Fig. A4.3a) (Ref. [23])

Detector	α [s^{-1}]	$\rho_\$$ [$]	$(\beta_{\mathrm{eff}}\,\Lambda)^{\mathrm{exp}}$ [s^{-1}]	C/E
BF-3 #1	294.78 ± 29.27	0.0228 ± 0.0070	288.20 ± 93.30	0.88 ± 0.29
BF-3 #2	226.96 ± 23.31	0.0158 ± 0.0054	223.44 ± 79.47	1.14 ± 0.41
BF-3 #3	258.58 ± 12.34	0.0194 ± 0.0027	253.66 ± 37.32	1.00 ± 0.15
Fiber #1	248.38 ± 39.04	0.0207 ± 0.0056	243.36 ± 72.07	1.04 ± 0.31
Fiber #2	229.95 ± 14.51	0.0200 ± 0.0032	225.45 ± 38.73	1.13 ± 0.19
Fiber #3	273.99 ± 26.24	0.0261 ± 0.0056	267.02 ± 62.56	0.95 ± 0.22

MCNP6.1 with JENDL-4.0: $\beta_{\mathrm{eff}} = 806 \pm 10$ [pcm], $\Lambda = 31.71 \pm 0.07$ [μs], $(\beta_{\mathrm{eff}}/\Lambda)^{\mathrm{cal}}_{\mathrm{J40}}$(MCNP6.1) $= 254.05 \pm 3.07$ [s^{-1}]; C/E $= (\beta_{\mathrm{eff}}/\Lambda)^{\mathrm{cal}}_{\mathrm{J40}}/ (\beta_{\mathrm{eff}}/\Lambda)^{\mathrm{exp}}$

Table 4.24 Measured prompt neutron decay constants α [s^{-1}] deduced by least-square fitting method, subcriticality $\rho_\$$ [$] (dollar units) by extended area ratio method, and $\beta_{\mathrm{eff}}/\Lambda$ [s^{-1}] by α-fitting method (# of HEU plates: 3000; subcritical core with spallation neutrons in Fig. A4.6a) (Ref. [23])

Detector	α [s^{-1}]	$\rho_\$$ [$]	$(\beta_{\mathrm{eff}}/\Lambda)^{\mathrm{exp}}$ [s^{-1}]	C/E
BF-3 #1	276.25 ± 7.17	0.0900 ± 0.0011	253.43 ± 7.25	1.00 ± 0.03
BF-3 #2	300.62 ± 7.03	0.0966 ± 0.0011	274.13 ± 7.15	0.93 ± 0.03
BF-3 #3	259.96 ± 6.00	0.0899 ± 0.0011	238.53 ± 6.18	1.07 ± 0.03
Fiber #1	283.79 ± 15.38	0.0973 ± 0.0025	258.62 ± 15.49	0.98 ± 0.06
Fiber #2	360.05 ± 64.85	0.1139 ± 0.0087	323.24 ± 63.19	0.79 ± 0.15
Fiber #3	269.83 ± 26.36	0.1392 ± 0.0039	236.85 ± 24.04	1.07 ± 0.11

MCNP6.1 with JENDL-4.0: $\beta_{\mathrm{eff}} = 806 \pm 10$ [pcm], $\Lambda = 31.71 \pm 0.07$ [μs], $(\beta_{\mathrm{eff}}/\Lambda)^{\mathrm{cal}}_{\mathrm{J40}}$(MCNP6.1) $= 254.05 \pm 3.07$ [s^{-1}]; C/E $= (\beta_{\mathrm{eff}}/\Lambda)^{\mathrm{cal}}_{\mathrm{J40}}/ (\beta_{\mathrm{eff}}/\Lambda)^{\mathrm{exp}}$

On the basis of the experimental results of α and $\rho_\$$, the value of $(\beta_{eff}/\Lambda)^{\text{exp}}$ was experimentally deduced by Eq. (4.52) for 14 MeV neutrons (Table 4.23) and spallation neutrons (Table 4.24), and compared with that of $(\beta_{eff}/\Lambda)^{\text{cal}}_{\text{J40}}$ obtained by MCNP6.1 together with JENDL-4.0. With the 14 MeV neutrons, since subcriticality in pcm units was found to be near critical, about 16.7 ± 4.5 pcm, the value of $(\beta_{\text{eff}}/\Lambda)^{\text{cal}}_{\text{J40}}$ was observed the same as that of α. The result was considered valid at a shallow subcritical state. Among optical fibers and BF-3 detectors, Fiber #1 revealed a fairly good agreement with the MCNP calculation within a relative difference of 4% in the C/E value (Table 4.23), which was attributable to the location of Fiber #1 near the center of the core (Fig. A4.1): the position dependence caused by the spatial effect was very small on the experimental results of α and $\rho_\$$. Also, with the spallation neutrons (Table 4.24), Fiber #1 demonstrated the same accuracy about 2% in the C/E value as with 14 MeV neutrons, although the subcriticality in pcm units was about 78.4 ± 2.2 pcm. The value of $(\beta_{\text{eff}}/\Lambda)^{\text{exp}}$ was found to be experimentally valid in the deduction of kinetics parameters, through a comparison between the experiments and the MCNP6.1 calculations with JENDL-4.0 (Tables 4.23 and 4.24).

4.4.2.3 Discussion

Experimental analyses of the ADS with spallation neutrons at KUCA have clearly demonstrated that the value of β_{eff} has a little effect on the evaluation of subcriticality in pcm units converted from that in dollar units, when compared with the results of numerical subcriticality in pcm units [2]. In the present study, considering that the values of β_{eff} are almost the same in the near-critical configurations, particular attention was directed to kinetic parameter Λ obtained by combining α and $\rho_\$$.

The values of β_{eff} by MCNP6.1 with JENDL-4.0 in the super-critical (3008 HEU fuel plates) and subcritical (3000 plates) configurations were 812 ± 10 and 806 ± 10 pcm as shown in Tables 4.21 and 4.22, respectively. Since the values of β_{eff} are almost the same and within the allowance of experimental uncertainty, the ratio of $(\beta_{\text{eff}}/\Lambda)^{\text{cal}}_{\text{super-critical}}$ and $(\beta_{\text{eff}}/\Lambda)^{\text{cal}}_{\text{subcritical}}$ in the near-critical configurations is approximated as follows:

$$(\beta_{\text{eff}}/\Lambda)^{\text{cal}}_{\text{super-critical}}/(\beta_{\text{eff}}/\Lambda)^{\text{cal}}_{\text{subcritical}} \approx \left(\Lambda_{\text{subcritical}}/\Lambda_{\text{super-critical}}\right)^{\text{cal}}. \qquad (4.54)$$

Here, Eq. (4.54) is defined as "Lambda ratio" that is a relative value of Λ in the near-critical configurations. From the results of Tables 4.23 and 4.24, assuming that the value of $(\beta_{\text{eff}}/\Lambda)^{\text{cal}}_{\text{subcritical}}$ in the subcritical state by MCNP6.1 is equal to that of $(\beta_{\text{eff}}/\Lambda)^{\text{exp}}$ by the experiment, Eq. (4.54) can finally be written as follows, by substituting $(\beta_{\text{eff}}/\Lambda)^{\text{cal}}_{\text{subcritical}}$ for $(\beta_{\text{eff}}/\Lambda)^{\text{exp}}$ and applying again the assumption with respect to β_{eff} in Eq. (4.54):

$$(\beta_{\text{eff}}/\Lambda)^{\text{cal}}_{\text{super-critical}}/\left(\beta_{eff}/\Lambda\right)^{\text{exp}} \approx \Lambda^{\text{exp}}/\Lambda^{\text{cal}}_{\text{super-critical}}. \qquad (4.55)$$

On the basis of the validity of $(\beta_{\text{eff}}/\Lambda)^{\text{exp}}$ mentioned in Sect. 4.4.2, Eq. (4.55) can be interpreted as a kind of Lambda ratio between the experiments and the calculations, with the combined use of $(\beta_{\text{eff}}/\Lambda)^{\text{cal}}_{\text{super-critical}}$ and $(\beta_{\text{eff}}/\Lambda)^{\text{exp}}$. Actually, the value of $(\beta_{\text{eff}}/\Lambda)^{\text{exp}}$ obtained by the ADS experiments at KUCA can be used as an index of Λ in the near-critical configurations, and is, indeed, expected to be applied for the validation of Λ, by stochastic and deterministic calculations, at near-critical configurations in a thermal neutron spectrum core.

4.5 Conclusion

A calculation methodology of $\beta^{\text{RR}}_{\text{source}}$ by the k-ratio method with an external neutron source has been proposed for the subcriticality estimation. Subcriticality measurements were carried at the KUCA-A cores to examine its applicability of the proposed calculation methodology by varying the subcriticality and the external neutron source. To confirm the validity of the proposed methodology, the eigenvalue calculations were firstly performed to estimate the multiplication factor k_{RR} with reaction rates by MCNPX. Further, $\beta^{\text{RR}}_{\text{eigen}}$ by the k-ratio method with reaction rates was compared with the reference ones obtained by MCNP6.1, respectively. For the estimation of subcriticality by the fixed-source calculations, $\beta^{\text{RR}}_{\text{source}}$ was observed to be dependent on the variation of subcriticality and external neutron source. Finally, the subcriticality with $\beta^{\text{RR}}_{\text{source}}$ was acquired well ranging between about 0.99 and 0.97 in k_{eff} and revealed the comparative tendency in the deep subcriticality through the estimation with the use of $\beta^{\text{RR}}_{\text{source}}$ obtained by the proposed calculation methodology.

In the methodology for estimating the value of β_{eff}, the Rossi-α method was applied to the neutron noise analyses at the PNS experiments with pulsed spallation neutrons. With the fitting curve obtained by neutron noise analyses, the signals from two detectors installed at the center of the core (optical fiber detector) and outside the core (BF$_3$ detector) and the correction factors, β_{eff} was measured and compared with the value calculated by MCNP6.1. The optical fiber detector located at the core center showed that the accuracy of the measured value, compared with the calculated one, was within a relative difference of about 13% in the subcritical range of the ADS operation around $k_{\text{eff}} = 0.93$. The result with BF$_3$ detector installed outside the core was not compared with the calculated one because of the low accuracy attributed to the uncorrelated probability. The applicability of the measurement methodology based on the Rossi-α method was demonstrated by the comparison between calculated and measured β_{eff} values in ADS experiments with spallation neutrons.

ADS experiments were carried out to evaluate $\beta_{\text{eff}}/\Lambda$ values by varying detector type, detector position, external neutron source, and subcriticality. The capability of λ-mode and ω-mode calculations was examined by comparing directly measured $\rho_\$$, α, and $\beta_{\text{eff}}/\Lambda$ in the PNS experiment with the target subcriticality of ADS ranging between 500 and 7500 pcm. The measurement of $\rho_\$$ and α indicated slight dependence on an external neutron source but not on any spatial effect except for the BF_3 detector located near the neutron source. In the experimental analyses, calculated $\rho_\$$ (λ-mode and ω-mode) showed good agreement with measured $\rho_\$$ within the whole range of subcriticality; however, the value of α by the λ-mode calculation showed a difference in the experiment at deep subcriticality. Conversely, α obtained by the ω-mode calculation agreed with the experiments. The calculated results of $\beta_{\text{eff}}/\Lambda$ were compared with the measured ones to examine their capability under subcriticality variation; consequently, an agreement was observed between the experiments and ω-mode calculations under a wide range of subcriticality. Notably, however, the λ-mode calculations differed from the experiments even under slight subcriticality, implying the necessity of introducing ω-mode calculations in ADS design for evaluating actual neutron multiplication factor in the subcritical system and kinetics parameters.

Main kinetics parameters, α and $\rho_\$$, were experimentally obtained from the KUCA core, and β_{eff} and Λ, were numerically validated by MCNP6.1, at the near-critical configurations: super-critical and subcritical states. The experimental value of $(\beta_{\text{eff}}/\Lambda)^{\text{exp}}$ was then available for use as an index of Λ in the near-critical configurations, with an attempt at the validation of Λ by the numerical calculations. From the results of experimental and numerical analyses, the importance of the experimental value of $(\beta_{\text{eff}}/\Lambda)^{\text{exp}}$ was emphasized for the verification of Λ, since the kinetics parameters were successfully obtained from the clean cores of near-critical configurations (super-critical and subcritical states) in the thermal spectrum core.

References

1. Gozani T (1962) A modified procedure for the evaluation of pulsed source experiments in subcritical reactors. Nukleonik 4:348
2. Yamanaka M, Pyeon CH, Misawa T (2016) Monte Carlo approach of effective delayed neutron fraction by k-ratio method with external neutron source. Nucl Sci Eng 184:551
3. Yagi T, Misawa T, Pyeon CH (2011) A small high sensitivity neutron detector using a wavelength shifting fiber. Appl Radiat Isot 69:176
4. Hendricks JS et al (2005) MCNPX user's manual, version 2.5.0. LA-UR-05-2675
5. Chadwick MB, Obložinský P, Herman M et al (2006) ENDF/B-VII.0: next generation evaluated nuclear data library for nuclear science and technology. Nucl Data Sheets 107:2931
6. Bretscher MM (1997) Perturbation independent methods for calculating research reactor kinetic parameters. ANL/RERTR/TM30

7. Goorley JT et al (2013) Initial MCNP6 release overview—MCNP6 version 1.0. LA-UR-13-22934
8. Takada H, Kosako K, Fukahori T (2009) Validation of JENDL high-energy file through analyses of spallation experiments at incident proton energies from 0.5 to 2.83 GeV. J Nucl Sci Technol 46:589
9. Spriggs GD (1993) Two Rossi-α techniques for measuring the effective delayed neutron fraction. Nucl Sci Eng 113:161
10. Kitamura Y, Pázsit I, Wright J et al (2005) Calculation of the pulsed Feynman- and Rossi-alpha formulae with delayed neutrons. Ann Nucl Energy 32:671
11. Degweker SB, Rana YS (2007) Reactor noise in accelerator driven systems-II. Ann Nucl Energy 34:463
12. Yamanaka M, Pyeon CH, Kim SH et al (2017) Effective delayed neutron fraction in accelerator-driven system experiments with 100 MeV protons at Kyoto University Critical Assembly. J Nucl Sci Technol 54:293
13. Shibata K, Iwamoto O, Nakagawa T et al (2011) JENDL-4.0: a new library for nuclear science and technology. J Nucl Sci Technol 48:1
14. Chiba G, Nagaya Y, Mori T (2011) On effective delayed neutron fraction calculations with iterated fission probability. J Nucl Sci Technol 48:1163
15. Yamanaka M, Pyeon CH, Endo T et al (2020) Experimental analyses of β_{eff}/Λ in accelerator-driven system at Kyoto University Critical Assembly. J Nucl Sci Technol 57:205
16. Watanabe K, Kawabata Y, Yamazaki A et al (2015) Development of an optical fiber type detector using a Eu:LiCaAlF6 scintillator for neutron monitoring in boron neutron capture therapy. Nucl Instrum Methods A 802:1
17. Alcouffe RE, Baker RS, Dahl JA et al (2008) PARTISN: a time-dependent, parallel neutral particle transport code system. LA-UR-08-07258
18. Endo T, Yamamoto A (2007) Development of new solid angle quadrature sets to satisfy eve- and odd- moment conditions. J Nucl Sci Technol 44:1249
19. Rearden BT, Jessee MA (2017) SCALE code system. ORNL/TM-2005/39 Version 6.2.3
20. Endo T, Chiba G, van Rooijen WFG et al (2018) Experimental analysis and uncertainty quantification using random sampling technique for ADS experiments at KUCA. J Nucl Sci Technol 55:450
21. Cullen DE, Clouse CJ, Procassini R et al (2003) Little RC. Static and dynamic criticality: are they different? UCRT-TR-201506
22. MacFarlane RE, Muir DW (1994) The NJOY nuclear data processing system, version91. LA-UR-08-07258
23. Pyeon CH, Yamanaka M, Endo T et al (2020) Neutron generation time in highly-enriched uranium core at Kyoto University Critical Assembly. Nucl Sci Eng 194:1116
24. Talamo A, Gohar Y, Sadovich S et al (2013) Correction factor for the experimental prompt neutron decay constant. Ann Nucl Energy 62:421
25. Simmons BE, King JS (1958) A pulsed technique for reactivity determination. Nucl Sci Eng 3:595

Chapter 5
Neutron Spectrum

Cheol Ho Pyeon

Abstract The subcritical multiplication factor is considered an important index for recognizing, in the core, the number of fission neutrons induced by an external neutron source. In this study, the influences of different external neutron sources on core characteristics are carefully monitored. Here, the high-energy neutrons generated by the neutron yield at the location of the target are attained by the injection of 100 MeV protons onto these targets. In actual ADS cores, liquid Pb–Bi has been selected as a material for the target that generates spallation neutrons and for the coolant in fast neutron spectrum cores. The neutron spectrum information is acquired by the foil activation method in the ^{235}U-fueled and Pb–Bi-zoned fuel region of the core, modeling the Pb–Bi coolant core locally around the central region. The neutron spectrum is considered an important parameter for recognizing information on neutron energy at the target. Also, the neutron spectrum evaluated by reliable methodologies could contribute to the accurate prediction of reactor physics parameters in the core through numerical simulations of desired precision. In the present chapter, experimental analyses of high-energy neutrons over 20 MeV are conducted after adequate preparation of experimental settings.

Keywords Subcritical multiplication factor · Reaction rates · Spectrum index · Spallation neutrons

5.1 Subcritical Multiplication Factor

5.1.1 Theoretical Background

In the Accelerator-Driven System (ADS) study on the subcritical system, neutron multiplication M and subcritical multiplication factor k_s [1] are theoretically expressed with the combined use of total fission reaction rates F in the core (fuel region) and external neutron source rates S at the target, respectively, as follows:

C. H. Pyeon (✉)
Institute for Integrated Radiation and Nuclear Science, Kyoto University, Osaka, Japan
e-mail: pyeon.cheolho.4z@kyoto-u.ac.jp

© The Author(s) 2021
C. H. Pyeon (ed.), *Accelerator-Driven System at Kyoto University Critical Assembly*,
https://doi.org/10.1007/978-981-16-0344-0_5

$$M = \frac{F + S}{S},$$
(5.1)

$$k_s = \frac{F}{F + S}.$$
(5.2)

Assuming that the fission reaction rates F do not vary along the wire radius in the thermal neutron field, the total fission reaction rates can be expressed approximately by a the [115] In wire length $(n, \gamma)^{116m}$In reaction rates R_{In} as follows:

$$F = \int_V \int_0^\infty \nu \Sigma_f(\underline{r}, E)\phi_s(\underline{r}, E)d\underline{r}dE = C_{Fission}^{Core} C_{3D \to 1D}^{Dimension} \int_a R_{In}(x, y_0, z_0)dx,$$
(5.3)

where V indicates the whole volume in the system, ν the average number of fission neutrons per fission reaction, Σ_f the fission cross sections, ϕ_s the neutron flux at position r with energy E in the presence of an external source, $C_{Fission}^{Core}$ the conversion coefficient of [115]In capture reactions to the [235]U fission ones, $C_{3D \to 1D}^{Dimension}$ the dimension coefficient of x-direction (1D) to x-y-z directions (3D) and a the [115]In wire along the fuel regions in the core. The external neutron source rates S can be expressed approximately by source reaction rates R_{source} of [115]In$(n, n')^{115m}$In reactions as follows:

$$S = \int_{r_s} \int_0^\infty s(\underline{r}, E)d\underline{r}dE = C_{Source}^{Target} \int_{r_s} R_{source}(\underline{r})d\underline{r},$$
(5.4)

where $s(r, E)$ indicates the external neutron source rate at position r with energy E, C_{Source}^{Target} the conversion coefficient of [115]In$(n, n')^{115m}$In reactions to neutron generation and r_s the position of the external neutron source. The validity of approximation and the applicability of the methodology mentioned above have been demonstrated previously [2].

Finally, using Eqs. (5.3) and (5.4), M and k_s in Eqs. (5.1) and (5.2), respectively, can be expressed as follows:

$$M = \frac{C_{Fission}^{Core} \cdot C_{3D->1D}^{Dimension} \int_a R_{In}(x, y_0, z_0)dx}{C_{Source}^{Target} \int_{r_s} R_{source}(\underline{r})d\underline{r}} + 1,$$
(5.5)

$$k_s = \frac{C_{Fission}^{Core} \cdot C_{3D->1D}^{Dimension} \int_a R_{In}(x, y_0, z_0)dx}{C_{Fission}^{Core} \cdot C_{3D->1D}^{Dimension} \int_a R_{In}(x, y_0, z_0) dx + C_{Source}^{Target} \int_{r_s} R_{source}(\underline{r})d\underline{r}}.$$
(5.6)

5.1.2 Characteristics of the Target

5.1.2.1 Dimension of Solid Targets

The combined use of the heavy (tungsten, lead and bismuth: W, Pb and Bi) [3] and the light nuclides (beryllium: Be; lithium) was considered useful in accomplishing the study objectives related to the neutron spectrum and the neutron yield. The neutron yield was obtained from the neutrons produced at the surface of the target. The two targets laid with the combined use of heavy and light nuclides are termed "two-layer" targets in this study, and their characteristics were numerically investigated by the MCNPX [4] and SRIM codes [5] with the JENDL/HE-2007 [6] library. The aim of these numerical analyses was to investigate the neutron spectrum in the high-energy region and to determine the thickness of solid targets so that incident protons could be fully stopped inside the target.

From the numerical results by MCNPX, the neutron spectrum was observed in the high-energy region of the solid target used: W, W–Be or Pb–Bi, as shown in Figs. A2.5 and A2.6, when 100 MeV protons were injected onto them. The neutron spectrum was somewhat similar, regardless of the kind of solid target used as shown in Fig. A2.5, although at each target it was comparatively different ranging from 1 to 10 MeV. Of particular interest here is the influence of the difference in the neutron spectrum, caused by the kind of target used, on neutron multiplication in the core. In the numerical simulations of neutron generation of the target, the neutron spectrum of W-Be target (two-layer target) was compared remarkably with other single targets, ranging from 85 to 100 MeV, as shown in Fig. A2.6, and the difference between the two spectra was attributed to the scattering reactions of Be to high-energy protons. The aim in using the two-layer target was to acquire the neutron spectrum in the high-energy region and the neutron yield of high-energy neutrons. Consequently, the proton beams actually penetrated the Be target, and conversely stopped inside the W target. The thickness of the solid targets was correctly determined by the numerical results of the range of high-energy protons with the use of the SRIM code. Lastly, the dimensions of the solid targets were determined as shown in Table A2.6, and since the size of the proton beam spot was 40 mm in diameter, the targets were adequately covered with the proton beams and satisfactorily penetrated and stopped fully in the solid targets.

5.1.2.2 Experimental Settings

The ADS experiments were carried out in the A-core (Fig. A2.1) with the combined use of fuel and polyethylene reflector rods. In the A-core, the fuel assembly shown in Fig. A2.2 is composed of 60 unit cells, and upper and lower polyethylene blocks about 25″ and 20″ long, respectively, in an aluminum (Al) sheath 2.1 × 2.1 × 60″. In the A-core, the neutron flux information was acquired from ^{115}In$(n, \gamma)^{116m}$In reactions using the indium (In) wire (1 mm diameter and 800 mm length), under the assumption that, in the thermal neutron region, the cross sections of ^{235}U(n, f) are proportional to those of ^{115}In$(n, \gamma)^{116m}$In shown in Fig. A2.7. The In wire was set along the vertical direction (14, 13–P, A$'$) at the axial center position shown in Fig. A2.1. An Al ($10 \times 10 \times 1$ mm, (15, H) in Fig. A2.1) and an In ($10 \times 10 \times 1$ mm, (15, H) in Fig. A2.1) foils were attached at the location of the target to monitor information on the generation of protons and spallation neutrons through ^{27}Al$(p, n + 3p)^{24}$Na and ^{115}In$(n, n')^{115m}$In reactions (threshold energy of 0.3 MeV), respectively. For an easy understanding of experimental analyses, the methodology of normalized reaction rates was introduced with the comparison between the experimental and calculated results at the location of the target and in the core region: the ^{115}In $(n, n')^{115m}$In and ^{115}In$(n, \gamma)^{116m}$In reaction rates normalized by ^{27}Al$(p, n + 3p)^{24}$Na and ^{115}In$(n, n')^{115m}$In ones, respectively, were estimated in the experiments, and interpreted as actual values of proton yield and neutron yield, respectively, in terms of the influence of the external neutron source. The main characteristics of proton beams were as follows: 100 MeV energy, 0.7 nA intensity, 20 Hz beam repetition, 100 ns beam width and 1.0×10^7 s^{-1} neutron generation. The irradiation time of all the foils and wire was about 3 h. The measured subcriticality, 2,900 pcm, of the core was obtained by the full insertion of control (C1, C2 and C3) and safety (S4, S5 and S6) rods, as shown in Fig. A2.1. The reactivity worth of all control and safety rods was evaluated by the rod drop method and the positive period method beforehand. The measured reaction rates varied according to the kind of solid target used, as shown in Fig. A2.8, under the subcritical level 2,900 pcm. The reaction rates were high with the combined use of W and Be targets, and the moderation (thermal) peak caused by the high-energy neutrons was observed in the polyethylene region, mostly in the two-layer target (W-Be target). Neutron multiplication has been obtained successfully by the ^{115}In$(n, \gamma)^{116m}$In reaction rates in the core region because the relation between ^{115}In$(n, \gamma)^{116m}$In and ^{235}U(n, f) reaction rates in the core region is apparently applicable to the subcritical multiplication analyses through the proportionality of ^{115}In$(n, \gamma)^{116m}$In and ^{235}U(n, f) cross sections in the thermal region [2].

5.1.2.3 Numerical Simulations

The numerical calculations were performed with the combined use of MCNPX and JENDL/HE-2007 for high-energy protons and high-energy neutrons, JENDL-4.0 [7] for transport, and JENDL/D-99 [8] for reaction rates. With MCNPX, the calculated reaction rates were obtained from the evaluation of volume tallies of activation foils, and since the effects of their reactivity are not negligible, they were included in the simulated geometry and transport calculations. The eigenvalue calculations were conducted for 1,000 active cycles of 100,000 histories. The subcriticalities in the eigenvalue calculations had statistical errors within 0.01 %$\Delta k/k$ (10 pcm), and the reaction rates in the fixed source calculations were within 3% as determined with the use of the total 1×10^8 histories. The precision of numerical subcriticality in the eigenvalue calculations was attained within the relative difference of 5% between the experimental and the numerical results.

The measured reaction rates of ^{115}In$(n, \gamma)^{116m}$In (wire: 1 mm diameter and 800 mm long) in the core were normalized by those of ^{115}In$(n, n')^{115m}$In (foil: $10 \times 10 \times 1$ mm) at the location of the target. The experimental errors in the activation wire and foil were estimated within 15% and 5%, respectively, including the statistical error of γ-ray counts and the full width at half maximum (FWHM) of the γ-ray spectrum peak. The calculated reaction rates of the ^{115}In wire and foil in the core were included in the simulated geometry and transport calculations, and deduced from tallies taken in the fixed source calculations. Also, the calculated reaction rates of the In foil at the target were obtained by the previously fixed source calculations, modeling the proton injection on the solid target.

The calculation/experiment (C/E) value of the experiments and the calculations of M in Eq. (5.5), as shown in Table 5.1, was good within an error of 7%, and the absolute value of M was large in the W-Be target, compared with that in the other two. Also, the values of neutron multiplication were differently compared with the W and Pb–Bi targets, indicating that neutron multiplication was mostly influenced by the neutron spectrum of the external neutron source, such as the W-Be target. For any target, F and S were numerically estimated with the use of conversion coefficients $C_{\text{Fission}}^{\text{Core}}$ ($\simeq 0.25$), $C_{3D \rightarrow 1D}^{\text{Dimension}}$ ($\simeq 1.05$) and $C_{\text{Source}}^{\text{Target}}$ ($\simeq 1.0 \times 10^5$) as shown in Eqs. (5.3) and (5.4), including the proportionality between ^{235}U fission and ^{115}In capture cross sections in the thermal neutron field, the dimension effect and the source conversion, respectively. These coefficients were applied to the evaluation of M because it is difficult to obtain F and S in Eqs. (5.1) and (5.2) directly by the experiments.

While the accuracy of M was attributable to the experimental validation of ^{115}In$(n, \gamma)^{116m}$In reaction rates, the actual influence of the kind of solid target used was considered significant: neutron multiplication increased over 30% in the two-layer target, compared with that in the W target. These experimental results clearly demonstrated the influence of the two-layer target on neutron multiplication, and the high-energy neutrons in the region ranging from 85 to 100 MeV contributed significantly to the neutron characteristics of ADS through the selection of the appropriate target.

The C/E values of the experiments and the calculations of k_s in Eq. (5.6) are shown in Table 5.2: the discrepancy between the experiments and the calculations was within the relative difference of 8%, as in M. The values of the measured and the calculated k_s demonstrated that the source term contributed largely to the estimation of k_s, since the external source was located inside the core (core target location

Table 5.1 Neutron multiplication M in Eq. (5.1) deduced from ^{115}In$(n, \gamma)^{116m}$In reaction rates in subcriticality 2,900 pcm (Ref. [3])

Target	Calculation	Experiment	C/E value
W	1.73 ± 0.01	1.85 ± 0.02	0.93 ± 0.01
W-Be	2.29 ± 0.01	2.36 ± 0.03	0.97 ± 0.01
Pb–Bi	1.95 ± 0.01	1.94 ± 0.02	1.01 ± 0.01

Table 5.2 Subcritical multiplication factor k_s in Eq. (5.2) deduced from ^{115}In$(n, \gamma)^{116m}$In reaction rates in subcriticality 2,900 pcm (Ref. [3])

Target	Calculation	Experiment	C/E value
W	0.42033 ± 0.00100	0.45874 ± 0.01003	0.92 ± 0.02
W–Be	0.56355 ± 0.00100	0.57662 ± 0.01510	0.98 ± 0.03
Pb–Bi	0.48830 ± 0.00100	0.48488 ± 0.01335	1.01 ± 0.03

(15, H) in Fig. A2.1). Moreover, the results demonstrated that the Pb–Bi target was confirmed fairly well for an upcoming target in actual ADS, as predicted in several experimental and numerical analyses, by its comparison with the W target, although its static core characteristics were almost the same as those of the W target.

5.1.3 Effects of Neutron Spectrum

5.1.3.1 Core Configuration

Starting from the reference core (Fig. A2.11) that reached a criticality, reaction rate experiments [9] (Cases II-1 to II-4 in Fig. 2A.16) in ADS were carried out in the KUCA A-core by varying subcriticality ranging from 2,483 to 11,556 pcm, as shown in Figs. A2.16a through A2.16d. The ADS cores comprise normal fuel assembly "F," partial fuel assembly "16", Pb–Bi-zoned fuel assembly "f," polyethylene moderator "p" and reflector rods. Normal fuel assembly "F" is composed of 36 cells of a highly-enriched uranium (HEU) fuel plate 1/16″ thick, polyethylene (p) plates 3/8″ thick, and upper and lower polyethylene blocks about 25″ and 20″ long, respectively, in an aluminum (Al) sheath 2.1 × 2.1 × 60″, as shown in Fig. A2.12. The numeral "16" corresponds to the number of unit cells in the partial fuel assemblies for reaching criticality in the core shown in Fig. A2.14. The Pb–Bi-zoned fuel assembly "f" is composed of 30 unit cells with two HEU plates 1/8″ thick, Pb–Bi plate 1/8″ thick, 30 unit cells of two HEU plates 1/8″ thick and polyethylene plates 1/8″ thick, as shown in Fig. A2.13.

The subcriticality in the pcm units was experimentally obtained with the combined use of control rod worth and its calibration curve measured by the rod drop method and the positive period method, respectively, in Case II-1 (Fig. A2.16a). For the deduction of subcriticality in Case II-1, the effective delayed neutron fraction β_{eff} and the neutron generation time Λ were obtained with 783 pcm and 4.64E-05 s, respectively, with MCNP6.1 and JENDL-4.0. In Cases II-2 to II-4 (Figs. A2.16b–d), the reference of subcriticality in the pcm units was attained with MCNP6.1 eigenvalue calculations and the JENDL-4.0 library, because the reactivity of control and safety rods was made by the substitution of fuel assembly rods for the polyethylene ones.

5.1.3.2 Experimental Settings

In reaction rate experiments, an In foil (10 × 10 × 1 mm) was attached at the boundary between (15, L) and (15, M) shown in Figs. A2.16a, d, in order to obtain the information on spallation neutrons with the use of ^{115}In$(n, n')^{115m}$In reactions (threshold energy 0.3 MeV). Six foils, including gold (^{197}Au; bare and cadmium: Cd covered), iron (^{56}Fe), aluminum (^{27}Al), ^{115}In and nickel (^{58}Ni), were used as activation foils (Table A2.22) to cover a wide range of threshold energy and acquire neutron spectrum information in the region of the Pb–Bi-zoned core affected by spallation neutrons. The foils were attached at the boundary between fuel assemblies (15, M) and (15, O) (Figs. A2.16a–d) with the dimensions and variations as shown in Fig. A2.19. Among the six foils, the ^{197}Au foil (Au-bare) was taken as a normalization factor of reactor power for ^{56}Fe, ^{27}Al, ^{115}In and ^{58}Ni foils. Furthermore, the Cd ratio was experimentally obtained by the combination of two Au foils: Au-bare and Au-Cd sandwiched between two Cd plates (10 mm diameter and 1 mm thick), as a spectrum index. For an easy understanding of neutron irradiation, neutron spectrum by the MCNP calculations was obtained at several significant positions around the Pb–Bi-zoned core (Fig. A2.16a), as shown in Fig. 5.1: at the location of the target, the Pb–Bi-zoned fuel assembly (15, M) and two positions (14-13, P-O) and (14-13, L-K) at the boundary between Pb–Bi-zoned and normal fuel regions.

The information on neutron flux distribution was acquired from ^{115}In$(n, \gamma)^{116m}$In reaction rate distribution with the use of the In wire (1 mm diameter and 680 mm long), which was set at the gap between the Pb–Bi-zoned and the normal fuel regions along (14-13, P-A) at a height of 700 mm from the bottom of the core. The reaction rates of the ^{115}In$(n, n')^{115m}$In foil at the Pb–Bi target (the boundary between (15, L) and (15, M)) were taken as the normalization factor of the source generating spallation neutrons.

Spallation neutrons were generated by bombarding 100 MeV proton beams from the FFAG accelerator onto the Pb–Bi target. To compensate for the drawback of

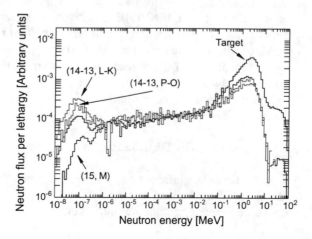

Fig. 5.1 Neutron spectra at target and several locations in Fig. A2.16a during injection of 100 MeV protons onto the Pb–Bi target (Ref. [9])

Table 5.3 Measured and calculated (MCNP6.1) subcriticalities [pcm] in Cases II-1 through II-4 (Ref. [9])

Case	Calculation	Experiment	C/E
II-1	2483 ± 12	2302 ± 6	1.08 ± 0.01
II-2	4812 ± 12	–	–
II-3	9895 ± 12	–	–
II-4	11556 ± 12	–	–

locating the original target outside the core at (15, A′), the Pb–Bi target was moved to (15, L) inside the core on the basis of experimental results obtained in a previous study [10]. Specifications of proton beams were as follows: 100 MeV energy, 1.0 nA intensity, 20 Hz pulsed frequency, 50 ns beam width and 40 mm diameter spot size at the location of the target.

5.1.3.3 Numerical Simulations

To validate the accuracy of calculated subcriticality, eigenvalue calculations were performed with MCNP6.1 [11] and JENDL-4.0 for transport, by comparison with measured subcriticality in Case II-1. Here, since the effect of neutron detectors, control (safety) rods and irradiation materials is not negligible, they were included in simulated geometry and transport calculations with MCNP6.1. The total number of histories used in the eigenvalue calculations was $1E + 08$ ($1E + 05$ histories; $1E + 03$ cycles) with a standard deviation of 8 pcm. A comparison between measured and calculated subcriticalities revealed an agreement with a relative difference of 8%, through the calculation/experiment (C/E) value, as shown in Table 5.3. Also, these results demonstrated that the precision of numerical subcriticality by the eigenvalue calculations was considered carefully to ensure the reliability of the fixed-source calculations with MCNP6.1 in the ADS core with spallation neutrons.

Reaction rate calculations with spallation neutrons were performed by the fixed-source calculations (total number of histories: $1E + 08$) with the combined use of MCNP6.1 and JENDL/HE-2007 for high-energy protons and spallation process, JENDL-4.0 for transport and JENDL/D-99 for reaction rates. Proton beams (100 MeV) were modeled as a spot size of 40 mm in diameter injected onto the Pb–Bi target at location (15, L). The reaction rates by the fixed-source calculations were numerically obtained by the evaluation of volume tallies of activation foils and the In wire with a statistical error within 5%.

5.1.3.4 Indium Wire Distribution

Thermal neutron flux information was acquired from ^{115}In$(n, \gamma)^{116m}$In reaction rates with the use of the In wire, assuming that, in the thermal neutron region, the cross sections of ^{115}In$(n, \gamma)^{116m}$In are proportional to those of ^{235}U(n, f), as shown in

Fig. A2.7. To investigate the accuracy of reaction rate analyses at different subcritical states, the ^{115}In$(n, \gamma)^{116m}$In reaction rate distributions along the (14-13, P-A) region in Figs. A2.16a–d were measured with subcriticality ranging from 2,483 to 11,556 pcm. From a comparison between the measured and the calculated ^{115}In$(n, \gamma)^{116m}$In reaction rate distributions in Cases II-1 and II-3, as shown in Figs. 5.2a, b, respectively, the calculated ^{115}In$(n, \gamma)^{116m}$In reaction rates reproduced successfully the measured ones at an acceptable accuracy with the MCNP fixed-source calculations in the subcritical states, demonstrating the allowance within the experimental statistical error.

To estimate the effect of spallation neutrons on the neutron multiplication in ADS, the subcritical multiplication factor k_s is acquired as Eq. (5.6), on the basis of a theoretical background [5]. A comparison between experiments and calculations of k_s

Fig. 5.2 Comparison between measured and calculated ^{115}In$(n, \gamma)^{116m}$In reaction rate distributions along (14-13, P-A) region (Ref. [9])

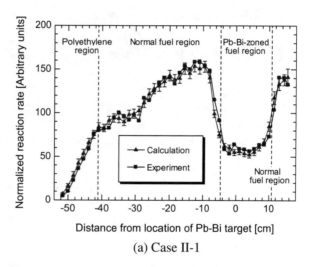

(a) Case II-1

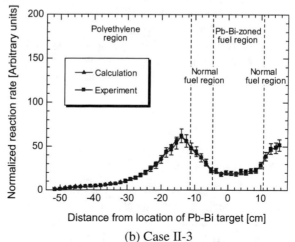

(b) Case II-3

Table 5.4 Comparison between measured and calculated (MCNP6.1) k_s values in Cases II-1 through II-4 (Ref. [9])

Case	Calculation	Experiment
II-1	0.9989 ± 0.0005	0.9996 ± 0.0008
II-2	0.9997 ± 0.0005	0.9991 ± 0.0009
II-3	0.9990 ± 0.0005	0.9991 ± 0.0009
II-4	0.9985 ± 0.0005	0.9989 ± 0.0005

deduced by Eq. (5.6) showed fairly good agreement within a relative difference of 1%, as shown in Table 5.4, and demonstrated an exact reproduction of ^{115}In$(n, \gamma)^{116m}$In and ^{115}In$(n, n')^{115m}$In reaction rates with the MCNP fixed-source calculations.

5.2 Threshold Energy Reactions

5.2.1 Foil Activation Method

The experimental results of reaction rates (RR) [s^{-1} cm^{-3}] were obtained by measuring total counts of the peak energy of γ-ray emissions. The value of RR was deduced from that of saturation activity D_∞ [s^{-1}] that is proportional to the reaction rates by using the following equations:

$$D_\infty = \frac{\lambda T_c C}{\varepsilon_D \varepsilon_E (1 - e^{-\lambda T_i}) e^{-\lambda T_w} (1 - e^{-\lambda T_c})}, \tag{5.7}$$

$$RR = D_\infty \frac{\rho}{M} = \frac{\lambda T_c C \rho}{\varepsilon_D \varepsilon_E (1 - e^{-\lambda T_i}) e^{-\lambda T_w} (1 - e^{-\lambda T_c}) M}, \tag{5.8}$$

where λ [s^{-1}] indicates the decay constant, $T_{-c}-$ [s] the measurement counting time, C [1/s] the counting rate, ε_D [%] the detection efficiency, ε_E [%] the emission rate, T_i [s] the irradiation time, T_w [s] the waiting time until the measurement starting after the irradiation, ρ [g cm^{-3}] the density and M [g] the mass of the foil. The experimental errors in the activation wire and foils were estimated within 15% and 5%, respectively, including the statistical errors of γ-ray counts and full width at half maximum of the γ-ray spectrum peak.

5.2.2 Activation Foils

The measured reaction rates of ^{197}Au (bare), ^{197}Au (Cd), ^{56}Fe, ^{27}Al, ^{115}In and ^{58}Ni foils irradiated in the subcritical states are shown in Table 5.5. A comparison between measured and calculated reaction rates is presented in Table 5.6 and Fig. 5.3. As shown in Table 5.6, in Case II-1 (2483 pcm), non-threshold and low-threshold energy

Table 5.5 Measured reaction rates of activation foils in Cases II-1 through II-4 (Ref. [9])

Reaction	Measured reaction rate ($s^{-1}\ cm^{-3}$)			
	Case II-1	Case II-2	Case II-3	Case II-4
$^{115}In(n, n')^{115m}In$ (target)	$(1.87 \pm 0.24) \times 10^5$	$(1.12 \pm 0.04) \times 10^5$	$(1.62 \pm 0.03) \times 10^5$	$(1.02 \pm 0.04) \times 10^5$
$^{197}Au(n, \gamma)^{198}Au$ (bare)	$(8.88 \pm 0.02) \times 10^6$	$(4.88 \pm 0.04) \times 10^6$	$(3.51 \pm 0.08) \times 10^6$	$(2.53 \pm 0.04) \times 10^6$
$^{197}Au(n, \gamma)^{198}Au$ (Cd)	$(7.84 \pm 0.09) \times 10^6$	$(4.46 \pm 0.04) \times 10^6$	$(3.11 \pm 0.04) \times 10^6$	$(2.30 \pm 0.03) \times 10^6$
$^{115}In(n, n')^{115m}In$ (Core)	$(8.60 \pm 0.13) \times 10^4$	$(4.27 \pm 0.22) \times 10^4$	$(4.27 \pm 0.03) \times 10^4$	$(2.86 \pm 0.06) \times 10^4$
$^{58}Ni(n, p)^{58}Co$	$(4.90 \pm 0.08) \times 10^4$	$(3.18 \pm 0.03) \times 10^4$	$(3.23 \pm 0.16) \times 10^4$	$(2.00 \pm 0.10) \times 10^4$
$^{56}Fe(n, p)^{56}Mn$	$(1.82 \pm 0.07) \times 10^3$	$(1.26 \pm 0.02) \times 10^3$	$(1.55 \pm 0.03) \times 10^3$	$(1.39 \pm 0.02) \times 10^3$
$^{27}Al(n, \alpha)^{24}Na$	$(1.54 \pm 0.05) \times 10^3$	$(1.11 \pm 0.02) \times 10^3$	$(1.62 \pm 0.03) \times 10^3$	$(1.10 \pm 0.01) \times 10^3$

Table 5.6 C/E values between measured and calculated reaction rates in Cases II-1 through II-4 (Ref. [9])

Reaction	Case II-1	Case II-2	Case II-3	Case II-4
^{197}Au$(n, \gamma)^{198}$Au (bare)	1.14 ± 0.09	0.96 ± 0.06	0.90 ± 0.06	0.75 ± 0.07
^{197}Au$(n, \gamma)^{198}$Au (Cd)	1.10 ± 0.09	0.81 ± 0.05	0.80 ± 0.06	0.71 ± 0.07
^{115}In$(n, n')^{115m}$In (Core)	0.86 ± 0.02	0.69 ± 0.04	0.52 ± 0.01	0.47 ± 0.02
^{58}Ni$(n, p)^{58}$Co	0.99 ± 0.04	0.74 ± 0.03	0.59 ± 0.03	0.67 ± 0.04
^{56}Fe$(n, p)^{56}$Mn	0.60 ± 0.03	0.54 ± 0.03	0.47 ± 0.02	0.40 ± 0.02
^{27}Al$(n, \alpha)^{24}$Na	0.47 ± 0.01	0.26 ± 0.02	0.19 ± 0.01	0.22 ± 0.02

Fig. 5.3 C/E value between measured and calculated reaction rates by varying subcriticality in Cases II-1 through II-4 (Ref. [9])

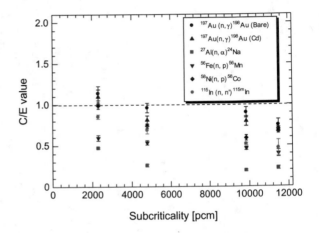

reaction rates showed good agreement between experimental and numerical reaction rates, within a relative difference of around 10%. Conversely, the numerical calculations for high-threshold reaction rates of ^{27}Al and ^{56}Fe foils revealed an underestimation of about 50% at most. Besides, by varying subcriticality, the deeper its level, the smaller the C/E value, as shown in Fig. 5.3. Notably, in ADS with spallation neutrons, the dependence of reaction rates on subcriticality was revealed in the accuracy of C/E, under subcriticality ranging from 2,483 to 11,556 pcm.

Compared with previous analyses of ADS with 14 MeV neutrons [12], a discrepancy between measured and calculated reaction rates was, by contrast, larger in ADS with spallation neutrons, and considered attributable mainly to the uncertainty of reaction rates in the high-energy thresholds and the difficulty in the exact simulation of defocused proton beams. In the reaction rate experiments at KUCA, the proton beams were transported through an un-vacuumed air space from the location of the original target to that of the Pb–Bi target, although the proton beams were in a vacuum until they reached the location of the original target. As a result, the proton beam spot was easily defocused at the location of the Pb–Bi target by the scattering reactions of high-energy protons in the air space (15, A-L; Figs. A2.16a–d).

5.3 Spectrum Index

5.3.1 Cd Ratio

In reactor physics experiments, the Cd ratio is generally interpreted as the neutron spectrum index, where the relation between absorption cross sections of Au and Cd is used in foil activation analyses. For thermal and resonance regions, 0.18 eV corresponds to the boundary energy (cut-off energy) between thermal and epithermal neutron regions. By defining the Cd ratio as that of activation reaction rates $R_{\text{Au-bare}}$ and $R_{\text{Au-Cd}}$ of Au foils (Au-bare and Au-Cd-sandwiched plates, respectively), the Cd ratio can be expressed as follows:

$$\text{Cd ratio} = \frac{R_{\text{Au-bare}}}{R_{\text{Au-Cd}}}. \tag{5.9}$$

On the basis of the definition in Eq. (5.9), the Cd ratio was evaluated by two Au foils (bare and Cd-sandwiched plates) in subcritical core configurations, as shown in Table 5.7. The C/E values of the Cd ratio were found in good agreement in terms of the relative difference error, demonstrating the accuracy of Au absorption reaction rates with MCNP simulations in both thermal and epithermal neutron regions.

5.3.2 In Ratio

As discussed in Sect. 5.3.1, the ratio of thermal neutrons to epithermal ones obtained by the Cd ratio is readily understood as a simple interpretation of the spectrum index being about 10%, by subtracting the unity value from the experimental results of the Cd ratio shown in Table 5.7. Moreover, among several spectrum indexes of fast reactors, the value of F8/F5, which is the ratio of fast fission and total fission reaction rates of ^{238}U and ^{235}U, respectively, is especially introduced as a quantitative evaluation of the neutron spectrum. From the point of view of the spectrum index, a new assumption suggested here is supportable by introducing the same concept as the Cd ratio and the value of F8/F5, regarding thermal and fast neutrons.

This section is devoted to a worthwhile discussion of a new index of the neutron spectrum in ADS through a new definition of ^{235}U fission ratio of thermal and fission

Table 5.7 Comparison between measured and calculated Cd ratios in Cases II-1 through II-4 (Ref. [9])

Case	Calculation	Experiment	C/E
II-1	1.15 ± 0.09	1.11 ± 0.06	1.04 ± 0.10
II-2	1.24 ± 0.09	1.09 ± 0.02	1.13 ± 0.09
II-3	1.27 ± 0.09	1.12 ± 0.04	1.13 ± 0.09
II-4	1.16 ± 0.09	1.10 ± 0.04	1.05 ± 0.09

neutrons in the specified Pb–Bi-zoned fuel region from (14-13, L; y_1) to (14-13, O; y_2) shown in Figs. A2.16a–d, with an interpretation similar to the Cd ratio mentioned in Sect. 5.3.1 as follows:

$$^{235}\text{U fission ratio} = \frac{\int_{y_1}^{y_2} R_{U-\text{fission}}^{\text{thermal}}(x_0, y, z_0)dy}{\int_{y_1}^{y_2} R_{U-\text{fission}}^{\text{fast}}(x_0, y, z_0)dy}, \tag{5.10}$$

where $R_{U-\text{fission}}^{\text{thermal}}$ and $R_{U-\text{fission}}^{\text{fast}}$ indicate $^{235}\text{U}(n, f)$ reaction rates in the thermal (less than 0.1 eV) and fast (more than 0.1 MeV) neutron regions, respectively.

The numerator in Eq. (5.10) is expressed as follows, with the assumption as discussed in Sect. 5.1.1:

$$\int_{y_1}^{y_2} R R_{U-\text{fission}}^{\text{thermal}}(x_0, y, z_0)dy = C_{\text{fission}}^{\text{thermal}} \int_{y_1}^{y_2} R_{In(n,\gamma)}^{\text{thermal}}(x_0, y, z_0)dy, \tag{5.11}$$

where $C_{\text{fission}}^{\text{thermal}}$ indicates the proportionality coefficient of cross sections between $^{115}\text{In}(n, \gamma)^{116m}\text{In}$ and $^{235}\text{U}(n, f)$ reactions in the thermal neutron region. Similarly, the denominator in Eq. (5.10), by a new assumption of the proportionality of the In inelastic scattering (threshold energy 0.4 MeV) and ^{235}U fast fission reactions, is written as follows:

$$\int_{y_1}^{y_2} R_{U-\text{fission}}^{\text{fast}}(x_0, y, z_0)dy = C_{\text{fission}}^{\text{fast}} \int_{y_1}^{y_2} R_{In(n,n')}^{\text{fast}}(x_0, y, z_0)dy, \tag{5.12}$$

where $C_{\text{fission}}^{\text{fast}}$ indicates the proportionality coefficient of cross sections between $^{115}\text{In}(n, n')^{115m}\text{In}$ and $^{235}\text{U}(n, f)$ reactions in the fast neutron region, although the assumption of proportionality in the fast neutron region is somewhat complicated from the viewpoint of the characteristics of cross sections. On the basis of Eqs. (5.11) and (5.12), the results of $C_{\text{fission}}^{\text{thermal}}$ and $C_{\text{fission}}^{\text{fast}}$ were found to be nearly constant around 1.11 and 6.25, respectively, with the MCNP fixed-source calculations.

Generally, in the experiments, while it is apparently difficult to measure $^{235}\text{U}(n, f)$ reaction rates directly, a convenient alternative is to introduce the proportionality of cross sections discussed in Sect. 5.1.2. Subsequently, special attention was directed to the property of In wire reaction rates to experimentally obtain neutron flux information on both thermal and fast energy regions simultaneously. Assuming that the ^{235}U fission ratio in Eq. (5.9) corresponds approximately to the ratio of $^{115}\text{In}(n, \gamma)^{116m}\text{In}$ and $^{115}\text{In}(n, n')^{115m}\text{In}$ reaction rate distributions, a new spectrum index of the In ratio is, by introducing the coefficients of $C_{\text{fission}}^{\text{thermal}}$ and $C_{\text{fission}}^{\text{fast}}$, defined as follows:

$$\text{In ratio} = \frac{C_{\text{fission}}^{\text{thermal}} \int_{y_1}^{y_2} R_{In(n,\gamma)}^{\text{thermal}}(x_0, y, z_0)dy}{C_{\text{fission}}^{\text{fast}} \int_{y_1}^{y_2} R_{In(n,n')}^{\text{fast}}(x_0, y, z_0)dy}. \tag{5.13}$$

Compared with the numerical results of ^{235}U fission ratio and In ratio, the assumption in Eqs. (5.11) and (5.12) is considered mostly valid with a relative difference of around 10%, as shown in Table 5.8. A comparison between measured and calculated In ratios revealed fairly good agreement within a relative difference of 5% in Cases II-1 through II-4 the ratio, as shown in Fig. 5.4, and was found constant especially in the Pb–Bi-zoned fuel regions, except at the boundaries between the fuel regions. Furthermore, the results in Fig. 5.4 demonstrated that fast neutrons were dominant over the Pb–Bi-zoned fuel region through the ratio of thermal and fast neutrons shown in Eq. (5.13). From the results in Table 5.8 and Fig. 5.4, the consistency between measured and calculated In ratios supported an interesting interpretation that the In ratio is useful in conveniently determining the ratio of thermal and fast neutrons as another neutron spectrum index of ADS, in addition to the Cd ratio, through two different In reaction rate distributions obtained by the In wire.

Table 5.8 Comparison between measured and calculated In ratios in Cases II-1 through II-4 (Ref. [9])

Case	^{235}U fission ratio in Eq. (5.10)	In ratio in Eq. (5.13)		
	Calculation	Calculation	Experiment	C/E
II-1	$(5.16 \pm 0.08) \times 10^{-3}$	$(4.62 \pm 0.08) \times 10^{-3}$	$(4.52 \pm 0.10) \times 10^{-3}$	1.02 ± 0.03
II-2	$(2.21 \pm 0.10) \times 10^{-3}$	$(2.15 \pm 0.10) \times 10^{-3}$	$(2.17 \pm 0.10) \times 10^{-3}$	0.99 ± 0.07
II-3	$(1.79 \pm 0.07) \times 10^{-3}$	$(1.60 \pm 0.07) \times 10^{-3}$	$(1.57 \pm 0.11) \times 10^{-3}$	1.02 ± 0.08
II-4	$(1.70 \pm 0.09) \times 10^{-3}$	$(1.54 \pm 0.09) \times 10^{-3}$	$(1.48 \pm 0.10) \times 10^{-3}$	1.04 ± 0.10

Fig. 5.4 Comparison between measured and calculated In ratios (Eq. (5.12)) in Case II-1 (Ref. [9])

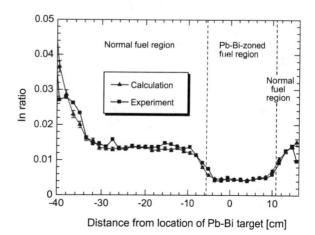

Distance from location of Pb-Bi target [cm]

5.4   Spallation Neutrons

5.4.1   Neutron Spectrum Analyses

5.4.1.1   Experimental Settings

High-energy protons were generated by the FFAG accelerator under the following parameters: 100 MeV energy, 30 pA intensity, 30 Hz repetition rate and 200 ns beam width. On the downstream of the FFAG beam line, the W was set at the location (15, A′; Fig. A2.1) of the original target (80 mm diameter and 10 mm thick); the thickness was determined on the basis of previous experimental and numerical analyses [13] for the injection of high-energy proton beams onto the W target. For the proton beam configuration modeled by numerical simulations, the size of the proton beam spot was requisite experimentally and precisely, when 100 MeV protons were injected onto the tungsten target where the spallation neutrons are generated. The Gafchromic film [14], which is very sensitive to the charged particles, was then used to evaluate the size of the proton beam spot injected onto the W target, since a graphic image on the film is acquired quickly after the irradiation of charged particles for a short time.

The reaction rates for threshold energy of high-energy neutrons and the continuous energy distribution of the spallation neutrons at the target were acquired by the foil activation method and the organic liquid scintillator, respectively [15]. The high-energy neutrons (spallation neutrons) of threshold reactions $^{209}Bi(n, xn)^{210-x}Bi$ ($n = 3$, to 12) over 15 MeV have been generated by the injection of high-energy protons over 100 MeV. Here, to obtain the reaction rates by high-energy neutrons at the target, ^{209}Bi was selected as an activation foil (Table 5.9) to cover threshold energies over 15 MeV, and ^{115}In was selected as a normalization factor for monitoring the spallation neutrons at the target to cover threshold reactions $^{115}In(n, n')^{115m}In$ over 0.3 MeV. Foil dimensions at the target were as follows: ^{209}Bi, 50 mm in diameter and 3 mm thick, ^{115}In, $10 \times 10 \times 1$ mm, and two foils were set around the target region as shown in Fig. 5.5. Additionally, nine other ^{115}In foils ($10 \times 10 \times 3$ mm) were placed in a circle (100 mm radius) around the target at 30° intervals on an acryl plate, to investigate the angular distribution of spallation neutrons as shown in Fig. 5.6. The irradiation time of ^{209}Bi and ^{115}In foils was four hours for measuring the neutron yield of spallation neutrons, and their reaction rates were measured by the high-purity germanium detector (ORTEC, GEM60P). Besides the previous study [12, 13, 16], the detection efficiency of the germanium detector was determined by the fitting line obtained from the energy calibration with the use of several γ-ray standard sources.

The continuous energy distribution of spallation neutrons was determined by the organic liquid scintillator (Nuclear Enterprises Ltd., NE213 Scintillator; 5″ in diameter and 5″ long) set directly facing the W target without any reactor components as shown in Fig. 5.7. The measurement circulation of the organic liquid scintillator was as indicated in Fig. 5.8. The main advantage of the measurement system is that the two signals (rise time and light output of γ-ray and neutron) acquired coincidently

Table 5.9 Threshold energy, half-life and γ-ray energy of ^{209}Bi$(n, xn)^{210-x}$Bi reactions ($x = 3$ to 12) (Ref. [15])

x	Threshold energy (MeV)	Half-life ($T_{1/2}$)	Emission γ-ray energy (keV)
3	14.42	38.3 y	569.7 (97.8) 1063.6 (74.9)
4	22.52	6.24 d	803.1 (98.9) 881.1 (66.2) 1718.7 (31.8)
5	29.62	15.31 d	703.4 (31.5) 1763.4 (32.5)
6	38.13	11.22 h	374.8 (81.8) 899.2 (98.5) 983.9 (59.1)
7	45.37	11.76 h	820.2 (29.6) 825.2 (14.6) 1847.3 (11.4)
8	54.24	1.67 h	422.1 (83.7) 658.5 (60.6) 961.7 (99.3)
9	61.69	1.77 h	786.4 (9.5) 935.7 (11.3)
10	70.89	36.45 m	419.7 (91.3) 462.3 (98.3) 1026.4 (100.0)
11	78.47	27.12 m	425.3 (22.0) 560.1 (22.0)
12	87.94	11.85 m	562.4 (79.0) 1063.4 (100.0)

(): Emission rate [%]

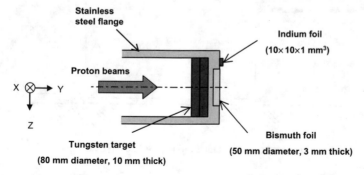

Fig. 5.5 Experimental setting of activation foils (Bi and In) for measuring the reaction rates at the target position (Ref. [15])

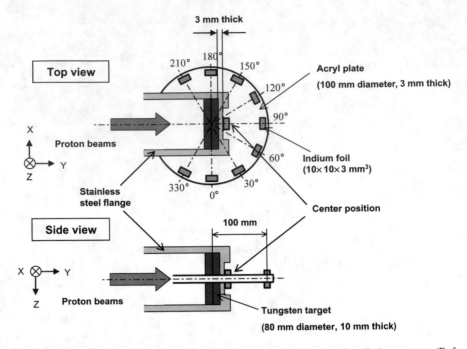

Fig. 5.6 Experimental setting of tungsten target for angular distribution of spallation neutrons (Ref. [15])

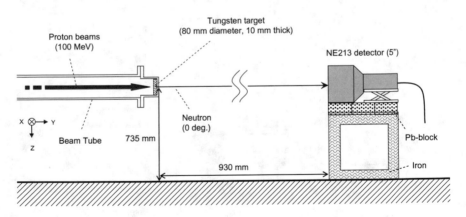

Fig. 5.7 Experimental setting of the organic liquid scintillator (NE213) (Ref. [15])

with the use of specific equipment (Laboratory Equipment Corp., Dual MCA) readily provide a two-dimensional graphic image (rise time and light output). This system allows the rise time information on two signals to accomplish easy discrimination between the γ-ray and the neutron, and the signals can be converted into fluorescence signals of charged particles (recoil protons generated in reaction with neutrons).

Fig. 5.8 Measurement circulation of the organic liquid scintillator (NE213) (Ref. [15])

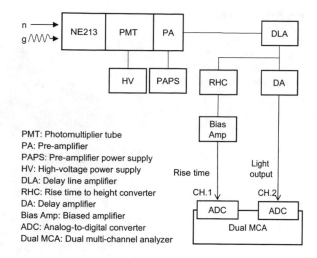

PMT: Photomultiplier tube
PA: Pre-amplifier
PAPS: Pre-amplifier power supply
HV: High-voltage power supply
DLA: Delay line amplifier
RHC: Rise time to height converter
DA: Delay amplifier
Bias Amp: Biased amplifier
ADC: Analog-to-digital converter
Dual MCA: Dual multi-channel analyzer

5.4.1.2 Experimental Analyses

The 10 mm-thick W target was determined on the basis of experimental and numerical analyses [13] from the viewpoint of the full stop of proton beams within the W target. On the other hand, the proton beam irradiation experiments were carried out to monitor the size of the proton beam spot with the use of a Gafchromic film, which is highly sensitive to charged particles. The Gafchromic film was irradiated for two minutes and set on the surface (downstream beam) outside the stainless steel flange (Fig. 5.5) without the W target, although the influence of the stainless steel flange was slightly found in the proton beam profile. It demonstrated that the proton strength was distributed by the downstream beams (Fig. 5.9), with a relative distribution within an approximately 40 mm diameter spot (Fig. 5.10). The experimental result of the Gafchromic film was very useful for modeling the size of the proton beam

Fig. 5.9 Measured result (Gafchromic film) of proton strength distribution of 100 MeV protons at downstream beam without the tungsten target (Ref. [15])

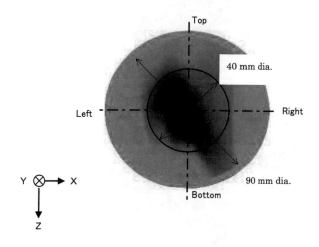

Fig. 5.10 Scanning relative result of proton strength distribution in Fig. 5.9 of 100 MeV protons at the downstream beam without the tungsten target (Ref. [15])

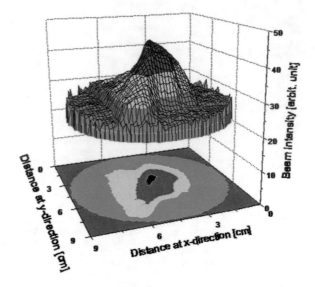

spot in numerical simulations, therefore, the irradiation experiments were important for evaluating the size of the proton beam spot for which the W target (80 mm in diameter) was considered sufficient to cover.

No value for the 100 MeV proton irradiation was observed in $3n$, $9n$ through $12n$ reactions of $^{209}\text{Bi}(n, xn)^{210-x}\text{Bi}$ ($x = 3–12$), since little activation was caused by insufficient irradiation with the low proton beam intensity of 30 pA and the long half-life (38.3 y) of $3n$ reactions. The MCNPX calculations with JENDL/HE-2007 for nuclear data and ENDF/B-VI for cross sections of Bi were executed by a total number of 2×10^8 histories within a statistical error of 1% to obtain the reaction rates of the irradiated ^{209}Bi foil. The spot size 40 mm diameter of proton beams was modeled in the MCNPX calculations on the basis of the experimental results in the Gafchromic film. For the irradiation experiments of the ^{209}Bi foil, a comparison (Table 5.10) between the experimental and the numerical values showed agreement around a relative difference of 10% in the calculation/experiment (C/E) values, excluding the $^{209}\text{Bi}(n, 8n)^{202}\text{Bi}$ reaction. Here, from the results of the ^{209}Bi foil irradiation, the high-energy neutrons up to 50 MeV generated by $^{209}\text{Bi}(n, xn)^{210-x}\text{Bi}$ reactions were confirmed to have been bombarded by the injection of 100 MeV protons.

Table 5.10 C/E values between measured and calculated reaction rates of $^{209}\text{Bi}(n, xn)^{210-x}\text{Bi}$ reactions ($x = 4–8$) for 100 MeV proton beams (Ref. [15])

Reaction	C/E value
$^{209}\text{Bi}(n, 4n)\,^{206}\text{Bi}$	1.00
$^{209}\text{Bi}(n, 5n)\,^{205}\text{Bi}$	0.94 ± 0.01
$^{209}\text{Bi}(n, 6n)\,^{204}\text{Bi}$	0.88 ± 0.01
$^{209}\text{Bi}(n, 7n)\,^{203}\text{Bi}$	0.95 ± 0.02
$^{209}\text{Bi}(n, 8n)\,^{202}\text{Bi}$	1.65 ± 0.03

The comparison between measured and calculated results of the reaction rates at 90 degrees is listed in Table 5.10. The highest reaction rate was observed at the center of the tungsten target as shown in Table 5.11, and the angular distribution (Fig. 5.11 and Table 5.11) of reaction rates appeared slightly polarized in the upper direction, when all reaction rates were normalized by that at position 210 degrees, which was the largest along the angular. The reaction rate at 210 degrees was larger than at other positions, whereas the effect of the acryl plate was considered insufficient in the measured reaction rates along the angular. Next, to investigate the effect of the acryl plate on measured reaction rates, numerical simulations were executed with the use of MCNPX and JENDL/HE-2007 with (w/) and without (w/o) the acryl plates. A comparison of the results (Fig. 5.12) in the presence or absence of acryl plates showed an apparent effect on the reaction rates: the high-energy neutron flux was attenuated

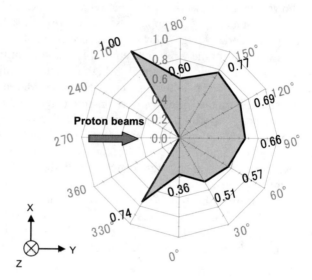

Fig. 5.11 Measured results of the angular distribution of indium reaction rates (Ref. [15])

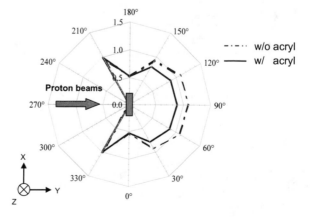

Fig. 5.12 Comparison between calculated results of angular distribution of indium reaction rates with and without the acryl plate (Ref. [15])

ahead at the target and influenced by the acryl plate. Thus, the spallation neutrons were considered significantly spherical in the angular distribution through the results in numerical simulations, although their angular distribution was observed actually reversed at the target. Subsequently, the neutron yield at the target was evaluated at $(9.73 \pm 0.12) \times 10^4$ s^{-1} over 20 MeV and $(1.03 \pm 0.04) \times 10^7$ s^{-1} over 0.3 MeV from the measured reaction rates of Bi and In foils, respectively.

The main characteristics of the measurements by the organic liquid scintillator are to acquire two signals of prompt (electrons) and delayed (protons, deuterons, or α-ray, etc.) fluorescence components and to discriminate γ-ray and neutron events caused by the two signals, respectively. Thus, the discrimination between the γ-ray and the neutron was possibly caused by the difference between their fluorescence intensities in a time-dependent manner.

A comparison between the combined (γ-ray and neutron) and the γ-ray events showed apparent discrimination between the γ-ray and the neutron in the experimental results (Fig. 5.13) of fluorescence distributions. The γ-ray events were found to be considerably large in low-fluorescence distribution and difficult to discriminate the two events of the γ-ray and neutron. The amount of fluorescence by high-energy neutrons was found to be small in high-fluorescence distribution because of very small counting rates in the high-channel region. Moreover, the spallation neutrons generated from the FFAG accelerator were considered to be a group of continuous energy neutrons with ambiguity in maximum energy, since an edge of the recoil proton corresponding to the neutron energy was not found in the measurements by the organic liquid scintillator. Thus, the neutron energy calibration [17, 18] of fluorescence to the light unit was conducted with the use of the results of ^{22}Na standard source (γ-ray energy; 1.274 MeV).

The neutron spectrum (Fig. 5.14) was obtained experimentally with the use of the SCINFUL-QMD code [19] for the matrix of response functions and with the UMG code [20] for the unfolding of experimental results (Fig. 5.13) together with the matrix by SCINFUL-QMD. As a reference of the neutron spectrum, the MCNPX

Fig. 5.13 Comparison between measurement results of the light output from the organic liquid scintillator before and after the discrimination of γ-ray and neutron (Ref. [15])

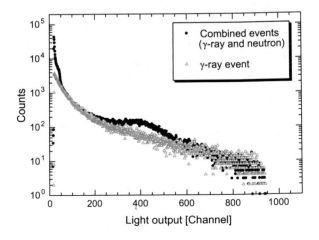

Fig. 5.14 Comparison
between measured
(Unfolding) and calculated
(MCNPX) results of neutron
spectrum (Ref. [15])

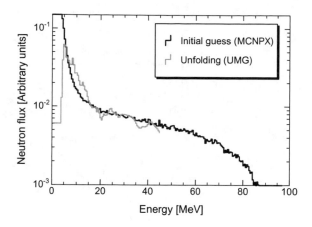

calculations were executed with the use of ENDF/B-VI.6 for cross section data and
of LAHET150 [21] for the high-energy neutron and proton libraries. A comparison
between the experiment (UMG) and the calculation (MCNPX) revealed an approximate reconstruction of the neutron spectrum in the experiment, ranging from 5 to
45 MeV neutrons, although the discrepancy was observed in some energy regions. In
the measurement system, the amount of fluorescence was insufficiently over 50 MeV
neutron, because of detector sensitivity in relating the intensity and the energy of
protons. Finally, the spallation neutrons up to 45 MeV were considered successfully
detected by the organic liquid scintillator, since the discrimination between the γ-ray
and the neutron was satisfactorily conducted.

5.4.2 Reaction Rates

5.4.2.1 Experimental Settings

At KUCA, the proton beam transport facility for injecting 100 MeV protons onto a
heavy metal target was used for experiments on ADS [22] equipped with a subcritical
core. The main specifications of proton beams were 100 MeV energy, 1 nA intensity,
30 Hz beam repetition, 100 ns beam width and 1×10^7 s^{-1} neutron yield, as shown
in Fig. 5.15. The heavy metal (Pb–Bi) target was set in the downstream of a stainless
steel flange, as shown in Fig. 5.16. The Pb–Bi target was 50 mm in diameter for
covering the proton beam shape and 18 mm thick for attaining the full stopping of
proton beams [3, 13] inside the Pb–Bi target. To monitor the size of the proton beam
spot, the Gafchromic film, which is highly sensitive to charged particles, was attached
to the surface of the stainless steel flange before setting the Pb–Bi target. Among
the main characteristics of the protons, the proton beam shape [23] was considered
essential for determining neutron multiplication [3, 9] in the subcritical core, and for
demonstrating adequate numerical precision of Monte Carlo calculations.

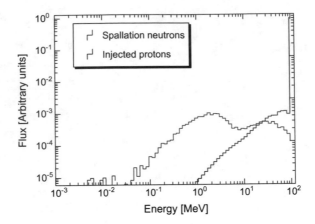

Fig. 5.15 Numerical results (PHITS3.0 [24]) of spectra of injected proton beams and spallation neutrons generated by injection of protons onto Pb–Bi target (Ref. [22])

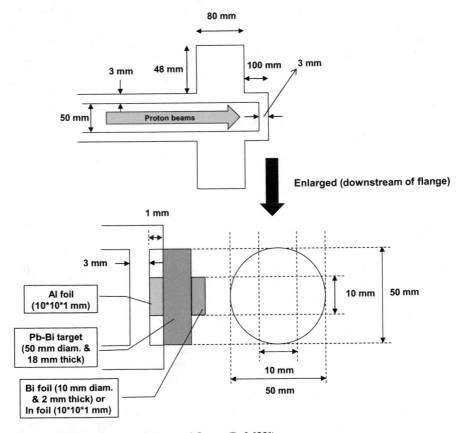

Fig. 5.16 Foil settings at stainless steel flange (Ref. [22])

Table 5.11 Measured results of indium reaction rates for the angular (Ref. [15])

Position (°)	Measured reaction rate (s^{-1} cm^{-3})
Center	11216 ± 96
0	162 ± 5
30	229 ± 6
60	254 ± 8
90	297 ± 9
120	311 ± 10
150	343 ± 11
180	269 ± 8
210	448 ± 14
330	331 ± 10

The ^{115}In$(n, n')^{115m}$In reaction rates of high-energy neutrons generated at the location of the Pb–Bi target were used for the monitoring of the source neutrons, showing the threshold energy as 0.3 MeV (Table 5.12). Also, the ^{27}Al$(p, n + 3p)^{24}$Na reaction rates were selected for monitoring high-energy protons (100 MeV). The number of neutrons per proton (^{115}In$(n, n')^{115m}$In/^{27}Al$(p, n + 3p)^{24}$Na) was then experimentally evaluated with the use of the two reaction rates.

High-energy neutrons over 20 MeV, generated by the interaction between the Pb–Bi target and 100 MeV protons, were evaluated with the use of the neutron threshold energies of ^{209}Bi$(n, xn)^{210-x}$Bi (Table 5.9). The ^{209}Bi foil was considered a very effective detector for easily acquiring the high-energy neutron spectrum information from the systematic threshold energies of high-energy neutrons ranging from 20 to 50 MeV (Table 5.9). Also, in the reaction rate analyses of ^{209}Bi$(n, xn)^{210-x}$Bi reactions, the ^{27}Al$(p, n + 3p)^{24}$Na reaction rates were taken as a normalization factor of the neutron spectrum for the injection of 100 MeV protons onto the Pb–Bi target.

5.4.2.2 Experimental Analyses

Monte Carlo calculations have been considered useful in accurately obtaining reaction rates in the ADS experimental analyses, with the combined use of major nuclear data libraries. In the present study, a series of numerical simulations was conducted with the PHITS3.0 [24] and the MCNP6.1 codes together with the JENDL libraries, the INCL model [25] and LAHET150, for the transport of neutrons and protons, as shown in Table 5.13. The ^{115}In$(n, n')^{115m}$In reaction rates of 0.3 MeV neutron threshold energy were obtained by JENDL/D-99 for cross sections under 20 MeV. For over 20 MeV, the ^{209}Bi$(n, xn)^{210-x}$Bi reaction rates were attained by ENDF/B-VI.8 [21, 26]. Finally, the ^{27}Al$(p, n + 3p)^{24}$Na reaction rates were acquired by the point-wise data of cross sections in JENDL/HE-2007. Reaction rate calculations performed by the fixed-source calculations (total number of histories: 1E + 08) were

Table 5.12 Threshold energy, half-life and emission γ-ray energy of activation foils (Ref. [22])

Reactions	Threshold energy (MeV)	Half-life ($T_{1/2}$) (h)	Emission γ-ray energy [keV] (Branching ratio (%))
^{115}In$(n, n')^{115m}$In	0.3	4.486	336.2 (37.0)
^{27}Al$(p, n + 3p)^{24}$Na	–	15.00	1368.6 (99.9)

numerically obtained by the evaluation of volume tallies of activation foils (In, Bi and Al foils) within a statistical error of 5%.

5.4.2.3 Number of Neutrons Per Proton

To determine the size of the proton beam spot, the Gafchromic film was attached to the stainless steel flange downstream of the proton beam path, because a modeling of the proton beam spot has a significant effect [3, 23] of high-energy neutron generation injected by 100 MeV protons onto the Pb–Bi target. By injecting 100 MeV protons in a few seconds, the configuration of the proton beam spot was found to be a triangle-like configuration, as shown in Fig. 5.17, and the actual size was decided by scanning the result of irradiation shown in Fig. 5.18.

On the basis of the experimental result of the proton beam spot (Fig. 5.17), reaction rates of neutrons (0.3 MeV threshold energy) and protons were numerically acquired by the PHITS3.0 code with JENDL-4.0 and the INCL model for neutron

Fig. 5.17 Measured result (Gafchromic film) of proton strength distribution of 100 MeV protons of 1 nA intensity at upstream beam of the Pb–Bi target (Ref. [22])

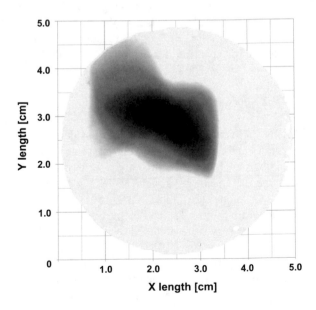

Fig. 5.18 Scanning relative result of proton strength distribution of 100 MeV protons in Fig. 5.17 (Ref. [22])

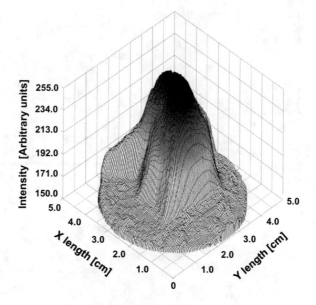

and proton transport, respectively, and JENDL/D-99 and JENDL/HE-2007 for reaction rates of neutrons and protons, respectively. The ratio of neutrons and protons (number of neutrons per proton) by PHITS3.0 was compared with the use of experimental (Table 5.14) and numerical results shown in Table 5.15, demonstrating that the calculation/experiment (C/E) value of the number of neutrons per proton showed agreement with a relative difference about 10%. Meanwhile, the MCNP code yielded a discrepancy between the experiments and the calculations about 30%, with the combined use of nuclear data libraries and data sets shown in Table 5.13.

From the results in Table 5.15, the PHITS3.0 code was successfully validated to obtain the reaction rate ratio of neutrons (under 20 MeV) and protons (100 MeV), together with the neutron spectrum by JENDL-4.0 and the ^{115}In$(n, n')^{115m}$In cross

Table 5.13 Combination of spectra and cross sections by PHITS3.0 and MCNP6.1 with nuclear data libraries (Ref. [22])

Spectrum (Transport)		
	PHITS3.0	MCNP6.1
Under 20 MeV (Neutrons)	JENDL-4.0	JENDL/HE-2007
Over 20 MeV (Neutrons)	INCL model	JENDL/HE-2007
100 MeV (Protons)	INCL model	LAHET150
Cross sections (Reaction rates)		
	PHITS3.0	MCNP6.1
^{115}In$(n, n')^{115m}$In	JENDL/D-99	
^{209}Bi$(n, xn)^{210-x}$Bi	ENDF/B-VI.8	
^{27}Al$(p, n + 3p)^{24}$Na	JENDL/HE-2007	

Table 5.14 Measured results of ^{115}In$(n, n')^{115m}$In and ^{27}Al$(p, n + 3p)^{24}$Na reaction rates (Ref. [22])

Reactions	Measured reaction rates (s^{-1} cm^{-3})
^{115}In$(n, n')^{115m}$In	$(2.89 \pm 0.03) \times 10^5$
^{27}Al$(p, n + 3p)^{24}$Na	$(1.58 \pm 0.02) \times 10^6$

Fig. 5.19 Cross sections of ^{27}Al$(p, n + 3p)^{24}$Na (JENDL/HE-207), ^{115}In$(n, n')^{115m}$In (JENDL/D-99) and ^{209}Bi$(n, xn)^{201-x}$Bi (ENDF/B-VI.8) reactions (Ref. [22])

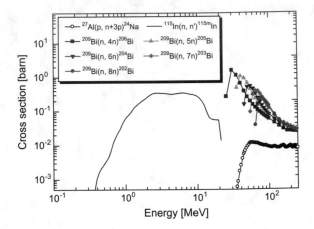

sections (Fig. 5.19) by JENDL/D-99, the proton spectrum by the INCL model of PHITS3.0 and the ^{27}Al$(p, n + 3p)^{24}$Na cross sections (Fig. 5.19) by JENDL/HE-2007.

5.4.2.4 Neutron Spectrum Over 20 MeV

Interestingly, another attempt with PHITS3.0 was applied to the reaction rate analyses of over 20 MeV neutrons generated by the injection of 100 MeV protons onto the Pb–Bi target. As shown in Fig. 5.16, the Al foil was attached to the downstream side of the stainless steel flange for injecting the proton beam, and the Pb–Bi target was set between the Al and Bi foils, which was attached to the backside of the Pb–Bi target. After the injection of proton beams, the γ-ray spectrum of ^{209}Bi$(n, xn)^{210-x}$Bi was experimentally obtained for numbers ranging from 4 to 8, as shown in Fig. 5.20. Experimental reaction rates were attained by Eq. (5.8), with the use of total counts of γ-rays in threshold energies, as shown in Table 5.9.

From the results in Table 5.16, the ratios of ^{209}Bi$(n, xn)^{210-x}$Bi and ^{27}Al$(p, n + 3p)^{24}$Na reaction rates by PHITS3.0 were found to be fairly good with a relative difference of about 10%, except for $x = 7$. Additionally, a comparison of the C/E value demonstrated significant validation of the INCL model of PHITS3.0 and verification of cross sections over 20 MeV threshold energy (ENDF/B-VI.8; Fig. 5.19) and 100 MeV protons (JENDL/HE-2007; Fig. 5.19). Also, the MCNP code showed good agreement with the experimental results shown in Table 5.16, demonstrating the accuracy of a relative difference of about 10%, except for $x = 7$.

Fig. 5.20 Experimental results of the γ-ray spectrum of ^{209}Bi$(n, xn)^{210-x}$Bi $(x = 4$–$8)$ (Ref. [22])

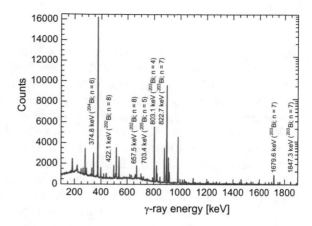

Table 5.15 Comparison between calculations and experiments of reaction rate ratio ^{115}In$(n, n')^{115m}$In/^{27}Al$(p, n + 3p)^{24}$Na (Ref. [22])

Ratio (number of neutrons per proton)	Experiment	C/E*	
		PHITS3.0	MCNP6.1
^{115}In$(n, n')^{115m}$In/ ^{27}Al$(p, n + 3p)^{24}$Na	0.185 ± 0.004	1.14 ± 0.03	1.34 ± 0.03

C/E*: Calculation/Experiment

Table 5.16 Comparison between calculations and experiments of reaction rate ratio ^{209}Bi$(n, xn)^{210-x}$Bi $(x = 4$ to $8)/^{27}$Al$(p, n + 3p)^{24}$Na (Ref. [22])

Reactions	Measured reaction rates (s^{-1} cm^{-3})	C/E*	
		PHITS3.0	MCNP6.1
^{27}Al$(p, n + 3p)^{24}$Na	$(1.64 \pm 0.01) \times 10^6$	–	–
^{209}Bi$(n, 4n)^{206}$Bi	$(8.22 \pm 0.09) \times 10^4$	1.01 ± 0.03	0.83 ± 0.03
^{209}Bi$(n, 5n)^{205}$Bi	$(4.86 \pm 0.25) \times 10^4$	1.06 ± 0.06	1.00 ± 0.06
^{209}Bi$(n, 6n)^{204}$Bi	$(2.23 \pm 0.02) \times 10^4$	0.83 ± 0.02	0.92 ± 0.04
^{209}Bi$(n, 7n)^{203}$Bi	$(1.42 \pm 0.04) \times 10^4$	0.69 ± 0.03	0.81 ± 0.04
^{209}Bi$(n, 8n)^{202}$Bi	$(0.35 \pm 0.01) \times 10^4$	0.99 ± 0.06	1.08 ± 0.07

C/E*: Calculation/Experiment

For experimental analyses of over 20 MeV neutrons and 100 MeV protons, a choice of the INCL model of PHITS3.0 or MCNP6.1 with LAHET150 still remained for examining the validity of ^{209}Bi$(n, 7n)^{203}$Bi reaction rate analyses. Meanwhile, the PHITS3.0 and the MCNP6.1 codes were significantly similar with respect to the reconstruction of reaction rates over 20 MeV neutrons and 100 MeV protons, through a comparative study on a suitable combination (Table 5.13) of the Monte Carlo codes and nuclear data libraries.

5.5   Conclusion

The ADS experiments with 100 MeV protons were carried out at KUCA to evaluate the accuracy of experiments and calculations in subcritical states, when the kind of solid target (W, W-Be and Pb–Bi) used was varied at the location of the target. The analyses of neutron multiplication and subcritical multiplication factor demonstrated a fairly good comparison between the experiments and the calculations.

Reaction rate experiments were carried out on ADS with spallation neutrons to examine the accuracy of reaction rates by the foil activation method at various subcriticalities. A comparison between measured and calculated ^{115}In$(n, \gamma)^{116m}$In reaction rate distributions in the subcritical cores proved the reliability of the MCNP simulations through analyses of k_s. The reaction rates by the activation foils for threshold energy were compared significantly with the discrepancy between experiments and calculations, and the dependence of reaction rates on subcriticality. Furthermore, the In ratio was proposed to examine neutron spectrum information on ADS by two different In reaction rate distributions, as another neutron spectrum index, and the validation of the In ratio was well supported by the comparison between experiments and calculations.

Neutron spectrum experiments on spallation neutrons were conducted in the ADS facility at KUCA to investigate the neutronic characteristics of spallation neutrons generated at the target. The reaction rates and the continuous energy distribution of spallation neutrons were measured by the foil activation method and by the organic liquid scintillator, respectively. For the reaction rate experiments of ^{209}Bi foil, the C/E values between the experiments and the calculations (MCNPX and ENDF/B-VI) were found well within the relative difference of 10% (^{209}Bi$(n, xn)^{210-x}$Bi; $x = 4$–7), except for some reactions ($x = 8$). For continuous energy distribution experiments, the spallation neutrons were observed up to 45 MeV with the use of the organic liquid scintillator and the numerical simulations (MCNPX with JENDL/HE-2007 and ENDF/B-VI.6). Moreover, a suitable combination of Monte Carlo codes and the nuclear data libraries was investigated for reaction rate analyses of high-energy neutrons and protons. The ratio of high-energy neutrons under 20 and 100 MeV protons was successfully reconstructed by combining JENDL-4.0 and the INCL model of the PHITS3.0 code, showing a relative difference of about 10% between experiments and calculations. For high-energy neutrons over 20 MeV, the Monte Carlo codes together with JENDL/HE-2007 for high-energy neutrons and LAHET150 for protons demonstrated significant agreement through ^{209}Bi$(n, xn)^{210-x}$Bi reaction rate analyses.

References

1. Kobayashi K, Nishihara K (2000) Definition of subcriticality using the importance function for the production of fission neutrons. Nucl Sci Eng 136:272

2. Shahbunder H, Pyeon CH, Misawa T et al (2010) Subcritical multiplication factor and source efficiency in accelerator-driven system. Ann Nucl Energy 37:1214
3. Pyeon CH, Nakano H, Yamanaka M et al (2015) Neutron characteristics of solid targets in accelerator-driven system with 100-MeV protons at Kyoto University Critical Assembly. Nucl Technol 192:181
4. X-5 Monte Carlo team (2003) MCNP—a general Monte Carlo N-particle transport code, version 5. LA-UR-03-1987
5. Ziegler JF, Biersack JP, Ziegler MD (1985) The stopping and range of ions in matter. Pergamon Press, New York
6. Takada H, Kosako K, Fukahori T (2009) Validation of JENDL high-energy file through analyses of spallation experiments at incident proton energies from 0.5 to 2.83 GeV. J Nucl Sci Technol 46:589
7. Shibata K, Iwamoto O, Nakagawa T et al (2011) JENDL-4.0: a new library for nuclear science and engineering. J Nucl Sci Technol 48:1
8. Kobayashi K, Iguchi T, Iwasaki S et al (2002) JENDL dosimetry file 99 (JENDL/D-99). JAERI Report 1344
9. Pyeon CH, Vu TM, Yamanaka M et al (2018) Reaction rate analyses of accelerator-driven system experiments with 100 MeV protons at Kyoto University Critical Assembly. J Nucl Sci Technol 55:190
10. Lim JY, Pyeon CH, Yagi T et al (2012) Subcritical multiplication parameters of the accelerator-driven system with 100 MeV protons at the Kyoto University Critical Assembly. Sci Technol Nucl Install 395878:9
11. Goorley JT, James MR, Booth TE et al (2013) Initial MCNP6 release overview - MCNP6 version 1.0. LA-UR-13-22934
12. Pyeon CH, Takemoto Y, Yagi T et al (2012) Accuracy of reaction rates in the accelerator-driven system with 14 MeV neutrons at the Kyoto University Critical Assembly. Ann Nucl Energy 40:229
13. Pyeon CH, Shiga H, Abe K et al (2010) Reaction rate analysis of nuclear spallation reactions generated by 150, 190 and 235 MeV protons. J Nucl Sci Technol 47:1090
14. Ashland (2017) Gafchromic film. http://gafchromic.com/, Accessed 16 Jun 2020
15. Pyeon CH, Azuma T, Takemoto Y et al (2013) Experimental analyses of spallation neutrons generated by 100 MeV protons at the Kyoto University Critical Assembly. Nucl Eng Technol 45:81
16. Pyeon CH, Shiga H, Misawa T et al (2009) Reaction rate analyses for an accelerator-driven system with 14 MeV neutrons in the Kyoto University Critical Assembly. J Nucl Sci Technol 46:965
17. Verbinski VV, Burrus WR, Love TA et al (1968) Calibration of an organic scintillator for neutron spectrometry. Nucl Instrum Methods 65:8
18. Dietzw G, Klien H (1982) Gamma-calibration of NE-213 scintillator counters. Nucl Instrum Methods 193:549
19. Satoh D, Sato T, Shingyo N et al (2006) SCINFUL-QMD: Monte Carlo based computer code to calculate response function and detection efficiency of a liquid organic scintillator for neutron energies up to 3 GeV. JAEA-Data/Code, 2006-23
20. Reginatto M, Wiegel B, Zimbal A et al (2004) The UMG-code package, ver. 3.3. NEA-1665/03
21. Lemmel HD et al (2001) ENDF/B-VI release 8 (Last release for ENDF/B-VI) The U.S. evaluated nuclear data library for neutron reaction data. IAEA-NDS-100
22. Pyeon CH, Yamanaka M, Lee B (2020) Reaction rate analyses of high-energy neutrons by injection of 100 MeV protons onto lead-bismuth target. Ann Nucl Energy 144:107498
23. Yamanaka M, Jang KW, Shin SH et al (2019) Proton beam characteristics with wavelength shifting fiber detector at Kyoto University Critical Assembly. Jpn J Appl Phys 58:036002
24. Sato T, Iwamoto Y, Hashimoto S et al (2018) Features of particle and heavy ion transport code system (PHITS) version 3.02. J Nucl Sci Technol 55:684

25. Boudard A et al (2013) New potentialities of the liege intranuclear cascade mode for reactions induced by nucleons and light charged particles. Phys Rev C 87:014606
26. Kim E, Nakamura T, Konno A et al (1998) Measurement of neutron spallation cross sections of ^{12}C and ^{209}Bi in the 20- to 150-MeV energy range. Nucl Sci Eng 129:209

Chapter 6
Nuclear Transmutation of Minor Actinide

Cheol Ho Pyeon

Abstract Integral experiments on critical irradiation of neuptium-237 (^{237}Np) and americium-241 (^{241}Am) foils are carried out in a hard spectrum core at KUCA with the use of the back-to-back fission chamber, and Monte Carlo calculations together with a reference nuclear data library are conducted for confirming the precision of numerical simulations. Subcritical irradiation of minor actinide (MA) by ADS is a very important step, before operating actual ADS facilities, in a critical assembly at zero power, such as KUCA, which is an exclusive facility for ADS that comprises a uranium-235 (^{235}U) fueled core and a 100 MeV proton accelerator. The first significant attempt is made to demonstrate the principle of nuclear transmutation of MA by ADS through the injection of high-energy neutrons into the KUCA core at a subcritical state. Here, the main targets of nuclear transmutation of MA by the ADS experiments are fission reactions of ^{237}Np and ^{241}Am, and capture reactions of ^{237}Np.

Keywords Nuclear transmutation · Minor actinides · BTB fission chamber · Neptunium-237 · Americium-241

6.1 Integral Experiments at Critical State

6.1.1 Critical Irradiation Experiments

6.1.1.1 Core Configuration

Critical irradiation experiments [1] were carried out in the A-core (reference core; Fig. 6.1) that has polyethylene (PE) moderator and reflector rods, and two fuel assemblies: normal "F" and special "f" (Fig. 6.2a, b, respectively). The normal fuel assembly "F" is composed of 60 unit cells, and upper and lower polyethylene blocks about 23″ and 21″ long, respectively, in an aluminum (Al) sheath (2.1 × 2.1 × 60″).

C. H. Pyeon (✉)
Institute for Integrated Radiation and Nuclear Science, Kyoto University, Osaka, Japan
e-mail: pyeon.cheolho.4z@kyoto-u.ac.jp

© The Author(s) 2021
C. H. Pyeon (ed.), *Accelerator-Driven System at Kyoto University Critical Assembly*,
https://doi.org/10.1007/978-981-16-0344-0_6

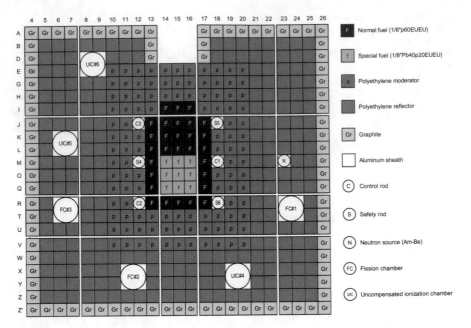

Fig. 6.1 Top view of the KUCA core (reference core) for critical irradiation experiments on MA (Ref. [1])

A unit cell is composed of two highly enriched uranium (HEU) fuel plates $2 \times 2''$ square and $1/8''$ thick ($1/16'' \times 2$), polyethylene plates $2 \times 2''$ square and $1/8''$ thick, for the normal fuel plate "F" (HEU-PE). The special fuel assembly "f" (HEU-Pb) is composed of 40 unit cells with two HEU fuel plates $1/8''$ ($1/16'' \times 2$) thick, lead (Pb) plate $2 \times 2''$ square and $1/8''$ thick, 20 unit cells of HEU and polyethylene plates as well as in the normal fuel assembly.

6.1.1.2 Back-to-Back Fission Chamber

Critical irradiation experiments for measuring the fission reaction rate ratio were carried out with the back-to-back (BTB) fission chamber (Fig. 6.3) set inside a special void element, containing two foils: a test foil neptunium-237 (^{237}Np) or americium-241 (^{241}Am) and a reference one uranium-235 (^{235}U). The reference foil ^{235}U was a normalization factor that has the fission reactions ^{235}U(n, f), to obtain two fission reaction rate ratios: ^{237}Np/^{235}U and ^{241}Am/^{235}U. The main characteristic of the BTB chamber is to obtain uniquely an original signal of fission event of one foil attached at the position of sample deposit, and a different signal of fission event of the other foil attached at the opposite side without any disturbance of the foil on the front side. The chamber has a structure of separating it from two electro-deposited foils that receive independently pulsed signals of fission events accumulated as electric pulses generated by fission fragments through the ionization process of filling gas

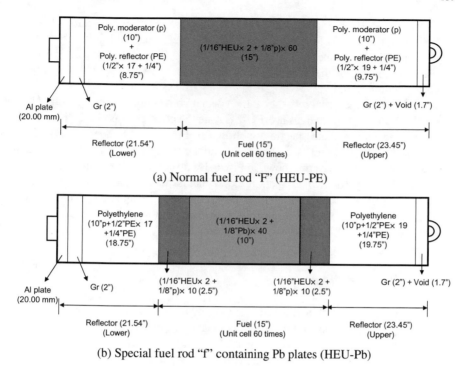

(a) Normal fuel rod "F" (HEU-PE)

(b) Special fuel rod "f" containing Pb plates (HEU-Pb)

Fig. 6.2 Schematic drawing of fuel assemblies in the KUCA A-core (Fig. 6.1) (Ref. [1])

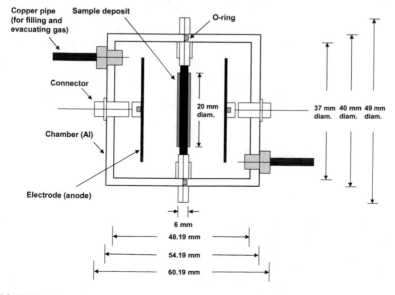

Fig. 6.3 BTB fission chamber (Ref. [1])

(97% argon and 3% nitrogen). For measuring capture reaction rate ratio, a test foil ^{237}Np was set in the BTB fission chamber, and a reference foil gold-197 (^{197}Au) was attached to the Al sheath containing the chamber. The reference foil ^{197}Au was a normalization factor that has the capture reactions ^{197}Au(n, γ) ^{198}Au, to obtain capture reaction rate ratio: ^{237}Np/^{197}Au.

The BTB fission chamber was set at the special void element position in HEU-Pb zone (15, O; Fig. 6.4a), whereas in HEU-PE zone the position was (15, K; Fig. 6.4b). The hard and intermediate spectra are shown in Fig. 6.5. As shown in Fig. 6.4a, b, one and five normal fuel assemblies were added to the cores, respectively, when comparing with reference core shown in Fig. 6.1. Further fuel assemblies were added to reach criticality for two irradiation experiments at the critical state shown in Table 6.1. A series of critical irradiation experiments on ^{237}Np and ^{241}Am foils was conducted under the following core conditions: irradiation time 1 h; reactor power 3.5 W; neutron flux (4.73 $\pm$ 0.24) $\times$ 10^7 s^{-1} cm^{-2}.

6.1.1.3 ^{237}Np and ^{241}Am Foils

Two ^{237}Np and ^{241}Am thin test foils (both 99.99% purity, 20 mm diameter, and 4 nm thick), electro-deposited on Al backing (28 mm diameter and 0.2 mm thick), had an isotopic mass, respectively, equal to 89 and 15 µg as shown in Table 6.2. Thin ^{235}U reference foils (20 mm diameter and 0.6 nm thick, 5 or 10 µg weight), used to determine the fission reaction rate ratios, had a 99.91% enrichment. Test and reference ^{237}Np and ^{197}Au foils (8 mm diameter and 0.05 mm thick) were used to determine the capture reaction rate ratio. After the irradiation at critical state, ^{237}Np and ^{197}Au capture reaction rates were deduced by the saturated radioactivity obtained from the γ-ray spectra of irradiated test and reference foils, respectively, with the use of a high-purity germanium (HPGe) detector. The γ-ray energy generated by the capture reactions was as follows: 984.45, 1025.87, and 1028.53 keV in ^{237}Np; 411.80 keV in ^{197}Au, as shown in Table 6.3.

6.1.2 Experimental Analyses

6.1.2.1 Experimental Analyses

During the irradiation experiments, the pulsed-height signals (voltage) induced by fission reactions were stored by the use of a multi-channel analyzer to discriminate the signals originated from fission products generated by fission reactions. In the HEU-Pb zone, pulsed-height distributions were clearly observed at two peaks of ^{237}Np (2.0 V and 5.0 V) and ^{235}U (1.6 V and 3.4 V) foils attributed to light and heavy fragments in fission products, as shown in Fig. 6.6, demonstrating the counts by their fission reactions. For ^{241}Am, one peak (0.1 V) was observed in the pulsed-height distribution shown in Fig. 6.7, and the counts by fission reactions were obtained by

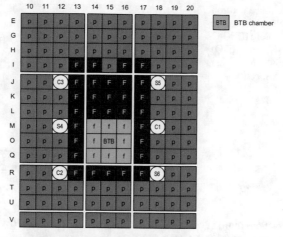

(a) Location of BTB fission chamber (15, O) in HEU-Pb zone

(b) Location of BTB fission chamber (15, K) in HEU-PE zone

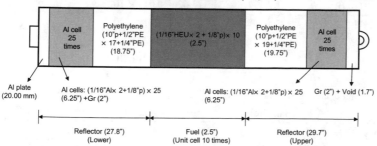

(c) Partial fuel rod "10" shown in Fig. 6.4(b)

Fig. 6.4 Core configuration of critical irradiation experiments on MA (Ref. [1])

Fig. 6.5 Calculated neutron
spectra (MCNP6.1 with
ENDF/B-VII.1) of HEU-PE
(15, K) and HEU-Pb (15, O)
fuel zones shown in
Figs. 6.2a, b, respectively
(Ref. [1])

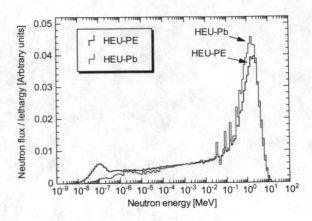

Table 6.1 Positions [mm] of control and safety rods at critical state in HEU-Pb and HEU-PE cores
in Figs. 6.4a, b, respectively (Ref. [1])

Measured fission reaction rate ratio	C1	C2	C3	S4	S5	S6
HEU-Pb zone						
^{237}Np/^{235}U (10 μg)	723.27	1200.00	1200.00	1200.00	1200.00	1200.00
^{241}Am/^{235}U (5 μg)	720.35	1200.00	1200.00	1200.00	1200.00	1200.00
HEU-PE zone						
^{237}Np/^{235}U (5 μg)	698.23	1200.00	1200.00	1200.00	1200.00	1200.00
^{241}Am/^{235}U (10 μg)	694.28	1200.00	1200.00	1200.00	1200.00	1200.00

1200.00 [mm]: Position of upper limit

Table 6.2 Number of atoms of ^{237}Np, ^{241}Am, and ^{235}U foils (Ref. [1])

Foil	Number of atoms
^{237}Np (HEU-Pb)	$(2.27 \pm 0.02) \times 10^{17}$
^{237}Np (HEU-PE)	$(1.48 \pm 0.02) \times 10^{17}$
^{241}Am (HEU-Pb)	$(3.62 \pm 0.01) \times 10^{15}$
^{241}Am (HEU-PE)	$(3.94 \pm 0.01) \times 10^{15}$
^{235}U (5 μg)	$(1.31 \pm 0.02) \times 10^{16}$
^{235}U (10 μg)	$(2.67 \pm 0.02) \times 10^{16}$

Table 6.3 Main characteristics of ^{237}Np and ^{197}Au capture reactions (Ref. [1])

Reaction	Half-life	γ-ray energy (keV)	Emission rate (%)
^{237}Np (n, γ) ^{238}Np	2.177 d	984.45	47.8
		1025.87	9.6
		1028.53	20.3
^{197}Au (n, γ) ^{198}Au	2.69517 d	411.80	95.5

Fig. 6.6 Measured pulsed heights of ^{237}Np and ^{235}U fission reaction rates in HEU-Pb zone (15, O; Fig. 6.4a) at critical state (Ref. [1])

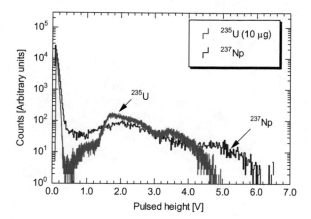

Fig. 6.7 Measured pulsed heights of ^{241}Am and ^{235}U fission reaction rates in HEU-Pb zone (15, O; Fig. 6.4a) at critical state (Ref. [1])

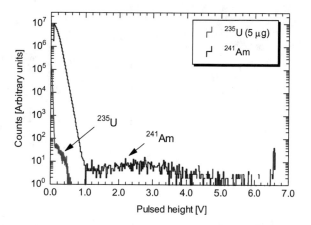

making lower-level discrimination about 1.0 V and integrating the count per channel over 1.0 V. Also, the difference in pulsed heights between 5 and 10 μg of ^{235}U for ^{241}Am fission reaction rates was clearly observed by comparing the HEU-Pb and HEU-PE zones shown in Figs. 6.7 and 6.8, respectively.

In addition to the fission reaction rates, capture reaction rates were obtained by measuring the γ-ray of ^{237}Np foils in HEU-Pb and HEU-PE zones with the HPGe detector after the irradiation experiments, as shown in Figs. 6.9 and 6.10, respectively. For the ^{237}Np capture reaction rates, three peaks (Table 6.3) were found to be clearly in the γ-ray measurements of two irradiation experiments in both HEU-Pb and HEU-PE zones, and ^{237}Np capture reaction rates were experimentally deduced through the saturated radioactivity.

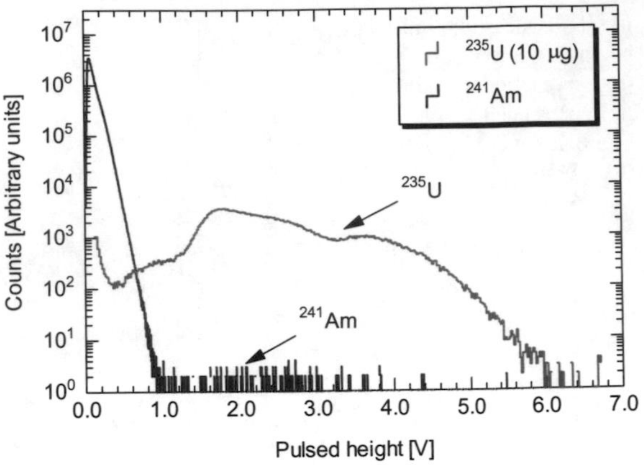

Fig. 6.8 Measured pulsed heights of ^{241}Am and ^{235}U fission reaction rates in HEU-PE zone (15, K; Fig. 6.4b) at critical state (Ref. [1])

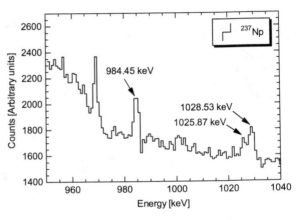

Fig. 6.9 Measured γ-ray spectrum of ^{237}Np capture reaction rates in HEU-Pb zone (15, O; Fig. 6.4a) at critical state (Ref. [1])

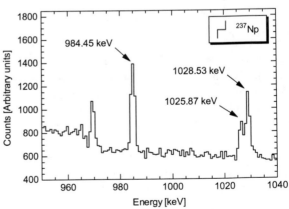

Fig. 6.10 Measured γ-ray spectrum of ^{237}Np capture reaction rates in HEU-PE zone (15, K; Fig. 6.4b) at critical state (Ref. [1])

6.1.2.2   Numerical Analyses

Numerical analyses were carried out with the use of MCNP6.1 [2] together with ENDF/B-VII.1 [3]; analyses of ^{197}Au capture reaction rates were done with JENDL/D-99 [4]. The comparison between measured and calculated results, as shown in Table 6.4, shows a good agreement demonstrating a relative difference within 5% in the C/E (calculation/experiment) value for ^{237}Np fission reaction rates ratio in the HEU-Pb core. For the ^{241}Am fission reaction rates ratio, the C/E discrepancy was about 10% at HEU-Pb and HEU-PE zones, demonstrating a difference between the neutron spectra of the two zones shown in Fig. 6.5. Here, the ^{237}Np capture reaction rate ratio revealed notably good agreement between the experiments and the calculations, as shown in Table 6.5, with a relative difference of about 5% in C/E values.

From the results in Tables 6.4 and 6.5, MCNP6.1 simulations with ENDF/B-VII.1 demonstrated a fairly agreement with the experimental data of ^{237}Np and ^{241}Am fission and capture reaction rates.

Table 6.4 Comparison of the results between experiments and calculations (MCNP6.1 with ENDF/B-VII.1) of fission reaction rate ratios at critical state (Ref. [1])

Fission reaction rate ratio	Calculation	Experiment	C/E
HEU-Pb zone			
^{237}Np/^{235}U (10 µg)	0.067 ± 0.001	0.071 ± 0.004	0.95 ± 0.06
^{241}Am/^{235}U (5 µg)	0.079 ± 0.003	0.069 ± 0.001	1.14 ± 0.05
HEU-PE zone			
^{237}Np/^{235}U (5 µg)	–	–	–
^{241}Am/^{235}U (10 µg)	0.036 ± 0.001	0.034 ± 0.001	1.08 ± 0.02

C/E: Calculation/Experiment

Table 6.5 Comparison of the results between experiments and calculations (MCNP6.1 with ENDF/B-VII.1) of capture reaction rate ratios at critical state (Ref. [1])

Capture reaction rate ratio	Calculation	Experiment	C/E
HEU-Pb zone			
^{237}Np/^{197}Au	2.09 ± 0.10	2.15 ± 0.33	0.97 ± 0.16
HEU-PE zone			
^{237}Np/^{197}Au	1.88 ± 0.08	2.02 ± 0.08	0.93 ± 0.05

C/E: Calculation/Experiment

6.1.3 Discussion

6.1.3.1 Fission Reaction Rate Ratio

The region-wise contribution of energy was examined by MCNP6.1 and ENDF/B-VII.1, for ^{237}Np and ^{241}Am fission reaction rates ratios to ^{235}U, with the use of reaction rates normalized by summarizing entire reaction rates over the whole energy. For the sake of comparison in Fig. 6.11, the energy distribution of ^{237}Np and ^{235}U fission reaction rates are shown for the critical irradiation in the HEU-Pb zone shown in Fig. 6.4a, demonstrating that ^{237}Np fission rate has high sensitivity to fission reactions around a few MeV region as well as in the neutron spectrum of the HEU-Pb zone shown in Fig. 6.5. Also, ^{235}U fission reactions were found to be dominant over the thermal and epi-thermal neutron regions for critical irradiation in the HEU-Pb zone, as shown in Fig. 6.11.

Fission reaction rates of ^{241}Am were acquired mainly in a few MeV region in HEU-Pb and HEU-PE zones shown in Figs. 6.12 and 6.13, respectively, although

Fig. 6.11 Region-wise contribution of energy of ^{237}Np and ^{235}U fission reaction rates by MCNP calculations in HEU-Pb zone at critical state (Ref. [1])

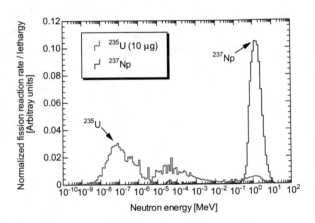

Fig. 6.12 Region-wise contribution of energy of ^{241}Am and ^{235}U fission reaction rates by MCNP calculations in HEU-Pb zone at critical state (Ref. [1])

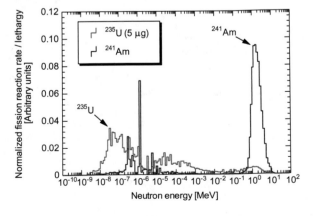

Fig. 6.13 Region-wise contribution of energy of ^{241}Am and ^{235}U fission reaction rates by MCNP calculations in HEU-PE zone at critical state (Ref. [1])

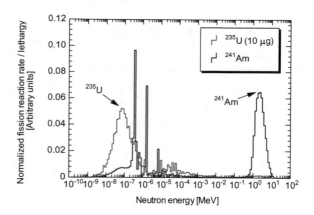

striking peaks of ^{241}Am fission reactions were observed in a wide range of the thermal and epi-thermal neutron regions. Moreover, by comparing the results in Figs. 6.12 and 6.13, a difference in the reaction rates energy distribution between HEU-Pb and HEU-PE zones was clearly observed over the entire energy regions, demonstrating the important effect of the neutron spectrum variation on ^{241}Am fission reaction rates, as shown in Table 6.4.

Through numerical analyses of the region-wise contribution of energy to ^{237}Np and ^{241}Am fission reactions, the effect of neutron spectrum variation (Fig. 6.5) between the HEU-Pb and HEU-PE zones was soundly confirmed on the irradiation of ^{237}Np and ^{241}Am foils at a critical state.

6.1.3.2   Capture Reaction Rate Ratio

Capture reaction rates of ^{237}Np were successfully obtained by the measurement of γ-ray spectra in HEU-Pb and HEU-PE zones after critical irradiation, and used for the evaluation of capture reaction rate ratio by comparison of ^{197}Au capture reaction rates. From the calculated capture reaction rate results in Fig. 6.14, two main peaks can be noticed around the thermal neutron region, and ^{237}Np capture reaction rates were found to be highly sensitive to the thermal neutron spectrum field even in the HEU-Pb zone.

As shown in Table 6.5, the selection of ^{197}Au as reference foil was experimentally meaningful for the evaluation of the ^{237}Np/^{197}Au capture reaction rate ratio, even if no significant differences were observed between the results in the HEU-Pb and HEU-PE zones.

Fig. 6.14 Region-wise contribution of energy of ^{237}Np capture reaction rates by MCNP calculations in HEU-Pb zone at critical state (Ref. [1])

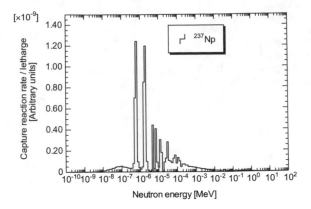

6.2 ADS Irradiation at Subcritical State

6.2.1 Experimental Settings

6.2.1.1 Core Configuration

Subcritical irradiation experiments [2] of MA by ADS were carried out in the solid-moderated and -reflected (A-core) at KUCA by the injection of high-energy protons onto lead-bismuth (Pb–Bi) target. As shown in Fig. 6.15, the ADS core consists of normal and partial fuel assemblies, and polyethylene moderators and reflectors. Normal fuel assembly "F" is composed of 60 unit cells, as shown in Fig. 6.2a. Numeral "12" (Fig. 6.16b) is a partial fuel assembly for reaching a critical mass, and represents the number of unit cells in a normal fuel assembly.

6.2.1.2 Proton Beams

High-energy neutrons generated by the interaction between high-energy protons and Pb–Bi target that set at the location of (15, H) shown in Fig. 6.15 were injected into the KUCA A-core. The main characteristics of proton beams were as follows: 100 MeV energy, 0.5 nA intensity (1 nA at most), 30 Hz pulsed frequency, and 100 ns beam width. The Pb–Bi target was 50 mm in diameter and 18 mm thick; the thickness was determined by taking into account the range and the full stoppage of 100 MeV protons inside the Pb–Bi target [6]. During the injection of 100 MeV protons (no vacuum; (15, A-H) in Fig. 6.15) onto the Pb–Bi target, the size of the proton beam spot was about 40 mm in diameter (circle line), as demonstrated by the scan of the Gafchromic film [7], at the front side (end of proton beams) of Pb–Bi target (15, H; Fig. 6.15), that is highly sensitive to the charged particles shown in Fig. 6.17. Furthermore, the neutron yield was $(1.33 \pm 0.04) \times 10^8$ s^{-1}, as measured through ^{115}In (n, n') ^{115m}In reaction rates with the use of the saturated activity that

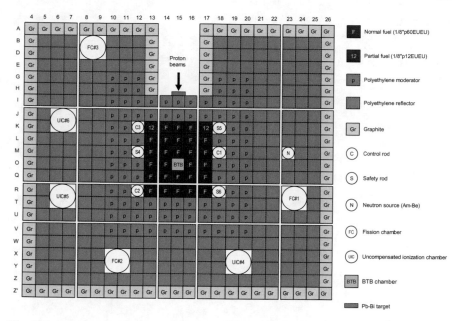

Fig. 6.15 Top view of the KUCA A-core for MA irradiation experiments by ADS (Ref. [5])

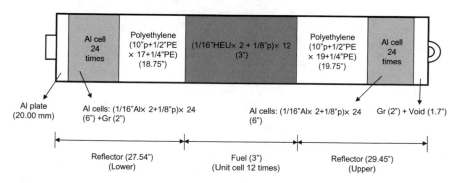

Fig. 6.16 Schematic drawing of "12" partial fuel rod (Fig. 6.15) (Ref. [5])

was deduced by the foil activation method [8] (over 0.3 MeV threshold energy; In foil; $10 \times 10 \times 1$ mm) obtained at the location of the Pb–Bi target.

6.2.1.3 Neutron Characteristics

For carrying out the ADS experiments, subcriticality state was made by inserting control rod C2 into the lower limit position (0.00 mm) and withdrawing other rods (C1, C3, S4, S5, and S6) fully (1200.00 mm) from the core. Subcriticality was then deduced experimentally by combining control rod C2 reactivity worth by the rod

Fig. 6.17 Scanning of Gafchromic film after proton irradiation at location of (15, H; Pb–Bi target) in Fig. 6.15 (Ref. [5])

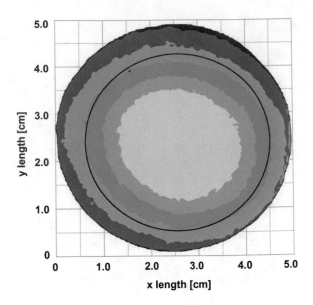

drop method and control rod C2 calibration curve by the positive period method: about 225 ± 10 pcm as a reference value, as shown in Table 6.6. Additionally, the supplemental result obtained by the α-fitting method [9] was 215 ± 9 pcm. During the injection of high-energy neutrons into the core, the reactor power and the neutron flux were experimentally obtained by the foil activation method [8] with the use of two gold foils (bare and cadmium-covered) irradiated at the location between (15, M) and (15, O) in Fig. 6.15 as follows: 1.35 ± 0.07 W and (1.82 ± 0.09) × 10⁷ s⁻¹cm⁻², respectively, for four hours of irradiation.

The neutron spectrum was numerically attained by the PHITS code [11], for the locations of the Pb–Bi target and the BTB fission chamber during the injection of high-energy neutrons, as shown in Fig. 6.18a, b, respectively. At the location of Pb–Bi target, high-energy neutrons showed a sharp peak around 2 MeV region and a unique distribution ranging between 10 and 100 MeV, as shown in Fig. 6.18a. In spite of small effect of high-energy neutrons over 10 MeV in ADS, as shown in Fig. 6.18b, no significant difference between the neutron spectra in ADS and critical cores was found at the location of the BTB fission chamber (15, O; Fig. 6.15).

Table 6.6 Measured core condition during injection of high-energy neutrons (Ref. [5])

Subcriticality [pcm] (Reference)	Subcriticality [pcm] (by α-fitting method)	Neutron flux (s⁻¹ cm⁻²)	Reactor power (W)
225 ± 10	215 ± 9	(1.82 ± 0.09) × 10⁷	1.35 ± 0.07

$\beta_{eff} = 810 \pm 10$ [pcm] and $\Lambda = (3.30 \pm 0.01) \times 10^{-5}$ [s] by MCNP6.1 (Ref. [3]) with JENDL-4.0 (Ref. [10])

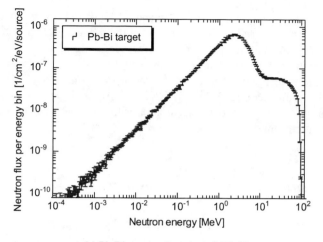

(a) Pb-Bi target at location of (15, H)

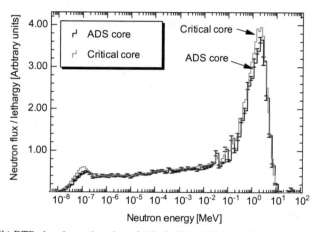

(b) BTB chamber at location of (15, O; Fig. 6.15) in ADS and critical cores

Fig. 6.18 Calculated neutron spectra (PHITS) at locations of (15, H) and (15, O) in Fig. 6.15 (Ref. [5])

6.2.1.4 ^{237}Np and ^{241}Am Foils

To measure fission and capture reactions of ^{237}Np and ^{241}Am (both 99.99% purity, 89 μg and 15 μg weight, respectively), the BTB fission chamber was used in subcritical irradiation experiments and set at the location of (15, O; Fig. 6.15) of a special void element. The main function of the BTB chamber is to obtain simultaneously two signals from specially installed test (^{237}Np or ^{241}Am) and reference (^{235}U; 99.91% enrichment, 10 μg weight) foils. For fission reaction rates, the two main signals from the test and reference foils come from electric ones of fission fragments caused by fission events. Capture reaction rates of ^{237}Np and ^{197}Au were deduced by the saturated radioactivity [8] obtained from two signals of the γ-ray emission of the ^{237}Np test foil in the BTB chamber and the ^{197}Au reference one that set at the location of (15, O) of the Al sheath (special void element), with the use of a germanium detector. From the two signals of test and reference foils, the fission reaction rate ratio was experimentally acquired by ^{237}Np/^{235}U or ^{241}Am/^{235}U, and the capture one by ^{237}Np/^{197}Au, and, finally, validated by comparison with previous experimental results [1] at a critical state.

6.2.2 Demonstration of Nuclear Transmutation

Fission reaction rates were experimentally obtained by the pulsed-height distributions (voltage) of ^{237}Np and ^{241}Am fission events, as shown in Figs. 6.19a, b, respectively, demonstrating original electric signals of test and reference foils. As shown in Fig. 6.19a, the fission events of two ^{237}Np and ^{235}U foils were clearly observed over entire pulsed heights. Moreover, for ^{241}Am and ^{235}U shown in Fig. 6.19b, discrimination between the signals of fission fragments and α-ray induced by ^{241}Am was made at 1.2 V, determining the small fission events by ^{241}Am over 1.2 V and the same fission events by ^{235}U over the entire pulsed heights, as confirmed in the critical irradiation [1]. Meanwhile, the difference between pulsed-height distributions of ^{235}U shown in Figs. 6.19a, b was mainly attributable to reproducibility of the gain in the BTB fission chamber when changing the ^{235}U sample installed. The fission reaction rate ratio determined by the experimental results was found to be a notable 0.048 ± 0.001 and 0.035 ± 0.003 for ^{237}Np/^{235}U and ^{241}Am/^{235}U, respectively, the same as at critical irradiation shown in Table 6.7, although no data of ^{237}Np/^{235}U were obtained at critical irradiation due to invalid data acquisition of two electric signals (^{237}Np and ^{235}U) actuated in the BTB fission chamber.

The capture reaction rate ratio was deduced by the saturated radioactivity on the basis of the γ-ray emission of ^{237}Np and ^{197}Au capture reactions (Fig. 6.20 and Table 6.8) was found to be a remarkable 1.88 ± 0.28, almost the same as at critical irradiation shown in Table 6.9.

From the results of fission and capture reaction rate ratios shown in Tables 6.7 and 6.9, respectively, fission and capture reaction events by ADS were successfully confirmed, and as a conclusion, nuclear transmutation of ^{237}Np and ^{241}Am by ADS was experimentally achieved and demonstrated in the KUCA core.

6.3 Conclusion

The integral experiments on irradiation of ^{237}Np and ^{241}Am in cores at critical conditions were carried out in the KUCA A-core with a neutron hard spectrum. The fission reaction rate ratios (^{237}Np/^{235}U and ^{241}Am/^{235}U) and the capture reaction rate ratio

Fig. 6.19 Measured pulsed heights of ^{237}Np, ^{241}Am, and ^{235}U fission reaction rates at subcritical state (Ref. [5])

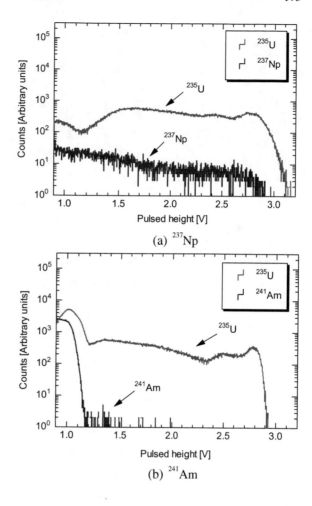

(a) ^{237}Np

(b) ^{241}Am

Table 6.7 Comparison of the results of fission reaction rate ratio at subcritical and critical states (Ref. [5])

Reaction rate ratio	Subcritical state	Critical state (Ref. [1])
^{237}Np/^{235}U	0.048 ± 0.001	–
^{241}Am/^{235}U	0.035 ± 0.003	0.034 ± 0.001

(^{237}Np/^{197}Au) were measured by a BTB (back-to-back) fission chamber. To investigate the behavior of ^{237}Np and ^{241}Am fission and capture reaction rates in the KUCA A-core, results were deeply analyzed by both the experimental signals from the BTB fission chamber and the numerical results of MCNP6.1 calculations with ENDF/B-VII.1. The calculated decomposition in energy of the fission and capture reaction rates allowed us to isolate the most important energy regions in the responses. From the results of experimental and numerical analyses, integral experiments on the irradiation of MA in cores at critical conditions were successfully carried out

Fig. 6.20 Measured γ-ray spectrum of ^{237}Np capture reaction rates at subcritical state (Ref. [5])

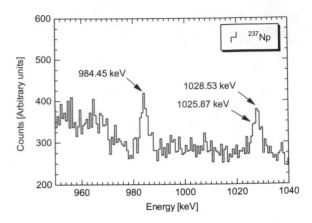

Table 6.8 Main characteristics of ^{237}Np and ^{197}Au capture reactions (Ref. [5])

Reaction	Half-life	γ-ray energy (keV)	Emission rate (%)
^{237}Np $(n, γ)$ ^{238}Np	2.177 d	984.45	27.8
		1025.87	9.6
		1028.53	20.3
^{197}Au $(n, γ)$ ^{198}Au	2.69517 d	411.80	95.5

Table 6.9 Comparison of the results of capture reaction rate ratio at subcritical and critical states (Ref. [5])

Reaction rate ratio	Subcritical state	Critical state (Ref. [1])
^{237}Np/^{197}Au	1.88 ± 0.28	1.88 ± 0.08

in the KUCA-A hard spectrum cores. In future studies, nuclear transmutation of MA by ADS is foreseen for implementation at KUCA, with the combined use of a hard spectrum core and 100 MeV protons, on the basis of measured and calculated methodologies obtained in the current at critical core conditions irradiation experiments on MA.

100 MeV proton beams were injected onto the Pb–Bi target, and subcritical irradiation experiments of MA (^{237}Np and ^{241}Am) by ADS were carried out with the use of high-energy neutrons generated by the interaction of 100 MeV protons and the Pb–Bi target. In the subcritical irradiation experiments, fission reaction rates of ^{237}Np and ^{241}Am were acquired by the electric signals of fission fragments obtained from the BTB fission chamber, and capture reaction rates of ^{237}Np were obtained by the measurement of the γ-ray spectrum after the irradiation. Here, for the first time, nuclear transmutation of ^{237}Np and ^{241}Am was soundly implemented by ADS that comprises a subcritical core and a 100 MeV proton accelerator with Pb–Bi target, and demonstrated at KUCA through the experimental results of the reaction rate ratio

obtained by combining the test (fission: ^{237}Np and ^{241}Am; capture: ^{237}Np) and the reference (fission: ^{235}U; capture: ^{197}Au) foils.

References

1. Pyeon CH, Yamanaka M, Sano T et al (2019) Integral experiments on critical irradiation of ^{237}Np and ^{241}Am foils at Kyoto University Critical Assembly. Nucl Sci Eng 193:1023
2. Goorley JT, James MR, Booth TE et al (2013) Initial MCNP6 release overview—MCNP6 Version 1.0. LA-UR-13-22934
3. Chadwick MB, Herman M, Oblozinsky P et al (2011) ENDF/B-VII.1 nuclear data for science and technology: cross sections, covariances, fission product yields and decay data. Nucl Data Sheets 112:2887
4. Kobayashi K et al (2002) JENDL dosimetry file 99 (JENDL/D-99). JAERI 1344
5. Pyeon CH, Yamanaka M, Oizumi A et al (2019) First nuclear transmutation of ^{237}Np and ^{241}Am by accelerator-driven system at Kyoto University Critical Assembly. J Nucl Sci Technol 56:684
6. Pyeon CH, Nakano H, Yamanaka M et al (2015) Neutron characteristics of solid targets in accelerator-driven system with 100 MeV protons at Kyoto University Critical Assembly. Nucl Technol 192:181
7. Ashland (2017) Gafchromi film. http://gafchromic.com/, Accessed 16 June 2020
8. Misawa T, Unesaki H, Pyeon CH (2010) Nuclear reactor physics experiments. Kyoto University Press, Kyoto, Japan, pp 61–100
9. Simmons BE, King JS (1958) A pulsed technique for reactivity determination. Nucl Sci Eng 3:595
10. Shibata K, Iwamoto O, Nakagawa T et al (2011) JENDL-4.0: a new library for nuclear science and technology. J Nucl Sci Technol 48:1
11. Sato T, Iwamoto Y, Hashimoto S et al (2018) Features of particles and heavy ion transport code system (PHITS) version 3.02. J Nucl Sci Technol 55:684

Chapter 7
Neutronics of Lead and Bismuth

Cheol Ho Pyeon

Abstract Cross-section uncertainties of Pb and Bi isotopes could consequently affect the precision of nuclear design calculations of preliminary analyses, before the actual operation of upcoming ADS, since Pb and Bi are composed partly of coolant material (lead-bismuth eutectic: LBE) in ADS facilities. The main characteristics of LBE in ADS are recognized as follows: chemically inactive; high boiling point mechanically; excellent neutron economy caused by large scattering cross sections. From the viewpoint of neutronics, LBE exerts considerable impact on nuclear design parameters for numerical simulations of neutron interactions of Pb and Bi isotopes. As a suitable way of investigating cross-section uncertainties, sample reactivity worth measurements in critical states are considered effective with the use of reference and test materials in a zero-power state, such as a critical assembly, because integral parameter information on cross sections of test materials can be acquired experimentally. For the required experimental study on Pb and Bi nuclear data uncertainties, the sample reactivity worth experiments are carried out at the KUCA core by the substitution of reference (aluminum) for test (Pb or Bi) materials, and numerical simulations are performed with stochastic and deterministic calculation codes together with major nuclear data libraries.

Keywords Sensitivity · Uncertainty · Sample reactivity worth · Lead · Bismuth

7.1 Sample Reactivity Worth Experiments

7.1.1 Core Configuration

7.1.1.1 Lead Sample Reactivity Worth

The lead (Pb) sample reactivity experiments [1, 2] were carried out in the A-core (Fig. 7.1) that has polyethylene moderator (polyethylene "p" in Fig. 7.1) and reflector

C. H. Pyeon (✉)
Institute for Integrated Radiation and Nuclear Science, Kyoto University, Osaka, Japan
e-mail: pyeon.cheolho.4z@kyoto-u.ac.jp

© The Author(s) 2021   177
C. H. Pyeon (ed.), *Accelerator-Driven System at Kyoto University Critical Assembly*,
https://doi.org/10.1007/978-981-16-0344-0_7

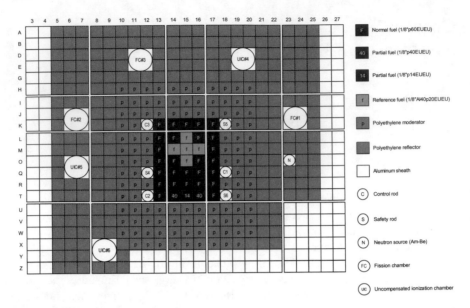

Fig. 7.1 Top view of the KUCA A-core in sample reactivity experiments (Reference core) (Ref. [1])

(conventional polyethylene) rods, and four different fuel assemblies: normal "F," partials "40" and "14," and reference "f" fuel assemblies (Figs. 7.2a–d, respectively). Normal fuel assembly "F" is composed of 60 unit cells, and upper and lower polyethylene blocks about 24″ and 21″ long, respectively, in an aluminum (Al) sheath. For the normal and partial fuel assemblies, a unit cell in the fuel region is composed of a highly enriched uranium (HEU) fuel plate 1/16″ thick and polyethylene plate 1/8″ thick. The numerals 40 and 14 correspond to the number of fuel plates in the partial fuel assembly used for reaching the criticality mass. The reference fuel assembly "f" is composed of 40 unit cells with an HEU fuel plate 1/16″ thick and Al plate 1/16″ thick, 20 unit cells of HEU and the polyethylene plate as in the normal fuel assembly, as shown in Fig. 7.2d.

7.1.1.2 Bismuth Sample Reactivity Worth

The bismuth (Bi) sample reactivity worth experiments [3] were carried out in the A-core (Fig. 7.3), which has polyethylene moderator and reflector rods, and four different fuel assemblies, including HEU, polyethylene moderator (p), polyethylene reflector (PE), graphite (Gr) and Al plate: normal "F," partials "40" and "14" and test "f" (Figs. 7.4a–d, respectively).

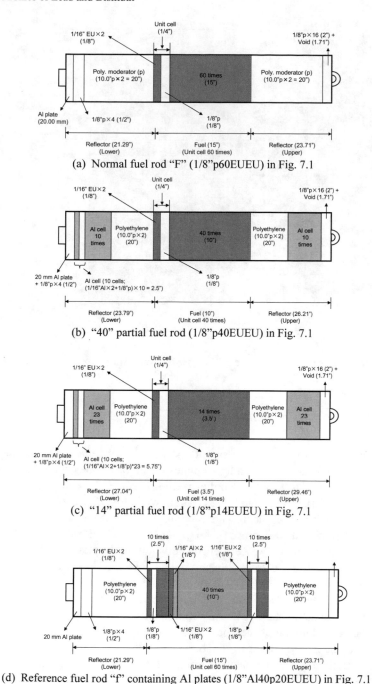

(a) Normal fuel rod "F" (1/8"p60EUEU) in Fig. 7.1

(b) "40" partial fuel rod (1/8"p40EUEU) in Fig. 7.1

(c) "14" partial fuel rod (1/8"p14EUEU) in Fig. 7.1

(d) Reference fuel rod "f" containing Al plates (1/8"Al40p20EUEU) in Fig. 7.1

Fig. 7.2 Schematic drawing of fuel assemblies in the KUCA A-core (Fig. 7.1) (Ref. [1])

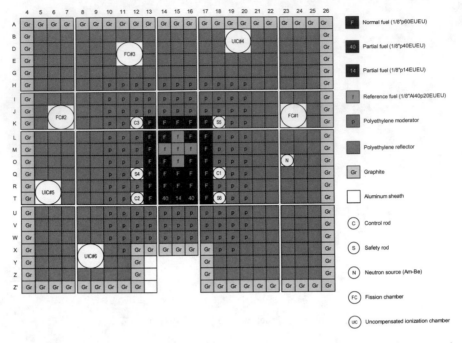

Fig. 7.3 Top view of the KUCA A-core in Bi sample reactivity worth experiments (Reference core) (Ref. [3])

7.1.2 Experimental Settings

In the sample reactivity experiments, a test-zoned fuel region was arranged for measuring the effects of substituting Al plates for Pb or Bi ones upon the criticality. In the test zone, five test fuel assemblies were set around the core at positions (14, M), (15, L), (15, M), (15, O) and (16, M), as shown in Figs. 7.1 and 7.3. The patterns of sample reactivity experiments were ranging between three and five, as shown in Fig. 7.5, substituting the reference fuel rods for Pb or Bi fuel rods. The test fuel rod was the same as in the reference fuel rod substituting Al plates for Pb or Bi ones shown in Fig. 7.6. The spectrum of experimental core at KUCA was compared with that of LBE core [4] in the JAEA ADS model, as shown in Fig. 7.7. The experimental core was a relatively hard spectrum one implemented in KUCA, though not to a fast spectrum core in actual ADS. The substitution was conducted in a total of 40 unit cells of the central region of fuel rods, such as changing Al plates in Figs. 7.2d and 7.4d into Pb and Bi ones in Figs. 7.6a, b, respectively. The sample reactivity caused by the substitution was experimentally obtained through the difference between the excess reactivities of Al reference core and Pb or Bi test core. In the experiments, the critical state was adjusted by maintaining the control rods (C1, C2 and C3) in certain positions shown in Tables 7.1 and 7.2; the excess reactivity was then deduced by the difference between the critical and super-critical states in the core. The experimental

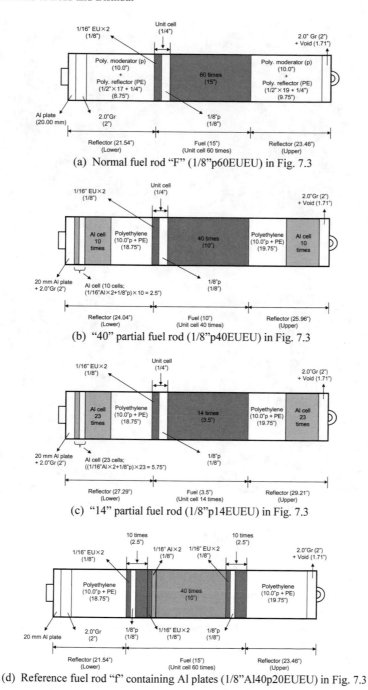

(a) Normal fuel rod "F" (1/8"p60EUEU) in Fig. 7.3

(b) "40" partial fuel rod (1/8"p40EUEU) in Fig. 7.3

(c) "14" partial fuel rod (1/8"p14EUEU) in Fig. 7.3

(d) Reference fuel rod "f" containing Al plates (1/8"Al40p20EUEU) in Fig. 7.3

Fig. 7.4 Schematic drawing of fuel assemblies (Fig. 7.3) in the A-core (Ref. [3])

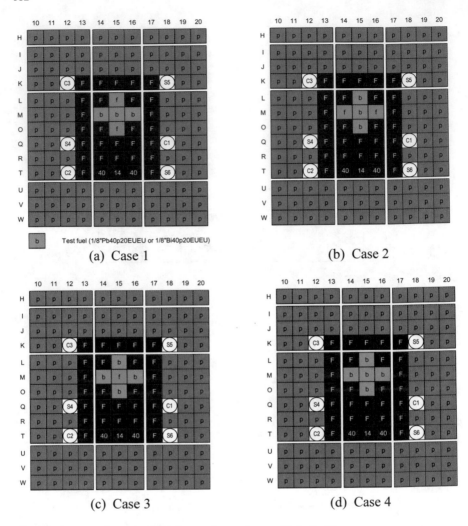

Fig. 7.5 Patterns of sample reactivity worth experiments (Refs. [1–3])

excess reactivity was obtained with the combined use of both the reactivity worth of each control rod evaluated by the rod drop method and its integral calibration curve obtained by the positive period method.

The estimated experimental error of excess reactivity measurement was less than 5%. In the Al reference and Pb or Bi test cores, the effective delayed neutron fraction (β_{eff}) was acquired by MCNP6.1 [5] (2,000 active cycles of 50,000 histories; 2 pcm statistical error) with JENDL-4.0, [6] and the values of 798 and 801 pcm were applied to these two cores, respectively, when the excess reactivity in dollar units was converted into that in pcm units.

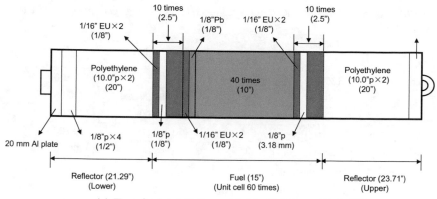

(a) Test fuel rod "b" containing Pb plates (Ref. [1])

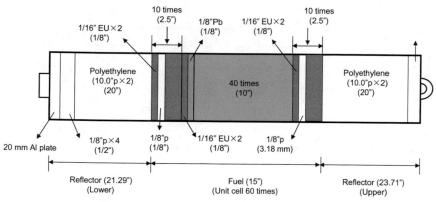

(b) Test fuel rod "b" containing Bi plates (Ref. [3])

Fig. 7.6 Schematic drawing of test fuel rods

Fig. 7.7 Comparison between the neutron spectra of HEU-Pb test zone in KUCA and LBE-cooled core in JAEA ADS model (Ref. [1])

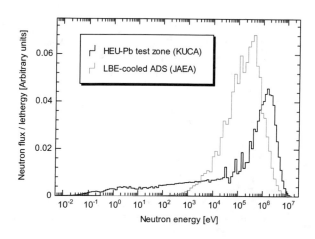

Table 7.1 Control rod positions at critical state in Al reference and Pb test cores (Cases 1 through 4) (Ref. [1])

	Rod position [mm]			
Core	C1	C2	C3	S4, S5, S6
Reference	1200.00	712.58	1200.00	1200.00
Case 1	1200.00	648.23	1200.00	1200.00
Case 2	1200.00	637.59	1200.00	1200.00
Case 3	1200.00	614.86	1200.00	1200.00
Case 4	1200.00	607.95	1200.00	1200.00

1200.00 [mm]: Position of upper limit

Table 7.2 Control rod positions at critical state in Al reference and Bi test cores (Cases 1 through 4) (Ref. [3])

	Rod position [mm]			
Core	C1	C2	C3	S4, S5, S6
Reference	1200.00	715.57	1200.00	1200.00
Case 1	1200.00	676.77	1200.00	1200.00
Case 2	1200.00	662.37	1200.00	1200.00
Case 3	1200.00	663.32	1200.00	1200.00
Case 4	1200.00	658.28	1200.00	1200.00

1200.00 [mm]: Position of upper limit

7.2 Monte Carlo Analyses

7.2.1 Evaluation Method

Experimental sample reactivity worth $\Delta\rho_{Al\to Pb}^{Exp}$ was deduced by the difference between two excess reactivities $\Delta\rho_{Excess}^{Exp,Al}$ and $\Delta\rho_{Excess}^{Exp,Pb}$ obtained by the positive period method in the reference and test cores, respectively, as follows, when the Al plates were substituted for the Pb (or Bi) ones:

$$
\Delta\rho_{Al\to Pb}^{Exp} = \rho_{Excess}^{Exp,Pb} - \rho_{Excess}^{Exp,Al} = \left(1 - \frac{1}{k_{Clean}^{Exp,Pb}}\right) - \left(1 - \frac{1}{k_{Clean}^{Exp,Al}}\right)
$$

$$
= \frac{1}{k_{Clean}^{Exp,Al}} - \frac{1}{k_{Clean}^{Exp,Pb}}, \tag{7.1}
$$

where $k_{Clean}^{Exp,Al}$ and $k_{Clean}^{Exp,Pb}$ indicate the effective multiplication factors deduced by the experimental excess reactivities obtained in super-critical cores (clean core) before

(Al) and after (Pb or Bi) substituting Al plates for Pb ones, respectively, under the condition of all the control and safety rods withdrawn.

In the MCNP analyses, numerical sample reactivity worth $\Delta\rho_{Al\to Pb}^{MCNP}$ was deduced by the difference between two excess reactivities $\Delta\rho_{Excess}^{MCNP,Al}$ and $\Delta\rho_{Excess}^{MCNP,Pb}$ in the reference and test cores, respectively, as follows, with the same method as that of experimental sample reactivity:

$$\Delta\rho_{Al\to Pb}^{MCNP} = \rho_{Excess}^{MCNP,Pb} - \rho_{Excess}^{MCNP,Al} = \left(\frac{1}{k_{Critical}^{MCNP,Pb}} - \frac{1}{k_{Clean}^{MCNP,Pb}}\right)$$
$$- \left(\frac{1}{k_{Critical}^{MCNP,Al}} - \frac{1}{k_{Clean}^{MCNP,Al}}\right), \tag{7.2}$$

where $k_{Clean}^{MCNP,Al}$ and $k_{Clean}^{MCNP,Pb}$ indicate the effective multiplication factors in super-critical cores before and after substituting Al plates for Pb ones, respectively. Also, $k_{Critical}^{MCNP,Al}$ and $k_{Critical}^{MCNP,Pb}$ need to be defined as the values of the effective multiplication factors in critical cores before and after substituting Al plates for Pb ones, since these numerical values always are not unity.

On the basis of the experimental methodology shown in Eq. (7.1), the numerical approach of sample reactivity worth $\Delta\rho_{Al\to Pb}^{Cal}$ can be generally expressed as follows, in case of substituting Al plates for Pb ones:

$$\Delta\rho_{Al\to Pb}^{Cal} = \rho_{Excess}^{Cal,Pb} - \rho_{Excess}^{Cal,Al} = \left(1 - \frac{1}{k_{Clean}^{Cal,Pb}}\right) - \left(1 - \frac{1}{k_{Clean}^{Cal,Al}}\right)$$
$$= \frac{1}{k_{Clean}^{Cal,Al}} - \frac{1}{k_{Clean}^{Cal,b}}, \tag{7.3}$$

where $k_{Clean}^{Cal,Al}$ and $k_{Clean}^{Cal,Pb}$ indicate the effective multiplication factors in super-critical cores.

Numerical sample reactivity $\Delta\rho_{Al\to Pb}^{MCNP}$ in Eq. (7.2) can be rewritten with the use of the concept of Eq. (7.3), as follows:

$$\Delta\rho_{Al\to Pb}^{MCNP} = \rho_{Excess}^{MCNP,Pb} - \rho_{Excess}^{MCNP,Al} = \left(\frac{1}{k_{Critical}^{MCNP,Pb}} - \frac{1}{k_{Clean}^{MCNP,Pb}}\right)$$
$$- \left(\frac{1}{k_{Critical}^{MCNP,Al}} - \frac{1}{k_{Clean}^{MCNP,Al}}\right)$$
$$= \Delta_{Critical,\,Al\to Pb}^{MCNP} + \left(\frac{1}{k_{Clean}^{MCNP,Al}} - \frac{1}{k_{Clean}^{MCNP,Pb}}\right), \tag{7.4}$$

where $\Delta_{\text{Critical},\text{Al}\to\text{Pb}}^{\text{MCNP}}$ indicates the difference between inverse values of eigenvalue calculations in the two critical states evaluated by MCNP6.1. Here, the first term in Eq. (7.4) is defined as "criticality bias" as follows:

$$\Delta_{\text{Critical},\text{Al}\to\text{Pb}}^{\text{MCNP}} = \frac{1}{k_{\text{Critical}}^{\text{MCNP,Al}}} - \frac{1}{k_{\text{Critical}}^{\text{MCNP,Pb}}}. \tag{7.5}$$

By introducing the evaluation methodology of the numerical sample reactivity worth shown in Eq. (7.3), $\Delta_{\text{Clean,J40}\to\text{xxx,yy}-\text{zzz}}^{\text{MCNP,Al}\to\text{Pb}}$ is investigated on the numerical sample reactivity worth as follows, when the nuclear data libraries and isotopes are varied in the MCNP calculations:

$$\Delta_{\text{Clean,J40}\to\text{xxx,yy}-\text{zzz}}^{\text{MCNP,Al}\to\text{Pb}} = \left(\frac{1}{k_{\text{Clean,J40,All}}^{\text{MCNP,Al}}} - \frac{1}{k_{\text{Clean,J40,All}}^{\text{MCNP,Pb}}} \right)$$
$$- \left(\frac{1}{k_{\text{Clean,xxx,yy}-\text{zzz}}^{\text{MCNP,Al}}} - \frac{1}{k_{\text{Clean,xxx,yy}-\text{zzz}}^{\text{MCNP,Pb}}} \right), \tag{7.6}$$

where J40 indicates the JENDL-4.0 library, *All* all the related isotopes and *xxx* a suitable choice of three nuclear data libraries: JENDL-3.3 [7], ENDF/B-VII.0 [8] and JEFF-3.1 [9], *yy* an isotope and *zzz* a mass of isotopes.

7.2.2 Lead Sample Reactivity Worth

7.2.2.1 Numerical Simulations

The numerical analyses were conducted with the use of MCNP6.1 together with the JENDL-3.3, JENDL-4.0, ENDF/B-VII.0 and JEFF-3.1 for transport. For actual experimental analyses, the capability of eigenvalue calculations by MCNP6.1 was useful to be discussed with the use of JENDL-4.0 in processing important data analyses. Also, JENDL-4.0, as a reference library, was compared with the other nuclear data libraries to reveal its uncertainty.

In the reference core shown in Fig. 7.1, criticality was reached by adjusting the position of control rod C2 and withdrawing control rods C1 and C3, and safety rods S4, S5 and S6 from the core, and excess reactivity was deduced with the combined use of control rod worth (C2) by the rod drop method and its calibration curve by the positive period method. The measured excess reactivity was within an uncertainty of 5%, and compared with the numerical one as shown in Table 7.3. The numerical excess reactivity was obtained by the MCNP6.1 eigenvalue calculations with JENDL-4.0 within a statistical error of 2 pcm through 2,000 active cycles of 25,000 histories and estimated with the use of the two eigenvalue calculations in critical and super-critical states. From a comparison between measured and calculated results shown

Table 7.3 Comparison between the results of measured and calculated excess reactivities in reference core shown in Fig. 7.1 (Ref. [1])

Calculation [pcm]	Experiment [pcm]	C/E*
98 ± 6	92 ± 5	1.07 ± 0.09

C/E*: calculation/experiment

Table 7.4 Comparison between the results of measured and calculated control rod worth in reference core shown in Fig. 7.1 (Ref. [1])

Rod	Calculation [pcm]	Experiment [pcm]	C/E
C1	1003 ± 6	980 ± 29	1.02 ± 0.03
C2	447 ± 6	442 ± 13	1.01 ± 0.03
C3	356 ± 6	364 ± 11	0.98 ± 0.03

(β_{eff} = 798 [pcm] and Λ = 3.394E-05 (s) by MCNP6.1 with JENDL-4.0)

in Table 7.3, the C/E (calculation/experiment) value revealed good agreement with a relative difference of 7%.

Furthermore, in the reference core, the measured control rod worth was compared with the calculated one, using the same method as for the excess reactivity. The results are as shown in Table 7.4. The MCNP eigenvalue calculations reproduced the experimental results of control rod worth accurately, with the C/E values within 2%, ranging between 350 and 1,000 pcm, regardless of the kind of control rod used (Table 7.4).

7.2.2.2 Eigenvalue Bias

On the basis of Eq. (7.2), numerical analyses of sample reactivity were conducted by the MCNP6.1 eigenvalue calculations with nuclear data libraries as in Sect. 7.2.2.1. In numerical simulations, sample reactivity worth was obtained by the two eigenvalue calculations in both critical and super-critical states: the difference between the inverse values of eigenvalue calculations in the two states. The calculated results of the sample reactivity were obtained by varying the nuclear data libraries, as shown in Table 7.5 (comparison between Eqs. (7.1) and (7.2)), and, through an estimation of C/E values, were compared at a high accuracy with the experimental results in almost all cases, regardless of the kind of nuclear data libraries used. From the calculated results in Tables 7.3 through 7.5, the precision of MCNP6.1 with JENDL-4.0 was considered fairly good in the eigenvalue calculations, and JENDL-4.0 was found reliable as a reference nuclear data library, comparing it with JENDL-3.3.

Prior to the MCNP numerical analyses, in the experiment results, interesting discussions were provided from two aspects. First, the positive reactivity effect was found in the sample reactivity experiments substituting Al plates for Pb ones, and was mainly attributable to the difference between the values of moderating ratio

Table 7.5 Comparison between the results of measured and calculated sample reactivities evaluated by Eqs. (7.1) and (7.2), respectively, by substituting Al plates for Pb ones as shown in Fig. 7.5 (Ref. [1])

Core	Experiment [pcm]	C/E			
		JENDL-4.0	JENDL-3.3	ENDF/B-VII.0	JEFF-3.1
Case 1	94 ± 5	0.93 ± 0.11	0.91 ± 0.11	0.88 ± 0.11	0.99 ± 0.11
Case 2	110 ± 6	0.85 ± 0.09	0.95 ± 0.09	1.01 ± 0.09	1.08 ± 0.09
Case 3	145 ± 7	0.97 ± 0.07	1.01 ± 0.07	1.06 ± 0.07	1.02 ± 0.07
Case 4	156 ± 8	0.94 ± 0.06	1.03 ± 0.06	1.02 ± 0.06	1.04 ± 0.06

($\xi \Sigma_s / \Sigma_a$: ξ, Σ_s and Σ_a indicate the average logarithmic energy decrement, macro cross sections of scattering and absorption, respectively.) in Al and Pb. Second, while the number of substitution of fuel rods was the same as in both Cases 1 and 2, a significant difference between sample reactivities was involved in the forward and adjoint functions of reactivity defined in the First-order perturbation theory with the variation of core sizes in horizontal and vertical directions shown in Fig. 7.1.

7.2.2.3 Criticality Bias

As discussed in Sect. 7.2, the ability of MCNP6.1 calculations was confirmed in terms of the general definition of sample reactivity by the MCNP approach. Here, the main objective of this study was to compare the experimental and numerical sample reactivities defined in Eqs. (7.1) and (7.3), respectively. By comparing Eqs. (7.1) and (7.3), as shown in Table 7.6, considering the uncertainties of C/E values, the accuracy of the numerical analyses by MCNP6.1 with JENDL-4.0 demonstrated a relative difference of about 5% and an overestimation by more than 50% with JENDL-3.3. By comparing JENDL-3.3 and JENDL-4.0, the calculated values with JENDL-4.0 improved more with a high accuracy of 30% in the C/E values than with the values calculated with JENDL-3.3. Regarding libraries ENDF/B-VII.0 and JEFF-3.1, the calculated sample reactivities were considered well within the relative difference of 10% as shown in Table 7.6.

Table 7.6 Comparison between the results of measured and calculated sample reactivities evaluated by Eqs. (7.1) and (7.3), respectively, by substituting Al plates for Pb ones as shown in Fig. 7.5 (Ref. [1])

Core	Experiment [pcm]	C/E			
		JENDL-4.0	JENDL-3.3	ENDF/B-VII.0	JEFF-3.1
Case 1	94 ± 5	1.13 ± 0.10	1.63 ± 0.13	0.79 ± 0.08	0.89 ± 0.09
Case 2	110 ± 6	1.07 ± 0.08	1.53 ± 0.10	0.85 ± 0.07	0.97 ± 0.07
Case 3	145 ± 7	1.12 ± 0.06	1.65 ± 0.08	0.94 ± 0.05	1.00 ± 0.05
Case 4	156 ± 8	1.13 ± 0.06	1.76 ± 0.08	0.94 ± 0.05	0.98 ± 0.05

Mention should be made, however, of the accuracy of the numerical analyses obtained by MCNP with the four libraries, especially the absolute values by JENDL-4.0, as shown in Table 7.6. Further investigation was needed to find the reason for the discrepancy of unit C/E values with JENDL-4.0. Consequently, in addition to the concept of eigenvalue bias mentioned in Sect. 7.2.2.2, a different evaluation, here termed "criticality bias," of sample reactivity worth was introduced to investigate C/E discrepancy, when the formulation of sample reactivity by MCNP6.1 defined in Eq. (7.2) is changed to that in Eq. (7.4). As shown in Eq. (7.5), the criticality bias $\Delta_{\text{Critical, Al}\to\text{Pb}}^{\text{MCNP}}$ by the MCNP approach was obtained by the difference between reactivity-like criticalities in critical cores by substituting of Al plates for Pb ones, and interpreted as a bias of reactivity induced by the difference between the experiments and the eigenvalue calculations. By the introduction of criticality bias in Eq. (7.5), a small discrepancy in C/E values (Table 7.6) was found in the numerical simulations.

On the basis of Eq. (7.5), criticality bias $\Delta_{\text{Critical, Al}\to\text{Pb}}^{\text{MCNP}}$ was compared with each nuclear data library as shown in Fig. 7.8 and Table 7.7. JENDL-4.0 revealed the bias around 20 pcm; JENDL-3.3 a further bias ranging between 50 and 100 pcm; ENDF/B-VII.0 a relatively small bias less than 20 pcm, compared with the JENDL libraries. Among the four libraries, JEFF-3.1 compared favorably with a small bias

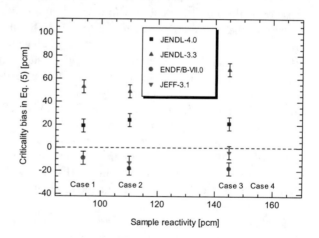

Fig. 7.8 Comparison between the values of criticality bias in critical states evaluated by Eq. (7.4) (Ref. [1])

Table 7.7 Comparison between the results of criticality bias by four nuclear data libraries corresponding to Fig. 7.8 (Ref. [1])

Core	Criticality bias [pcm]			
	JENDL-4.0	JENDL-3.3	ENDF/B-VII.0	JEFF-3.1
Case 1	19.04 ± 5.66	53.04 ± 5.66	−8.98 ± 5.65	−8.97 ± 5.65
Case 2	24.05 ± 5.66	49.04 ± 5.66	−17.95 ± 5.65	−12.96 ± 5.65
Case 3	21.04 ± 5.66	68.05 ± 5.66	−17.95 ± 5.65	−3.99 ± 5.65
Case 4	30.06 ± 5.66	99.04 ± 5.66	−12.96 ± 5.65	−9.97 ± 5.65

around 10 pcm, and resulted in a markedly high accuracy of C/E values, as shown in Table 7.6.

7.2.2.4 Discussion

Special attention was paid to the second term in Eq. (7.4) to investigate the difference between JENDL libraries mentioned in Sect. 7.2.2.3. The second term in Eq. (7.4) was significantly demonstrated in actual sample reactivity by the MCNP analyses, as well as by the experiments, and the bias between JENDL-4.0 and the other libraries were studied with a new definition, as shown in Eq. (7.6): contribution of individual isotope to sample reactivity. In the analyses of differences defined in Eq. (7.6), JENDL-4.0 was selected as the reference library, and the sample reactivities in clean cores were obtained, as shown in Figs. 7.9a, d along with four cases in Fig. 7.5, respectively, when the libraries and isotopes were varied separately: core composition materials of Pb isotopes, ^{27}Al, ^{235}U and ^{238}U in the fuel rod of the core.

A comparison between the two JENDL libraries showed a significant effect on the reactivity resulting from large differences among all Pb isotopes (^{204}Pb, ^{206}Pb, ^{207}Pb and ^{208}Pb), regardless of the magnitude of sample reactivity: especially from those of ^{206}Pb and ^{207}Pb; contrary to that among the others (^{27}Al, ^{235}U and ^{238}U). Regarding the discussion between the two JENDL libraries, the reason for total difference was attributable mainly to those of all Pb isotopes through the analyses of differences in Eq. (7.6). As discussed in previous studies [11, 12], this fact provided valuable knowledge that an improvement of the inelastic scattering cross sections around a few MeV neutron energy region of ^{206}Pb and ^{207}Pb had been pointed out importantly in the difference between JENDL-3.3 and JENDL-4.0 libraries through the analyses of the Pb void reactivity in the JAEA ADS model [10] and of the Pb reflector effect on SEG experiments through JENDL-4.0 benchmarks [12]. From the results of ENDF/B-VII.0, a small effect of the difference was compared inversely with that in JENDL-4.0 about 20 pcm in all cases, with regard to Pb isotopes and ^{27}Al, but not to ^{235}U and ^{238}U, although the total difference between JENDL-4.0 and ENDF/B-VII.0 was slight.

Furthermore, while a difference about 20 pcm was found in ^{238}U and ^{27}Al of Cases 2 and 4, respectively, the difference between JENDL-4.0 and JEFF-3.1 was considered notably minor within the allowance of relative errors.

On the basis of these observations, a library update from JENDL-3.3 to JENDL-4.0 was demonstrated by the fact that the difference between Pb isotopes of the two JENDL libraries was dominant in the comparative study, through the numerical analyses of sample reactivity by the MCNP approach. Moreover, JENDL-4.0 revealed a slight difference from ENDF/B-VII.0 in all the Pb isotopes to ^{27}Al, and from JEFF-3.1 in ^{238}U to ^{27}Al.

Fig. 7.9 Comparison
between the values of
difference in Eq. (7.6) in
JENDL-4.0 and other nuclear
data libraries (Ref. [1])

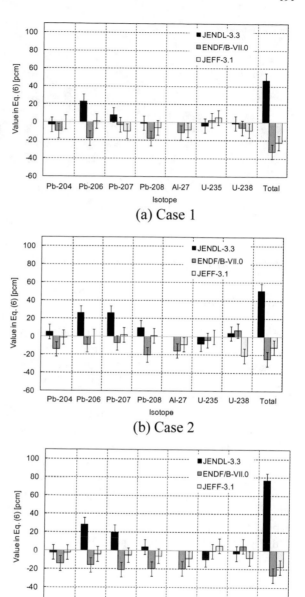

(a) Case 1

(b) Case 2

(c) Case

Fig. 7.9 (continued)

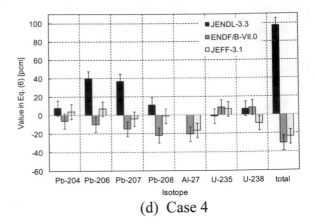

(d) Case 4

7.2.3 Bismuth Sample Reactivity Worth

7.2.3.1 Eigenvalue Calculations

Numerical analyses were conducted with the use of the Monte Carlo code MCNP6.1 together with the JENDL-4.0 nuclear data library for transport. For actual experimental analyses, the capability of eigenvalue calculations by MCNP6.1 was useful in the discussion with the use of JENDL-4.0 for processing important data analyses, and JENDL-4.0 has already been compared with other nuclear data libraries in a previous study [4], while demonstrating a reference library.

In the reference core shown in Fig. 7.3, criticality was reached by adjusting the position of control rod C2 and withdrawing control rods C1 and C3, and safety rods S4, S5 and S6 from the core; excess reactivity was then deduced from the combined use of control rod worth of C2 by the rod drop method and its calibration curve by the positive period method. The measured excess reactivity was attained within an uncertainty of 3%, and compared with the numerical one as shown in Table 7.8. Here, effective delayed neutron fraction (β_{eff}) was attained by MCNP6.1 with JENDL-4.0 in both reference (Al: 798 ± 3 pcm) and test (Bi: 801 ± 3 pcm)

Table 7.8 Comparison between the results of measured and calculated (Eq. (7.2); MCNP6.1) excess reactivities in reference (Al) and test (Bi) cores (Ref. [3])

Core	Calculation [pcm]	Experiment [pcm]	C/E
Reference (Al)	88 ± 6	87 ± 1	1.01 ± 0.07
Case 1	129 ± 6	143 ± 3	0.90 ± 0.05
Case 2	164 ± 6	165 ± 3	0.99 ± 0.04
Case 3	156 ± 6	163 ± 3	0.96 ± 0.04
Case 4	166 ± 6	171 ± 3	0.97 ± 0.04

($\beta_{\mathrm{eff}} = 798 \pm 3$ [pcm] and $\Lambda = (3.39 \pm 0.01)$E-05 [s] by MCNP6.1 with JENDL-4.0; C/E: calculation/experiment)

Table 7.9 Comparison between the results of measured and calculated (MCNP6.1) control rod worth in reference core (Al reference core) (Ref. [3])

Rod	Calculation [pcm]	Experiment [pcm]	C/E
C1	1005 ± 10	945 ± 32	1.06 ± 0.04
C2	454 ± 10	438 ± 1	1.04 ± 0.02
C3	360 ± 11	350 ± 8	1.03 ± 0.04

($\beta_{eff} = 798 \pm 3$ [pcm] and $\Lambda = (3.39 \pm 0.01)$E-05 (s) by MCNP6.1 with JENDL-4.0; C/E: calculation/experiment)

cores. Since two values were almost same within statistical errors, the β_{eff} in reference core was used, when converting measured values in dollar units into ones in pcm units. The numerical excess reactivity was obtained by the MCNP6.1 eigenvalue calculations with JENDL-4.0 within a statistical error of 6 pcm through 2,000 active cycles of 25,000 histories. By comparing the measured and calculated results shown in Table 7.8, the C/E (calculation/experiment) value revealed good agreement within a relative difference of 4%, except in Case 1.

Furthermore, the measured control rod worth in the reference core was compared with the calculated one by the same method used for excess reactivity, as shown in Table 7.9. As shown in Tables 7.8 and 7.9, the MCNP eigenvalue calculations with JENDL-4.0 revealed accurate reproduction of the experimental results of excess reactivity and control rod worth, respectively, with the C/E values within 6%, ranging widely between 87 and 945 pcm.

7.2.3.2 Criticality Bias

As discussed in Sect. 7.2.3.1, the accuracy of MCNP6.1 calculations was confirmed in terms of the general definition of sample reactivity worth by the MCNP approach. Here, the actual objective of this subsection was to compare the difference between experimental and numerical sample reactivity worth defined in Eqs. (7.1) and (7.2), respectively, and to confirm the precision of MCNP calculations of the Bi sample reactivity worth experiments.

Special mention is made of the accuracy of numerical results by MCNP, especially of the absolute values shown in Table 7.10, and additional investigation was requisite to find the reason for the discrepancy between the results of experiments

Table 7.10 Comparison of measured and calculated Bi sample reactivity worth in Eqs. (7.1) and (7.4), respectively, and criticality bias in Eq. (7.5) (Ref. [3])

Core	Experiment in Eq. (7.1) [pcm]	Calculation in Eq. (7.4) [pcm]	Criticality bias in Eq. (7.5) [pcm]
Case 1	56 ± 3	41 ± 8	24 ± 6
Case 2	78 ± 3	76 ± 8	6 ± 6
Case 3	76 ± 3	68 ± 8	33 ± 6
Case 4	84 ± 3	78 ± 8	37 ± 6

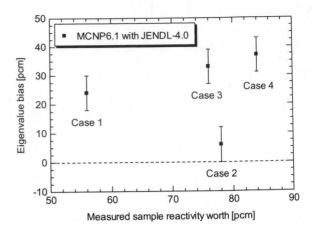

Fig. 7.10 Comparison between the values of criticality bias at critical state evaluated by Eq. (7.5) (Ref. [3])

and calculations in Eq. (7.4) shown in Table 7.10. Then, as suggested in Sect. 7.2.1, "criticality bias" of sample reactivity worth was useful in the investigation of the discrepancy, when a formulation of sample reactivity worth by MCNP6.1 defined in Eq. (7.2) is changed into that by Eq. (7.4). The criticality bias $\Delta^{MCNP}_{Critical,Al \rightarrow Bi}$ defined in Eq. (7.5) was around 30 pcm, as shown in Table 7.10 and Fig. 7.10. From these results, the criticality bias of 37 pcm at most was confirmed as being included in the sample reactivity worth about 80 pcm, even in the analyses of MCNP calculations, although the absolute value of sample reactivity worth was small.

7.3 Sensitivity Coefficients

7.3.1 Theoretical Background

7.3.1.1 Sensitivity Coefficients

The sensitivity coefficient S of the integral reactor physics parameter R is defined by the ratio of the rate of change in R and a certain parameter x as follows:

$$S = \frac{dR}{R} \bigg/ \frac{dx}{x}. \tag{7.7}$$

The effective multiplication factor k_{eff} can be expressed by a balance equation of neutrons as follows:

$$\mathbf{A}\phi = \frac{1}{k_{eff}}\mathbf{F}\phi, \tag{7.8}$$

where **A** and **F** indicate operators of transport and fission terms, respectively, and ϕ the forward neutron flux. Multiplying Eq. (7.8) by adjoint neutron flux ϕ^* and integrating over whole volume and energy, the following equation is obtained:

$$\frac{1}{k_{\text{eff}}} = \frac{\langle \phi^* \mathbf{A} \phi \rangle}{\langle \phi^* \mathbf{F} \phi \rangle},$$

(7.9)

where brackets <> indicate an integration over the whole volume and energy.

Assuming that the value of k_{eff} is a function x, taking the logs of both sides in Eq. (7.9) and differentiating Eq. (7.9) with respect to x, the following equation is obtained, on the basis of theoretical considerations [13–16]:

$$-\frac{d}{dx} \log k_{\text{eff}} = \frac{d}{dx} \log(\langle \phi^* \mathbf{A} \phi \rangle) - \frac{d}{dx} \log(\langle \phi^* \mathbf{F} \phi \rangle)$$

$$\Leftrightarrow -\frac{1}{k_{\text{eff}}} \frac{d}{dx} k_{\text{eff}} = \frac{1}{\langle \phi^* \, \mathbf{A} \, \phi \rangle} \frac{d}{dx} \langle \phi^* \mathbf{A} \phi \rangle - \frac{1}{\langle \phi^* \mathbf{F} \phi \rangle} \frac{d}{dx} \langle \phi^* \mathbf{F} \phi \rangle$$

$$= \frac{\langle \frac{\partial}{\partial x} \phi^* \mathbf{A} \phi \rangle}{\langle \phi^* \mathbf{A} \phi \rangle} - \frac{\langle \frac{\partial}{\partial x} \phi^* \mathbf{F} \phi \rangle}{\langle \phi^* \mathbf{F} \phi \rangle} + \frac{\langle \phi^* \frac{\partial}{\partial x} \mathbf{A} \phi \rangle}{\langle \phi^* \mathbf{A} \phi \rangle} - \frac{\langle \phi^* \frac{\partial}{\partial x} \mathbf{F} \phi \rangle}{\langle \phi^* \mathbf{F} \phi \rangle}$$

$$+ \frac{\langle \phi^* \mathbf{A} \frac{\partial}{\partial x} \phi \rangle}{\langle \phi^* \mathbf{A} \phi \rangle} - \frac{\langle \phi^* \mathbf{F} \frac{\partial}{\partial x} \phi \rangle}{\langle \phi^* \mathbf{F} \phi \rangle}.$$

(7.10)

With the use of an operator **B**, Eq. (7.8) can be expressed as follows:

$$\left(\mathbf{A} - \frac{1}{k_{\text{eff}}} \mathbf{F} \right) \phi = \mathbf{B} \phi = 0.$$

(7.11)

Here, assuming that parameter x, operator **B** and neutron flux ϕ are changed into $x + \delta x$, $\mathbf{B} + \delta \mathbf{B}$ and $\phi + \delta \phi$, respectively, in a critical state, the following equations are obtained:

$$(\mathbf{B} + \delta \mathbf{B})(\phi + \delta \phi) = 0.$$

(7.12)

Neglecting second-order perturbation terms, Eq. (7.11) can be expressed as follows:

$$\mathbf{B} \delta \phi + \delta \mathbf{B} \phi = 0.$$

(7.13)

Introducing the generalized adjoint flux Γ^*, the following equation is obtained with the use of adjoint operator $\mathbf{B}^*$ and a certain adjoint source term q^*, defined as reactivity in these analyses:

$$\mathbf{B}^* \Gamma^* = q^*.$$

(7.14)

Multiplying Eq. (7.13) by the generalized adjoint flux Γ^* on the left side, and integrating over the whole volume and energy, the following equations are obtained with the use of theoretical consideration [16]:

$$\langle \Gamma^* \mathbf{B} \delta \phi \rangle + \langle \Gamma^* \delta \mathbf{B} \phi \rangle = 0, \tag{7.15}$$

$$q^* = \frac{A^* \phi^*}{\langle \phi^* A \phi \rangle} - \frac{F^* \phi^*}{\langle \phi^* F \phi \rangle}. \tag{7.16}$$

From the formation of q^* in Eq. (7.16), q^* is interpreted as an adjustment term for numerically obtaining Γ^* in Eq. (7.14), on the basis of the Generalized Perturbation Method [14].

Finally, with the use of Eqs. (7.11) through (7.16), the sensitivity coefficient in Eq. (7.7) can be expressed as follows, on the basis of the first-order perturbation approximation [17]:

$$S = \frac{\langle \phi^* \frac{\partial A}{\partial x} \phi \rangle}{\langle \phi^* A \phi \rangle} - \frac{\langle \phi^* \frac{\partial F}{\partial x} \phi \rangle}{\langle \phi^* F \phi \rangle} + \left\langle \Gamma^* \frac{d\mathbf{B}}{dx} \phi \right\rangle - \left\langle \Gamma^* \frac{d\mathbf{B}^*}{dx} \phi \right\rangle. \tag{7.17}$$

7.3.1.2 Difference Between Nuclear Data Libraries

With the use of the sensitivity coefficient described in Sect. 7.3.1.1, reactivity change by a data library variation was evaluated by multiplying a relative value of cross sections between data libraries by the sensitivity coefficient.

For example, the sensitivity in JENDL-4.0 is expressed as follows:

$$S_{\rho, \sigma_{n,i,g}^{J40}} = \frac{\sigma_{n,i,g}^{J40}}{\rho_{J40}} \cdot \frac{d\rho}{d\sigma}, \tag{7.18}$$

where ρ_{J40} indicates the calculated sample reactivity by JENDL-4.0, σ the microscopic cross section, n the kind of nuclides, i the kind of reactions and g the energy group. Equation (7.18) can be rewritten as follows:

$$\frac{d\rho}{\rho_{J40}} = \frac{d\sigma}{\sigma_{n,i,g}^{J40}} \cdot S_{\rho, \sigma_{n,i,g}^{J40}}. \tag{7.19}$$

A variation $\Delta \rho_{n,i,g}^{\text{Lib}}$ of sample reactivity by some library (Lib) is evaluated by comparing with that by JENDL-4.0 as follows:

$$\Delta \rho_{n,i,g}^{\text{Lib}} = \left(\frac{\sigma_{n,i,g}^{\text{Lib}} - \sigma_{n,i,g}^{J40}}{\sigma_{n,i,g}^{J40}} \cdot S_{\rho, \sigma_{n,i,g}^{J40}} \right) \cdot \rho_{J40}. \tag{7.20}$$

7.3.2 Lead Isotopes

7.3.2.1 Numerical Approach

The numerical analyses were conducted with the combined use of SRAC2006 and MARBLE code systems: collision probability calculations (PIJ [18]), eigenvalue calculations (CITATION [19]), sensitivity coefficient calculations (SAGEP [20]) of SRAC2006 and uncertainty calculations (UNCERTAINTY [21]) of MARBLE shown in Fig. 7.11, coupled with JENDL-3.3, JENDL-4.0, ENDF/B-VII.0 and JEFF-3.1 nuclear data libraries. The cross-section data set in 107-energy-group processed by the NJOY code [22] is pre-installed with the use of each data library in the SRAC2006 and the MARBLE code systems, to conduct numeral analyses of the thermal neutron spectrum core, such as the KUCA core. For the experimental analyses, the accuracy of deterministic (diffusion-based) calculations by CITATION was useful in the discussion with the use of JENDL-4.0 in processing important data analyses. Also, JENDL-4.0, as a reference library in this study, was compared with other nuclear

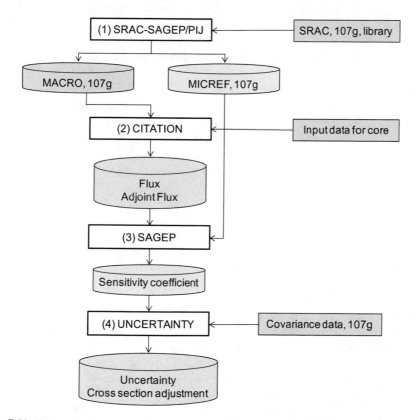

Fig. 7.11 Calculation flow of sensitivity and uncertainty analyses (Ref. [2])

Table 7.11 Comparison between measured and calculated sample reactivities, substituting reference (Al) for test-zoned (Pb) rods in Cases 1 through 4 shown in Fig. 7.5 (Ref. [2])

Core	Experiment [pcm]	Calculation* [pcm]
Case 1	94 ± 5	67
Case 2	110 ± 6	72
Case 3	145 ± 7	140
Case 4	156 ± 8	178

*CITATION in 107-enery-group and 3-D (x-y-z) with JENDL-4.0

data libraries to reveal its uncertainty. Finally, covariance data of cross sections were obtained by NJOY99 with the use of cross-section data contained in JENDL-4.0.

7.3.2.2 Diffusion-Based Eigenvalue Calculations

The measured excess reactivity was within an uncertainty of 5%, and compared with the numerical one as shown in Table 7.11. The numerical excess reactivity was deduced, for a clean core (all control and safety rod withdrawal) in a super-critical state, by the result of diffusion-based eigenvalue calculations (CITATION) in 107-energy-group and x-y-z dimensions (3-D) with JENDL-4.0. The numerical error was within an absolute value of about 30 pcm, compared with the experimental result, as shown in Table 7.10.

Among the four cases shown in Table 7.11, CITATION reproduced the experimental results of sample reactivity with an error of about 20 pcm in Case 4, which was the maximum value in a series of Pb sample reactivity experiments. The experimental result of Case 4 was selected as a representative one in sensitivity and uncertainty analyses, because of the maximum value of experiments and an acceptable accuracy of deterministic calculations by CITATION.

7.3.2.3 Sensitivity Coefficients

Sensitivity coefficients in Eq. (7.17) of sample reactivity were analyzed by the SAGEP code, for cross-section data of inelastic scattering, elastic scattering and capture reactions in Pb isotopes ($^{204, 206, 207, 208}$Pb), as shown in Figs. 7.12a–c, respectively. The sensitivity coefficients of inelastic scattering reactions (Fig. 7.12a) of all Pb isotopes were found to be dominant over the high energy (MeV) region with the other two reactions. The sensitivity coefficients were relatively highly positive in ^{208}Pb mostly around 1 MeV for the elastic scattering reactions (Fig. 7.12b); conversely, they were negative in all Pb isotopes for the capture reactions (Fig. 7.12c). Furthermore, as shown in Fig. 7.12c, the capture cross sections of ^{207}Pb were highly sensitive in the thermal neutron region, since the neutron spectrum of the KUCA core revealed extensive thermalization ranging between 0.01 and 100 eV shown in Fig. 7.13.

Fig. 7.12 Sensitivity
coefficients of sample
reactivity for Pb isotopes
(Ref. [2])

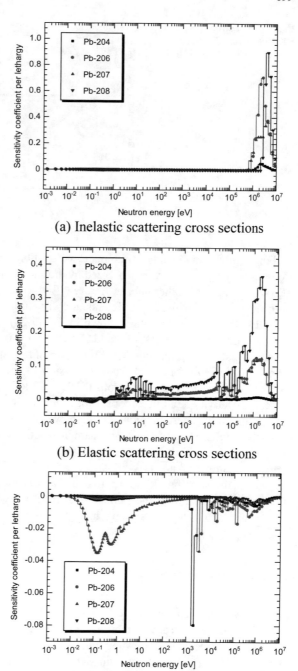

(a) Inelastic scattering cross sections

(b) Elastic scattering cross sections

(c) Capture cross sections

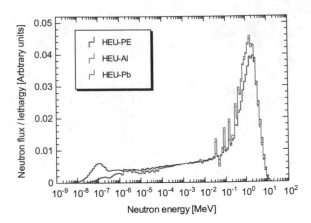

Fig. 7.13 Neutron spectra of HEU-PE, -Al and -Pb fuel zones in KUCA (Reg. [2])

For a comparative study of the nuclear data libraries, the contributions of reactions and energy regions were analyzed by the sample reactivity difference between JENDL-4.0 and one other nuclear library (JENDL-3.3, END/F-VII.0 or JEFF-3.1), with the use of sensitivity coefficients and variation of sample reactivity between JENDL-4.0 and another library shown in Eq. (7.20). As shown in Fig. 7.14a, the comparison between JENDL-3.3 and JENDL-4.0 was large, about 30 and 20 pcm, in inelastic scattering reactions of ^{206}Pb and ^{207}Pb, respectively. This tendency was taken into account for a well-known revision of inelastic scattering reactions of $^{206, 207}$Pb isotope cross sections from JENDL-3.3 to JENDL-4.0. The energy breakdown of reactivity and microscopic cross sections for inelastic scattering reactions were found in the energy region ranging between 1 and 5 MeV shown in Fig. 7.14b. From the results, a large difference of sample reactivity between two JENDL libraries was attributable mainly to the contribution of inelastic scattering reactions of two $^{206, 207}$Pb isotopes. Another comparison between ENDF/B-VII.0 and JENDL-4.0 revealed mainly a difference of 6 pcm in inelastic scattering reactions of ^{208}Pb shown in Fig. 7.14c, and almost the same with JEFF-3.1, as well as with ENDF/B-VII.0, except for ^{208}Pb isotopes, as shown in Fig. 7.14d.

7.3.3 Bismuth Isotope

Sensitivity coefficients of k_{eff} (Case 4) in Eq. (7.17) were analyzed by the SAGEP code, for cross-section data of inelastic scattering, elastic scattering and capture reactions in Bi isotope (^{209}Bi) shown in Figs. 7.15, 7.16 and 7.17, respectively, compared as ^{27}Al, ^{235}U and ^{238}U that are mainly core components. The sensitivity coefficients of inelastic scattering reactions (Fig. 7.15a) of ^{209}Bi were found to be dominant over the high-energy (MeV) region, like those of ^{27}Al shown in Fig. 7.15b. The sensitivity coefficients of ^{209}Bi elastic scattering reactions revealed an increasing tendency between epi-thermal and fast neutron energy regions shown in Fig. 7.16a, although

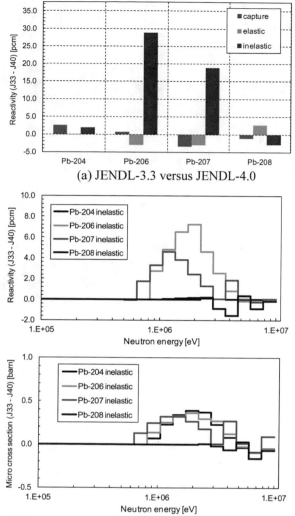

(a) JENDL-3.3 versus JENDL-4.0

(b) Energy breakdown of reactivity (upper) and difference of microscopic cross sections (lower) for inelastic scattering cross sections of Pb isotopes between JENDL-3.3 and JENDL-4.0

Fig. 7.14 Contribution of reactivity by reactions of Pb isotopes between nuclear data libraries (Ref. [2])

the coefficients of ^{209}Bi and ^{27}Al were compared with the mostly same distribution around 1 MeV of the elastic scattering reactions (Figs. 7.16a, b, respectively). For the sensitivity coefficients of capture reactions, ^{209}Bi was found at a highly negative peak around 10^3 eV regions shown in Fig. 7.17a, whereas ^{27}Al and ^{235}U showed a locally strong depression around the thermal neutron region, as shown in Fig. 7.17b.

Fig. 7.14 (continued)

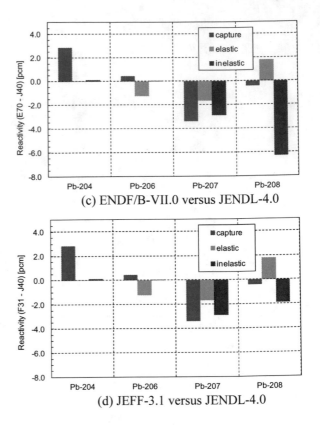

(c) ENDF/B-VII.0 versus JENDL-4.0

(d) JEFF-3.1 versus JENDL-4.0

Nonetheless, from all the results of sensitivity coefficients, absolute values of [209]Bi were markedly very small in vertical axes shown in Figs. 7.15, 7.16 and 7.17, as compared with the values of [27]Al and [235]U, demonstrating that the impact of [209]Bi cross sections was considered minor in the sensitivity coefficient analyses of k_{eff} in the Bi sample reactivity worth experiments at KUCA.

7.4 Uncertainty Quantification

7.4.1 Theoretical Background

7.4.1.1 Uncertainty

As for the cross-section uncertainty analyses of nuclear data [23], the uncertainty of reactor physics parameters v can be expressed as follows:

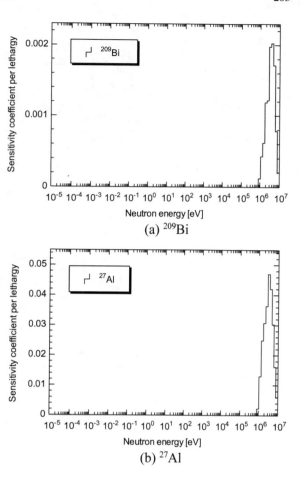

Fig. 7.15 Sensitivity coefficients of inelastic scattering reactions of ^{209}Bi and ^{27}Al in JENDL-4.0 (Ref. [3])

$$v = \mathbf{G}_{\mathrm{tar}}\,\mathbf{M}\,(\mathbf{G}_{\mathrm{tar}})^{\mathrm{t}} = \sum_i \sum_j s_i c_{i,j} s_j \equiv \sum_i \sum_j v_{i,j}\,(1 \le i, j \le p), \qquad (7.21)$$

where $\mathbf{G}_{\mathrm{tar}}$ $(1 \times p)$ indicates the sensitivity vector of reactor physics parameters, $\mathbf{M}$ $(p \times p)$ the covariance matrix of nuclear reaction parameters, s_i the sensitivity coefficient, $c_{i,j}$ the covariance, $v_{i,j}$ the factor of uncertainty and p the number of nuclear reactions including the nuclides. Thus, the contribution of uncertainty u_i in each nuclear reaction can be defined as follows:

$$u_i \equiv \sum_i v_{i,j}. \qquad (7.22)$$

Generally, since sensitivity coefficient s_i and covariance $c_{i,j}$ are dominant in the energy group, the factor of uncertainty is finally expressed with the use of the maximum number of energy group G as follows:

Fig. 7.16 Sensitivity coefficients of elastic scattering reactions of ^{209}Bi and ^{27}Al in JENDL-4.0 (Ref. [3])

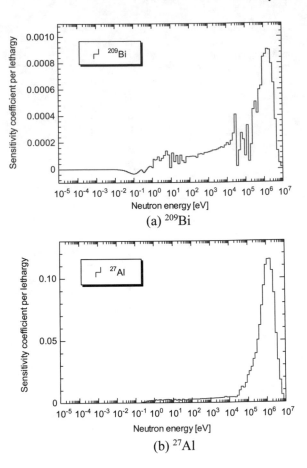

(a) ^{209}Bi

(b) ^{27}Al

$$v_{i, j} = \sum_{g} \sum_{g'} s_g^i c_{g, g'}^{i, j} s_{g'}^j \left(1 \leq g, g' \leq G\right). \tag{7.23}$$

7.4.1.2 Cross-Section Adjustment Method

In the cross-section adjustment method [24], probability P(**T**) with a certain cross-section set **T** is obtained as follows, assuming that a set of nuclear cross sections provides a true value in normal distribution around a true value **T$_0$** of the nuclear cross-section set with dispersion **M**:

$$P(\mathbf{T}) = P(\mathbf{T_0}) \propto \exp\left\{-(\mathbf{T} - \mathbf{T_0})^t \mathbf{M}^{-1}(\mathbf{T} - \mathbf{T_0})/2\right\}. \tag{7.24}$$

Substituting the values of **T$_0$**, **T** and **M** in Eq. (7.24) for those of experiments **R$_e$**, the true value of experiments **R$_{e0}$** and covariance **V$_e$** of experiment errors, Eq. (7.18)

Fig. 7.17 Sensitivity coefficients of capture reactions of ^{209}Bi, ^{27}Al, ^{235}U and ^{238}U in JENDL-4.0 (Ref. [3])

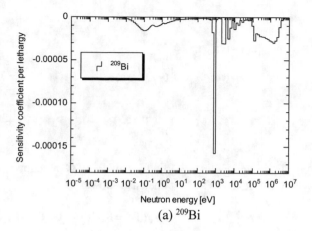

(a) ^{209}Bi

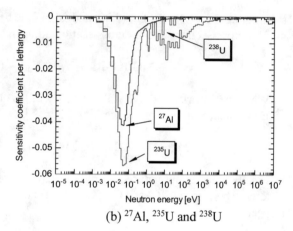

(b) ^{27}Al, ^{235}U and ^{238}U

can be expressed as follows:

$$P(\mathbf{R_0}) \propto \exp\left\{-(\mathbf{R_e} - \mathbf{R_{e0}})^t \mathbf{V_e}^{-1}(\mathbf{R_e} - \mathbf{R_{e0}})/2\right\}. \tag{7.25}$$

The values $\mathbf{R_e}$ are distributed around true values $\mathbf{R_{e0}}$ of experiment with covariance $\mathbf{V_e}$ of experimental value, giving true values $\mathbf{T_0}$ of a set of nuclear cross sections. Also, the values $\mathbf{R_c}(\mathbf{T_0})$ of experiment with the true value of the nuclear data cross-section set are distributed around true value $\mathbf{R_{e0}}$ with covariance $\mathbf{V_e} + \mathbf{V_m}$, giving true value $\mathbf{T_0}$ of a set of nuclear cross sections, as follows:

$$P(\mathbf{R_0}|\mathbf{T_0}) \propto \exp\left\{-(\mathbf{R_e} - \mathbf{R_c}(\mathbf{T_0}))^t (\mathbf{V_e} + \mathbf{V_m})^{-1}(\mathbf{R_e} - \mathbf{R_c}(\mathbf{T_0}))/2\right\}, \tag{7.26}$$

where $\mathbf{V_m}$ indicates the covariance of calculation value.

Using Eqs. (7.24) through (7.26), the following equations are obtained with the consideration of mathematical formulation [24]:

$$P(\mathbf{T_0}|\mathbf{R_0}) = P(\mathbf{R_0}|\mathbf{T_0}) \cdot \frac{P(\mathbf{T_0})}{P(\mathbf{R_0})}$$

$$= (\text{const.}) \cdot \exp\left[(-J)\Big/ \exp\left\{-(\mathbf{R_e} - \mathbf{R_c(T_0)})^t (\mathbf{V_e} + \mathbf{V_m})^{-1} (\mathbf{R_e} - \mathbf{R_c(T_0)})/2\right\}\right],$$
(7.27)

$$J = (\mathbf{T} - \mathbf{T_0})^t \mathbf{M}^{-1} (\mathbf{T} - \mathbf{T_0}) + (\mathbf{R_e} - \mathbf{R_c(T_0)})^t (\mathbf{V_e} + \mathbf{V_m})^{-1} (\mathbf{R_e} - \mathbf{R_c(T_0)}).$$
(7.28)

Introducing the sensitivity coefficient $\mathbf{G}$ as shown in Eq. (7.21), the relation between $\mathbf{R_c}$ and $\mathbf{G}$ is obtained as follows:

$$\mathbf{R_c(T_0)} = \mathbf{R_c(T)} - \mathbf{G(T} - \mathbf{T_0}),$$
(7.29)

substituting Eq. (7.29) for Eq. (7.28) and taking the derivative of Eq. (7.28), a set of nuclear cross-sections $\mathbf{T'}$ after cross-section adjustment is expressed as follows:

$$\mathbf{T'} = \mathbf{T} + \mathbf{MG^t}(\mathbf{GMG} + \mathbf{V_e} + \mathbf{V_m})^{-1}(\mathbf{R_e} - \mathbf{R_c(T_0)}).$$
(7.30)

When the covariance of $(\mathbf{T} - \mathbf{T_0})$ in Eq. (7.29) is obtained, applying to the cross-section adjustment, the covariance $\mathbf{M'}$ of $\mathbf{T'}$ can be expressed as follows:

$$\mathbf{M'} = \mathbf{M} - \mathbf{MG^t}(\mathbf{GMG} + \mathbf{V_e} + \mathbf{V_m})^{-1}\mathbf{GM}.$$
(7.31)

Finally, uncertainty induced by the errors of cross sections is evaluated by the difference between $\mathbf{GMG^t}$ and $\mathbf{GM'G^t}$ before and after the cross-section adjustment, respectively.

7.4.2 Lead Isotopes

7.4.2.1 Uncertainty

The uncertainty analyses by the UNCERTAINTY code of the MARBLE system were conducted with the use of JENDL-4.0 covariance data (107-energy-group) generated by NJOY99. Since the covariance data of H, C and Al nuclides consisted mainly of core components that were not prepared in JENDL-4.0, the uncertainty analyses were executed for several reactions of U and Pb isotopes composed of the reference and the test zones in fuel assemblies of the KUCA A-core, including capture, elastic scattering, inelastic scattering, fission and $(n, 2n)$ reactions. As shown in Table 7.12, the results of uncertainty in reactivity induced by covariance data were large about the total reactivity of 33.1 pcm, compared with an experimental error around 8 pcm of sample reactivity. The value of total uncertainty was acquired by

Table 7.12 Reaction-wise uncertainty contribution [pcm] to changes in sample reactivity induced by covariance data of JENDL-4.0 (Ref. [2])

Isotopes	Reactions					
	Capture	Elastic	Inelastic	Fission	$(n, 2n)$	Total
^{235}U	19.4	1.9	4.1	9.7	0.1	22.2
^{238}U	2.6	0.0	0.3	0.1	0.0	2.6
^{204}Pb	0.1	−0.4	1.6	–	0.0	1.7
^{206}Pb	−1.0	−4.9	20.0	–	−0.8	19.4
^{207}Pb	0.9	−2.6	9.0	–	1.5	8.8
^{208}Pb	−0.6	2.2	9.0	–	3.1	11.9
					Total	33.1

a square root of the sum of squares for reaction-wise contributions, ignoring the covariance between different nuclides in the sum of squares. Among the nuclides, the reaction-wise contribution was dominant over the capture (19.4 pcm) and the inelastic scattering (20.0 pcm) reactions of ^{235}U and ^{206}Pb, respectively, shown in Table 7.12. A large contribution was attributable to the sensitivity coefficients of ^{235}U capture and fission reactions (Fig. 7.18). Also, the reaction-wise contribution of $^{207, 208}$Pb inelastic scattering reactions was observed to obtain meaningful values (9.0 pcm).

For additional study on uncertainty, close attention was paid to the reliability of Pb isotope covariance data of JENDL-4.0 through a comparison between JENDL-4.0 and another library, such as JENDL-3.3, ENDF/B-VII.0 or JEFF-3.1. For a comparison with JENDL-3.3 shown in Fig. 7.19a, contributions of the inelastic scattering cross sections of $^{206, 207}$Pb isotopes were found to remarkably exceed the standard deviation evaluated by JENDL-4.0. This tendency was demonstrated mainly with the energy breakdown of reactivity and the difference of microscopic cross sections, with respect to the Pb isotope inelastic scattering reactions, as shown in Fig. 7.19b. This

Fig. 7.18 Sensitivity coefficients of ^{235}U fission and capture cross sections (JENDL-4.0) (Ref. [2])

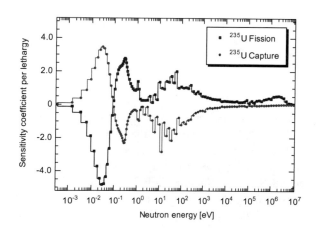

Fig. 7.19 Reactivity
contributions of Pb isotope
reactions induced by
uncertainties (The error bars
indicate the standard
deviation evaluated by
JENDL-4.0.) (Ref. [2])

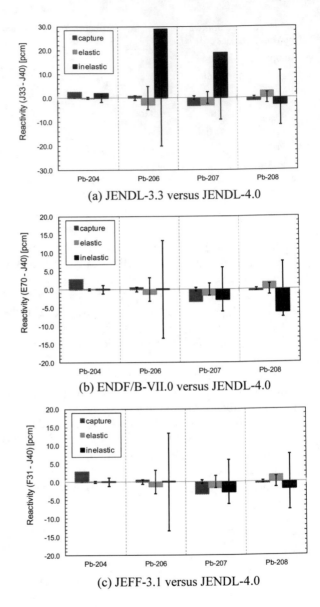

(a) JENDL-3.3 versus JENDL-4.0

(b) ENDF/B-VII.0 versus JENDL-4.0

(c) JEFF-3.1 versus JENDL-4.0

was also the same tendency as the sensitivity coefficients discussed in Sect. 7.3. With ENDF/B-VII.0 and JEFF-3.1, the tendency was not found to be greatly different from JENDL-4.0 as shown in Figs. 7.19c, d, respectively, except for the capture reactions of [204, 207]Pb isotopes. Although most cross-section data of Pb isotopes are the same in both ENDF/B-VII.0 and JEFF-3.1, a notable difference in [208]Pb inelastic scattering cross sections was observed between the two libraries, through a comparison with the standard deviation by JENDL-4.0.

7.4.2.2 Cross-Section Adjustment Method

As discussed in Sect. 7.4.2.1, the uncertainty induced by covariance data was compared with that of sample reactivity obtained by the experiments. In this section, the effect of decreasing uncertainty induced by the nuclear data was investigated by the cross-section adjustment method shown in Eqs. (7.30) and (7.31), on calculated reactivity. Here, the uncertainty induced by the analyses was assumed to be null, in order to estimate the maximum effect of decreasing uncertainty on the calculated reactivity. The cross-section adjustment with U and Pb isotopes was considered useful analyses in that the effects of covariance data (U and Pb isotopes) were significant, although the covariance data of H, C and Al isotopes could give inadequate results of the effect on the evaluation of uncertainty.

As shown in Table 7.13, the effect of decreasing uncertainty on the calculated reactivity was significant in ^{235}U and Pb isotopes. Generally, the effect of decreasing uncertainty regards as becoming large, when errors induced by both experimental and numerical analyses are compared with smaller errors of uncertainty induced by covariance data. In the analyses, the cross-section adjustment method was useful for decreasing the uncertainty, demonstrating a large uncertainty over 30 pcm induced by nuclear data of JENDL-4.0 toward experimental uncertainty of 7 pcm. As a representative example, the C/E value of sample worth reactivity in Case 4 shown in Table 7.11 was greatly improved over 10% shown in Table 7.14, applying the cross-section adjustment method to the uncertainty analyses. Additionally, the C/E values of sample reactivity in Cases 1 through 3 shown in Table 7.10 were remarkably improved to around 5% error with the use of the results of Case 4.

Table 7.13 Reaction-wise uncertainty contribution [pcm] to changes in sample reactivity induced by nuclear data of JENDL-4.0 (Ref. [2])

Isotopes	Reactions					
	Capture	Elastic	Inelastic	Fission	$(n, 2n)$	Total
^{235}U	4.3	0.4	0.9	2.1	0.1	4.9
^{238}U	0.6	0.0	0.1	0.0	0.0	0.6
^{204}Pb	0.0	−0.1	0.4	–	0.0	0.4
^{206}Pb	−0.2	−1.1	4.4	–	-0.2	4.5
^{207}Pb	0.2	−0.6	2.0	–	0.3	2.1
^{208}Pb	−0.1	0.5	2.5	–	0.7	2.6
					Total	7.5

Table 7.14 Results of C/E values of sample worth reactivity in Case 4 before and after application of cross-section adjustment method (Ref. [2])

Core	Before	After
Case 4	1.14	1.01

Table 7.15 Reaction-wise contribution [pcm] to changes in Bi sample reactivity worth (Case 4) induced by covariance data of JENDL-4.0 (Ref. [3])

Isotopes	Reactions					
	Capture	Elastic	Inelastic	Fission	$(n, 2n)$	Total
^{235}U	19.4	1.9	4.1	9.7	0.1	22.2
^{238}U	2.6	0.0	0.3	0.1	0.0	2.6
^{209}Bi	–	–	10.0	–	–	10.0
					Total	24.4

7.4.3 Bismuth Isotope

7.4.3.1 Uncertainty

Uncertainty analyses by the UNCERTAINTY code of the MARBLE system were conducted with the use of JENDL-4.0 covariance data (107-energy-group). Since the covariance data of H, C and Al nuclides consisted mainly of core components that are not provided in JENDL-4.0, the uncertainty was analyzed for several reactions of ^{209}Bi, ^{235}U and ^{238}U, including capture, elastic scattering, inelastic scattering, fission and $(n, 2n)$ reactions, comprising reference and test fuel assemblies of the KUCA A-core. Nonetheless, among the covariance data of ^{209}Bi, inelastic scattering reactions were only prepared in JENDL-4.0. As shown in Table 7.15, the results of uncertainty induced by covariance data were large, with total reactivity of 24.4 pcm, compared with the experimental error around 3 pcm of sample reactivity worth (Table 7.8). The value of total uncertainty was acquired by the square root of the sum of squares of reaction-wise contributions, disregarding the covariance between different nuclides in the sum of squares. Among the nuclides, the reaction-wise contribution was dominant mainly over the capture (19.4 pcm) and fission (9.7 pcm) reactions of ^{235}U, and reasonable in the inelastic scattering reactions (10.0 pcm) of ^{209}Bi. In other words, non-negligible contribution of ^{209}Bi inelastic scattering reactions was observed in the uncertainty analyses of Bi sample reactivity worth.

7.4.3.2 Comparative Study on Bi and Pb

Bi sample reactivity worth experiments were considered successfully carried out from the viewpoint of the reproducibility of previous Pb sample reactivity worth experiments, since the measured excess reactivity of the Al reference core was compared with the Bi and Pb experiments under the same condition, as shown in Table 7.16. With the combined use of experimental and numerical results, a comparative study on Bi and Pb sample reactivity worth was instrumental in examining the neutron characteristics of Pb–Bi coolant material in the actual ADS experimental facility.

In terms of the absolute values of sample reactivity worth shown in Table 7.16, the difference between Bi and Pb sample reactivity worth clearly emphasized the

Table 7.16 Comparison of the results of measured excess reactivities between Bi and Pb test cores (Ref. [3])

Core	Bi sample [pcm]	Pb sample [pcm] (Ref. [1])
Reference (Al)	87 ± 1	92 ± 5
Case 1	143 ± 3	186 ± 7
Case 2	165 ± 3	202 ± 8
Case 3	163 ± 3	237 ± 9
Case 4	171 ± 3	248 ± 9

Table 7.17 Comparison of reaction-wise contributions [pcm] between Bi and Pb (Ref. [6]) sample reactivity worth (Case 4) induced by covariance data of JENDL-4.0 (Ref. [3])

Isotopes	Reactions					
	Capture	Elastic	Inelastic	Fission	(n, 2n)	Total
^{204}Pb	0.1	−0.4	1.6	–	0.0	1.7
^{206}Pb	−1.0	−4.9	20.0	–	−0.8	19.4
^{207}Pb	0.9	−2.6	9.0	–	1.5	8.8
^{208}Pb	−0.6	2.2	11.3	–	3.1	11.9
^{209}Bi	–	–	10.0	–	–	10.0

significance of the characteristics of the actual ADS facility attributed to the reactivity effect. Interestingly, on the basis of the neutronics of Pb–Bi, an ADS with a Pb–Bi coolant core could exactly be analyzed by nuclear design calculations. Additionally, from the results of the uncertainty of Bi and Pb isotopes shown in Table 7.17, the impact of Bi induced by nuclear covariance data was considered small compared with that of the total contribution of Pb isotopes, and invaluable in understanding the reason for choosing Pb–Bi as coolant material in ADS.

7.5 Conclusion

The Pb sample reactivity worth experiments were carried out at KUCA to examine the uncertainties of cross sections of Pb and other isotopes. The comparison between the experiments and the calculations by MCNP6.1 with JENDL-3.3, JENDL-4.0, ENDF/B-VII.0 and JEFF-3.1 libraries revealed as follows: The library update from JENDL-3.3 to JENDL-4.0 demonstrated that the difference between Pb isotopes was dominant in the comparative study, through the experimental analyses of sample reactivity by the MCNP approach. Moreover, JENDL-4.0 revealed a slight difference from ENDF/B-VII.0 in all the Pb isotopes and ^{27}Al, and from JEFF-3.1 in ^{238}U and ^{27}Al. For the Bi sample reactivity worth, the comparison between the experiments and the calculations by MCNP6.1 with JENDL-4.0 revealed the importance of the effect of criticality bias on the precision of numerical simulations.

Sensitivity and uncertainty analyses of Pb isotope cross sections were conducted with the combined use of sample reactivity experiments carried out at KUCA and numerical simulations by the SRAC2006 and MARBLE code systems. The experimental sample reactivity was compared with the calculated one by the deterministic approach with the covariance data of JENDL-4.0 as follows: A series of sensitivity and uncertainty analyses demonstrated the reliability of Pb isotope cross-section data of JENDL-4.0, such as the uncertainty of the covariance data, compared with JENDL-3.3, ENDF/B-VII.0 and JEFF-3.1 libraries. Additionally, the numerical results revealed the applicability of sensitivity and uncertainty analyses to the thermal neutron spectrum cores, such as the KUCA core, demonstrating the improvement of calculation results induced by the cross-section adjustment.

For the Bi isotope, sensitivity coefficients of the Bi isotope were relatively small with the comparison of ^{27}Al, ^{235}U and ^{238}U comprising of fuel plates and core components. Uncertainty induced by Bi cross sections demonstrated a reasonable result of the Bi sample reactivity worth. From the results of Bi isotope uncertainty, the comparative study on Bi and Pb sample reactivity worth was instrumental in emphasizing the neutronics and the impact of Pb–Bi coolant material in ADS.

References

1. Pyeon CH, Fujimoto A, Sugawara T et al (2016) Validation of Pb nuclear data by Monte Carlo analyses of sample reactivity experiments at Kyoto University Critical Assembly. J Nucl Sci Technol 53:602
2. Pyeon CH, Fujimoto A, Sugawara T et al (2017) Sensitivity and uncertainty analyses of lead sample reactivity experiments at Kyoto University Critical Assembly. Nucl Sci Eng 185:460
3. Pyeon CH, Yamanaka M, Oizumi A et al (2018) Experimental analyses of bismuth sample reactivity worth at Kyoto University Critical Assembly. J Nucl Sci Technol 55:1324
4. Tsujimoto K, Sasa T, Nishihara K et al (2004) Neutronics design for lead-bismuth cooled accelerator-driven system for transmutation of minor actinide. J Nucl Sci Technol 41:21
5. Goorley JT, James MR, Booth TE et al (2013) Initial MCNP6 release overview—MCNP6 version 1.0. LA-UR-13-22934
6. Shibata K, Iwamoto O, Nakagawa T et al (2011) JENDL-4.0: a new library for nuclear science and technology. J Nucl Sci Technol 48:1
7. Shibata K, Kawano T, Nakagawa T et al (2002) Japanese evaluated nuclear data library version 3 revision-3: JENDL-3.3. J Nucl Sci Technol 39:1125
8. Chadwick MB, Oblozinsky P, Herman M et al (2006) ENDF/V-II.0: next generation evaluated nuclear data library for nuclear science and technology. Nucl Data Sheet 107:2931
9. Koning A, Forrest R, Kellett M et al (2006) The JEFF-3.1 nuclear data library—JEFF report 21, OECD/NEA No. 6190
10. Iwamoto H, Nishihara K, Sugawara T et al (2013) Sensitivity and uncertainty analysis for an accelerator-driven system with JENDL-4.0. J Nucl Sci Technol 50:856
11. Tsiboulia A, Khomyakov Y, Koscheev V et al (2002) Validation of nuclear data for Pb and Bi using critical experiments. J Nucl Sci Technol suppl 2:1010
12. Chiba G, Okumura K, Sugino K et al (2011) JENDL-4.0 benchmarking for fission reactor applications. J Nucl Sci Technol 48:172

13. Usachev LN (1964) Perturbation theory for the breeding ratio and for other number ratios pertaining to various reactor processes. J Nucl Energy 18:571
14. Gandini A (1967) A generalized perturbation method for bi-linear functionals of the real and adjoint neutron fluxes. J Nucl Energy 21:755
15. Cecchini GP, Salvatores M (1971) Advances in the generalized perturbation theory. Nucl Sci Eng 46:304
16. Kobayashi K (1996) Reactor physics. Corona Publishing Co., Ltd., Tokyo, Japan [in Japanese]
17. Cacuci DG (2004) On the neutron kinetics and control of accelerator driven systems. Nucl Sci Eng 148:55
18. Okumura K, Kugo T, Kaneko K et al (2007) SRAC2006: a comprehensive neutronic calculation code system. JAERI-Data/Code 2007-004
19. Fowler TB, Vondy DR (1969) Nuclear reactor core analysis code: Citation. ORNL-TM-2496, Rev. 2
20. Hara A, Takeda T, Kikuchi Y (1984) SAGEP: two-dimensional sensitivity analysis code based on generalized perturbation theory. JAERI-M 84-027
21. Hazama T, Chiba G, Numata K et al (2006) Development of fine and ultra-fine group cell calculation code SLAROM-UF for fast reactor analysis. J Nucl Sci Technol 43:908
22. Muir DW, Bicourt RM, Kahler AC (2012) The NJOY nuclear data processing system, version 2012. LA-UR-12-27079
23. Broadhead BL, Rearden BT, Hopper CM et al (2004) Sensitivity- and uncertainty-based criticality safety validation techniques. Nucl Sci Eng 146:340
24. Cacuci DG, Inoescu-bujor M (2010) Best-estimate model calibration and prediction through experimental data assimilation: I mathematical framework. Nucl Sci Eng 165:18

Chapter 8
Sensitivity and Uncertainty of Criticality

Masao Yamanaka

Abstract Excess reactivity and control rod worth are generally considered important reactor physics parameters for experimentally examining the neutron characteristics of criticality in a core, and for maintaining safe operation of the reactor core in terms of neutron multiplication in the core. For excess reactivity and control rod worth at KUCA, as well as at the Fast Critical Assembly in the Japan Atomic Energy Agency, special attention is given to analyzing the uncertainty induced by nuclear data libraries based on experimental data of criticality in representative cores (EE1 and E3 cores). Also, the effect of decreasing uncertainty on the accuracy of criticality is discussed in this study. At KUCA, experimental results are accumulated by measurements of excess reactivity and control rod worth. To evaluate the accuracy of experiments for benchmarks, the uncertainty originated from modeling of the core configuration should be discussed in addition to uncertainty induced by nuclear data, since the uncertainty from modeling has a potential to cover the eigenvalue bias more than uncertainty by nuclear data. Here, to investigate the uncertainty of criticality depending on the neutron spectrum of cores, it is very useful to analyze the reactivity of a large number of measurements in typical hard (EE1) and soft (E3) spectrum cores at KUCA.

Keywords Sensitivity · Uncertainty · Criticality · Solid-moderated and solid-reflected core

M. Yamanaka (✉)
Institute for Integrated Radiation and Nuclear Science, Kyoto University, Osaka, Japan
e-mail: yamanaka.masao.6z@kyoto-u.ac.jp

© The Author(s) 2021
C. H. Pyeon (ed.), *Accelerator-Driven System at Kyoto University Critical Assembly*,
https://doi.org/10.1007/978-981-16-0344-0_8

8.1 Experimental Settings

8.1.1 Core Configuration

The experiments of reactivity measurement [1, 2] were carried out in the A cores (EE1 and E3 cores in Figs. 8.1a, b, respectively) that have polyethylene moderator and reflector rods, and different fuel assemblies: "F" and "f" (Figs. 8.2a, b, respectively). In EE1 core (Fig. 8.1a), fuel assembly "F" (1/8″P60EUEU) in Fig. 8.2a is composed of 60 unit cells, and upper and lower polyethylene blocks about 25″ and 20″ long, respectively, in an aluminum (Al) sheath (2.1″ × 2.1″ × 60″). For the fuel assembly, a unit cell in the fuel region is composed of two highly enriched uranium (HEU) fuel plates 1/8″ (2 × 1/16″) thick and a polyethylene moderator plate 1/8″ thick. In E3 core (Fig. 8.1b), another fuel assembly "f" (3/8″P36EU) in Fig. 8.2b is composed of 36 unit cells with an HEU fuel plate 1/16″ thick, polyethylene plates 3/8″ thick, and upper and lower polyethylene blocks about 23″ and 21″ long, respectively, in the Al sheath as in fuel assembly "F". The neutron spectrum in EE1 core is compared with that in E3 core as shown in Fig. 8.3, demonstrating representatively hard (EE1 core) and soft (E3 core) neutron spectra in the KUCA A-core.

8.1.2 Reactivity Measurements

For measuring the excess reactivity, the critical state was adjusted by maintaining a certain position of C3 rod in EE1 core and C1 rod in E3 core, and by withdrawing fully the other control (C1 and C2 in EE1; C2 and C3 in E3) and safety (S4, S5, and S6) rods from the core, respectively. Furthermore, excess reactivities in EE1 and E3 cores were measured by the positive period method, when C3 and C1 rods, respectively, were, under the critical state, fully withdrawn from the core.

For measuring the control rod worth, in cases of C1 and C2 rods in EE1 core, criticality was maintained by C3 rod, and the control rod worth of C1 or C2 rod was acquired by the rod drop method, after full insertion of C1 or C2 rod into the core. For C3 rod in EE1 core, criticality was adjusted by C1 rod, and the control rod worth of C3 rod was obtained by the rod drop method, after the full insertion of C3 rod into the core. Almost the same procedures were followed for the E3 core, with the use of the rod drop method, and the control rod worth of C1, C2, and C3 rods was obtained experimentally.

To estimate the experimental uncertainty of excess reactivity and control rod worth, the dimensions of the HEU plate comprising of core components are considered manufacturing tolerances among the significant factors taken into account, as shown in Table 8.1. In addition to the dimensions of the HEU plate, as previously demonstrated at the Fast Critical Assembly in the Japan Atomic Energy Agency [3], other important uncertainty factors include mechanical reproducibility of control rod position, measurement errors of doubling time by the positive period method, core

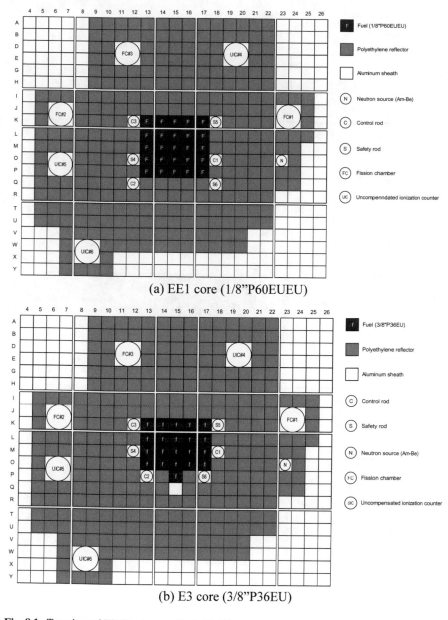

Fig. 8.1 Top view of KUCA A-core (Refs. [1, 2])

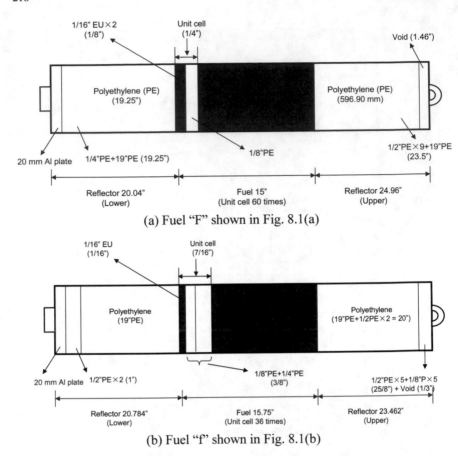

(a) Fuel "F" shown in Fig. 8.1(a)

(b) Fuel "f" shown in Fig. 8.1(b)

Fig. 8.2 Description of fuel rods in A-core shown in Fig. 8.1 (Refs. [1, 2])

Fig. 8.3 Neutron spectra of EE1 and E3 cores shown in Figs. 8.1a, b, respectively (Refs. [1, 2])

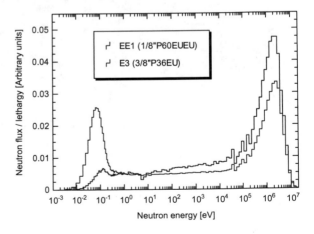

Table 8.1 Experimental uncertainties of HEU fuel plate (Refs. [1, 2])

Cause of uncertainty	Error (%)
Composition	3.0
Length of both sides (one side: 0.2%)	0.3
Thickness	3.1

Table 8.2 Measured results [pcm] of excess reactivity and control rod worth with standard deviation obtained from experimental data in EE1 and E3 cores (Refs. [1, 2])

Reactivity	EE1 (error: %)	E3 (error: %)
Excess	210 ± 6 (3.1)	264 ± 2 (0.8)
C1 rod	838 ± 11 (1.4)	551 ± 6 (1.2)
C2 rod	144 ± 3 (2.4)	429 ± 4 (1.0)
C3 rod	521 ± 8 (1.7)	329 ± 4 (1.4)

EE1: β_{eff}: 831 pcm and Λ: 3.027E-05 s by MCNP6.1 with JENDL-4.0

E3: β_{eff}: 805 pcm and Λ: 4.771E-05 s by MCNP6.1 with JENDL-4.0

temperature, delayed neutron parameters induced by numerical analyses, fuel composition uncertainty caused by heterogeneity distribution and deformation inside the Al sheath. Among these factors, the mechanical reproducibility of control rod position, in this study, was considered of great impact caused by tolerance and instability of control rod inside the Al sheath in a horizontal direction than the experimental uncertainty of excess reactivity and control rod worth, because the other uncertainties are relatively minor. Furthermore, heterogeneous arrangement of fuel plates in an axial direction was attributable to random selection of fuel plates for making fuel cells with a record number of fuel plates. Therefore, the experimental uncertainty of excess reactivity and control rod worth was finally determined to be about 3% at most by averaging the experimental data, and estimating its standard deviation, as observed in operations of the previous decade in the KUCA A-core, as shown in Table 8.2.

8.2 Criticality

8.2.1 Numerical Simulations

8.2.1.1 Stochastic Calculations

Excess reactivity was numerically deduced by the MCNP6.1 code [4] with the JENDL-4.0 [5] and the ENDF/B-VII.0 [6] libraries through the difference between the critical and super-critical states in the core; control rod worth was numerically obtained by the difference between critical and subcritical states. For the evaluation of excess reactivity and control rod worth, in the critical core, the effective delayed neutron fraction (β_{eff}) was acquired by MCNP6.1 (2,000 active cycles of 50,000 histories; 5 pcm statistical error) with JENDL-4.0. The values of β_{eff} at critical state of EE1 and E3 cores were 831 and 805 pcm, respectively; those of the neutron generation time (Λ) were 3.027E-05 and 4.771E-05 s, respectively. Atomic number densities of core components comprise fuel elements, moderator (reflector), control rod, and Al sheath, for analyzing experimental values of excess reactivity and control rod worth.

8.2.1.2 Deterministic Calculations

Numerical analyses by deterministic calculations were performed by combining the SRAC2006 [7] and the MARBLE [8] code systems: collision probability calculations (PIJ [7]) and eigenvalue calculations (CITATION [9]) of SRAC2006, sensitivity coefficient calculations (SAGEP [10]) and uncertainty calculations (UNCERTAINTY [8, 11]) of MARBLE, coupled with the JENDL-4.0 nuclear data library. Experimental analyses of uncertainty were conducted with the use of covariance data of cross sections contained in JENDL-4.0, including uncertainty of excess reactivity and control rod worth induced by the nuclear data, and the effects of decreasing uncertainty on the accuracy of excess reactivity and control rod worth. Here, in a series of deterministic calculations, the CITATION code was notably executed for obtaining sensitivity coefficients by the SAGEP code based on the diffusion calculations.

8.2.2 Sensitivity and Uncertainty

8.2.2.1 Numerical Reactivity

The experimental values of excess reactivity (positive value) $\Delta\rho_{\text{Excess}}^{\text{Exp}}$ and control rod worth (negative value) $(-\Delta\rho_{\text{Rod}}^{\text{Exp}})$ were deduced by effective multiplication factors $k_{\text{Clean}}^{\text{Exp}}$ and $k_{\text{Rod}}^{\text{Exp}}$ in super-critical (clean core) and subcritical states (rod insertion core) obtained by the positive period method and the rod drop method, respectively, as follows:

$$\Delta\rho_{Excess}^{Exp} = 1 - \frac{1}{k_{Clean}^{Exp}}, \tag{8.1}$$

$$-\Delta\rho_{Rod}^{Exp} = 1 - \frac{1}{k_{Rod}^{Exp}}. \tag{8.2}$$

In MCNP analyses, numerical value $\Delta\rho_{\text{Excess}}^{\text{MCNP}}$ or $(-\Delta\rho_{\text{Rod}}^{\text{MCNP}})$ was deduced by the difference between two effective multiplication factors $k_{\text{Clean}}^{\text{MCNP}}$ or $k_{\text{Rod}}^{\text{MCNP}}$ and $k_{\text{Critical}}^{\text{MCNP}}$ in the super-critical or subcritical and critical cores, respectively, as follows:

$$\Delta \rho_{\text{Excess}}^{\text{MCNP}} = \frac{1}{k_{\text{Critical}}^{\text{MCNP}}} - \frac{1}{k_{\text{Clean}}^{\text{MCNP}}}, \tag{8.3}$$

$$-\Delta \rho_{\text{Rod}}^{\text{MCNP}} = \frac{1}{k_{\text{Critical}}^{\text{MCNP}}} - \frac{1}{k_{\text{Rod}}^{\text{MCNP}}}. \tag{8.4}$$

In MCNP calculations, $k_{\text{Critical}}^{\text{MCNP}}$ needs to be defined as the value of the effective multiplication factor in the critical core, since the numerical value is not always a unit.

On the basis of the experimental methodology shown in Eqs. (8.1) and (8.2), a numerical approach of excess reactivity $\Delta \rho_{\text{Excess}}^{\text{CITATION}}$ and control rod worth $(-\Delta \rho_{\text{Rod}}^{\text{CITATION}})$ by deterministic calculations (CITATION) is generally expressed, respectively, as follows, as were experimental values:

$$\Delta \rho_{\text{Excess}}^{\text{CITATION}} = 1 - \frac{1}{k_{\text{Clean}}^{\text{CITATION}}}, \tag{8.5}$$

$$-\Delta \rho_{\text{Rod}}^{\text{CITATION}} = 1 - \frac{1}{k_{\text{Rod}}^{\text{CITATION}}}, \tag{8.6}$$

where $k_{\text{Clean}}^{\text{CITATION}}$ and $k_{\text{Rod}}^{\text{CITATION}}$ indicate the effective multiplication factors in the super-critical and subcritical cores, respectively.

Of the two numerical values by CITATION and MCNP, as mentioned in Sect. 8.2.2.1, the CITATION calculations needed to conduct a series of sensitivity and uncertainty analyses by SAGEP and UNCERTAINTY, respectively. Meanwhile, the MCNP calculations were requisite to assess the precision of eigenvalue calculations, such as eigenvalue bias [12]. That is why two numerical values were introduced by the stochastic (Eqs. (8.3) and (8.4)) and the deterministic (Eqs. (8.5) and (8.6)) approaches, and compared differently with the experimental values shown in Eqs. (8.1) and (8.2).

8.2.2.2 Sensitivity Coefficient

Sensitivity coefficient S of the integral reactor physics parameter (effective multiplication factor) R is defined by the ratio of the rate of change in R and a certain parameter x as follows:

$$S = \frac{dR}{R} \bigg/ \frac{dx}{x}. \tag{8.7}$$

The effective multiplication factor k_{eff} is expressed by a balance equation of neutrons as follows:

$$\mathbf{A}\phi = \frac{1}{k_{\text{eff}}} \mathbf{F}\phi, \tag{8.8}$$

where $\mathbf{A}$ and $\mathbf{F}$ indicate operators of transport and fission terms, respectively, and ϕ the forward neutron flux. Multiplying Eq. (8.8) by adjoint neutron flux ϕ^* and integrating over whole volume and energy, the following equation is obtained:

$$\frac{1}{k_{\mathrm{eff}}} = \frac{\langle \phi^* \mathbf{A} \phi \rangle}{\langle \phi^* \mathbf{F} \phi \rangle},$$ (8.9)

where brackets $<>$ indicate integration over the whole volume and energy.

With the use of an operator $\mathbf{B}$, Eq. (8.8) is expressed as follows:

$$\left(\mathbf{A} - \frac{1}{k_{\mathrm{eff}}} \mathbf{F} \right) \phi = \mathbf{B} \phi = 0.$$ (8.10)

Here, assuming that parameter x, operator $\mathbf{B}$, and neutron flux ϕ are changed into $x + \delta x$, $\mathbf{B} + \delta \mathbf{B}$, and $\phi + \delta \phi$, respectively, in a critical state, the following equations are obtained:

$$(\mathbf{B} + \delta \mathbf{B})(\phi + \delta \phi) = 0 .$$ (8.11)

Neglecting second-order perturbation terms, Eq. (8.10) is expressed as follows:

$$\mathbf{B} \, \delta \phi + \delta \mathbf{B} \, \phi = 0 .$$ (8.12)

Introducing the generalized adjoint flux Γ^*, the following equation is obtained with the use of adjoint operator $\mathbf{B}^*$ and a certain adjoint source term q^*, defined as reactivity in these analyses:

$$\mathbf{B}^* \, \Gamma^* = q^* .$$ (8.13)

Considering the theoretical background [13–16] and using Eqs. (8.10) through (8.13), the sensitivity coefficient in Eq. (8.7) is finally expressed by applying the first-order perturbation approximation [17], as follows:

$$S = \frac{\langle \phi^* \left(-\frac{\partial \mathbf{B}}{\partial x} \right) \phi \rangle}{\langle \phi^* \mathbf{F} \phi \rangle} k_{eff} .$$ (8.14)

Finally, applicability of sensitivity analyses to k_{eff} was investigated for a thermal spectrum core, such as the KUCA core, with the use of SAGEP that had been originally developed for conducting the sensitivity analyses of fast reactors.

8.2.2.3 Uncertainty

In analyzing the cross-section uncertainty of nuclear data [18], the uncertainty of reactor physics parameter v is expressed as follows:

$$v = \mathbf{G}_{tar}\,\mathbf{M}\,(\mathbf{G}_{tar})^{\mathrm{t}} = \sum_i \sum_j s_i\,c_{i,j}\,s_j \equiv \sum_i \sum_j v_{i,j} \quad (1 \le i,\, j \le p)\ ,$$

$$(8.15)$$

where $\mathbf{G}_{tar}$ ($1 \times p$) indicates the sensitivity vector of reactor physics parameters, $\mathbf{M}$ ($p \times p$) the covariance matrix of nuclear reaction parameters, s_i the sensitivity coefficient, $c_{i,j}$ the covariance, $v_{i,j}$ the factor of uncertainty and p the number of nuclear reactions including the nuclides. Thus, the contribution of uncertainty u_i in each nuclear reaction can be defined as follows:

$$u_i \equiv \sum_i v_{i,j}. \qquad (8.16)$$

Generally, since sensitivity coefficient s_i and covariance $c_{i,j}$ are dominant in the energy group, the factor of uncertainty is finally expressed with the use of the maximum number of energy group G as follows:

$$v_{i,j} = \sum_g \sum_{g'} s_g^i c_{g,g'}^{i,j} s_{g'}^j \left(1 \le g,\, g' \le G\right), \qquad (8.17)$$

where g and g' indicate the energy groups.

8.2.3 Results and Discussion

8.2.3.1 Eigenvalue Calculations

In MCNP simulations, excess reactivity and control rod worth were obtained by two eigenvalue calculations in critical and super-critical, and, critical and subcritical states, respectively: the difference between the inverse values of eigenvalue calculations in the two states. Here, MCNP eigenvalue calculations were made with 2,000 active cycles of 50,000 histories, resulting in a standard deviation within 10 pcm. The numerical results of excess reactivity and control rod worth were obtained by MCNP6.1 with JENDL-4.0 in EE1 and E3 cores, as shown in Table 8.3a, b, respectively. Moreover, estimation of C/E (calculation/experiment) values revealed an accuracy of around 5% error with the use of the experimental and numerical results of excess reactivity and control rod worth, as shown in Eqs. (8.1) and (8.2), respectively, excluding small values of excess reactivity and C2 control rod worth in EE1 core. From the calculated results in Table 8.3a, b, the difference between numerical analyses by MCNP6.1 with JENDL-4.0 and ENDF/B-VII.0 was found within an error of 3% of the C/E value.

The ability of MCNP6.1 calculations was confirmed at a critical state in terms of the eigenvalue bias by the MCNP approach, as shown in Table 8.4. In EE1 core, eigenvalue bias by MCNP6.1 with JENDL-4.0 demonstrated a relatively small value

Table 8.3 Comparison between measured and calculated (MCNP6.1) results of excess reactivity and control rod worth in EE1 and E3 cores (Ref. [1])

(a) EE1

Reactivity	Exp. [pcm]	Cal. [pcm] (C/E value)	
		JENDL-4.0	ENDF/B-VII.0
Excess	210 ± 6	198 ± 18 (0.94 ± 0.09)	196 ± 18 (0.93 ± 0.09)
C1 rod	838 ± 11	857 ± 18 (1.02 ± 0.02)	843 ± 18 (1.01 ± 0.02)
C2 rod	144 ± 3	134 ± 18 (0.93 ± 0.13)	135 ± 18 (0.94 ± 0.13)
C3 rod	521 ± 8	500 ± 18 (0.96 ± 0.04)	491 ± 18 (0.94 ± 0.04)

(b) E3

Reactivity	Exp. [pcm]	Cal. [pcm] (C/E value)	
		JENDL-4.0	ENDF/B-VII.0
Excess	264 ± 2	246 ± 16 (0.93 ± 0.06)	241 ± 16 (0.91 ± 0.06)
C1 rod	551 ± 6	526 ± 16 (0.95 ± 0.03)	536 ± 16 (0.97 ± 0.03)
C2 rod	429 ± 4	400 ± 16 (0.93 ± 0.04)	388 ± 16 (0.90 ± 0.04)
C3 rod	329 ± 4	314 ± 16 (0.95 ± 0.05)	315 ± 16 (0.96 ± 0.05)

Table 8.4 Numerical results by MCNP6.1 eigenvalue calculations in critical state (Ref. [1])

EE1		E3	
JENDL-4.0	ENDF/B-VII.0	JENDL-4.0	ENDF/B-VII.0
1.00246 ± 0.00009	1.00461 ± 0.00009	1.00443 ± 0.00008	1.00578 ± 0.00008

of about 250 pcm, and with ENDF/B-VII.0 a value about 460 pcm (Table 8.4). Similarly, in E3 core, the value with JENDL-4.0 was small, about 440 pcm, compared with that with ENDF/B-VII.0, about 580 pcm. From the analyses of eigenvalue bias, a significant index of the accuracy of experimental analyses was acquired by MCNP calculations in EE1 and E3 cores at KUCA; also, as in previous studies [12, 19], JENDL-4.0 was considered reliable as a reference nuclear data library in a series of sensitivity and uncertainty analyses of excess reactivity and control rod worth in the KUCA A core.

The numerical values of k_{eff} in excess reactivity and control rod worth shown in Eqs. (8.5) and (8.6), respectively, were deduced, for a clean core (withdrawal of all control and safety rods) and a subcritical core (control rod insertion) in super-critical and subcritical states, respectively, through the results of diffusion-based eigenvalue calculations (CITATION) in the 107-energy-group and x-y-z dimensions (3-D) with JENDL-4.0. Also, on the basis of accuracy of experimental analyses by CITATION (Table 8.5), sensitivity analyses of k_{eff} in excess reactivity and control rod worth were notably conducted by diffusion-based calculations (SAGEP).

Table 8.5 Comparison between measured and calculated (CITATION with JENDL-4.0) results of k_{eff} (excess reactivity: super-critical state; control rod worth: subcritical state) in EE1 and E3 cores (Ref. [1])

Reactivity	EE1		E3	
	Experiment	Calculation	Experiment	Calculation
Excess	1.00210	1.00030	1.00265	1.00067
C1 rod	0.99169	0.99525	0.99452	0.99513
C2 rod	0.99856	0.99903	0.99573	0.99673
C3 rod	0.99482	0.99447	0.99672	0.99762

8.2.3.2 Sensitivity Coefficients

Sensitivity coefficients Eq. (8.7) of k_{eff} in excess reactivity and control rod worth were analyzed by the SAGEP code for assessing cross-section data of inelastic scattering, elastic scattering and capture reactions of ^{27}Al, boron isotopes ($^{10, 11}B$), carbon (^{12}C), hydrogen (^1H), oxygen (^{16}O), and uranium isotopes ($^{234, 235, 236, 238}U$) comprising the core components.

For excess reactivities in EE1 and E3 cores, the sensitivity coefficients of elastic scattering reactions were relatively highly positive, mostly 1 MeV, in ^{27}Al, ^{12}C, and ^1H, as shown in Fig. 8.4a, and b, respectively. Sensitivity coefficients were dominant over the high-energy (MeV) region of the inelastic scattering reactions of ^{27}Al in k_{eff} (excess reactivities) at EE1 and E3 cores shown in Fig. 8.5a, b, respectively. In thermal neutron region shown in Fig. 8.6a, b, the capture cross sections of ^{27}Al, ^1H, and ^{235}U were highly sensitive at EE1 and E3 cores, respectively. Also, the sensitivity coefficients of ^{27}Al, ^1H, and ^{235}U were remarkably higher in E3 core than in EE1 core ranging between 0.01 and 100 eV shown in Fig. 8.6b, because E3 core is a relatively soft-spectrum core shown in Fig. 8.3. In a series of sensitivity analyses shown in Figs. 8.4 through 8.6, effects of Al on sensitivities were observed in entire reactions and energy regions, and attributable to containing Al itself comprising of U-Al alloy (HEU) fuel plates and Al sheath of fuel assembly.

Further study of the sensitivity coefficients was made of k_{eff} (worth of C1 control rod) at EE1 and E3 cores, since the worth of C1 control rod was mostly larger in both EE1 and E3 cores, compared with other reactivities shown in Table 8.3. Also, the worth of C1 control rod was selected to investigate directly the effect of the boron isotope component of the control rod. As shown in Fig. 8.7a, b, the sensitivity coefficients of ^{27}Al, ^{10}B, ^1H, and ^{235}U were negative in the capture reactions; among these, the capture cross sections of ^{27}Al and ^{235}U were highly sensitive in the thermal neutron region, as well as for capture reactions in the excess reactivity shown in Fig. 8.6a, b. Moreover, the sensitivity coefficient of ^{27}Al was remarkably large in the E3 core with a thermal neutron spectrum, as shown in Fig. 8.7b, as was that of ^{235}U. Finally, in C1 control rod worth, sensitivity coefficient of ^{10}B was found relatively of negligible significance due to an insertion of C1 control rod, although that of ^{27}Al comprising of core components (U-Al alloy fuel plates and Al sheath) was large.

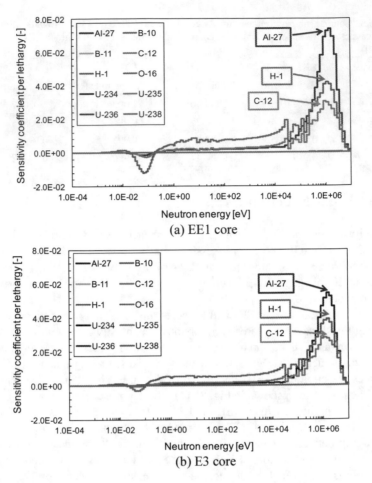

Fig. 8.4 Sensitivity coefficients for elastic scattering reactions in excess reactivity (Ref. [1])

8.2.3.3 Uncertainty

The uncertainty analyses by the UNCERTAINTY code of the MARBLE system were conducted with the use of JENDL-4.0 covariance data (107-energy-group). Since the covariance data of ^{27}Al, ^{12}C, and ^{1}H isotopes consisted mainly of core components that are not contained in JENDL-4.0, uncertainty analyses were executed for various reactions of $^{10,\,11}$B, ^{16}O, and $^{235,\,238}$U isotopes comprising of control and fuel rods in the KUCA A-core, including capture, elastic scattering, inelastic scattering, fission, and (n, $2n$) reactions covered in the SAGEP code. As shown in Table 8.6, the uncertainty of excess reactivity in EE1 core induced by covariance data was large, total uncertainty 135.8 pcm, compared with an experimental error of 6 pcm (Table 8.2). The value of total uncertainty was acquired through the square root of the sum of squares for reaction-wise contributions. Among the isotopes, uncertainty was

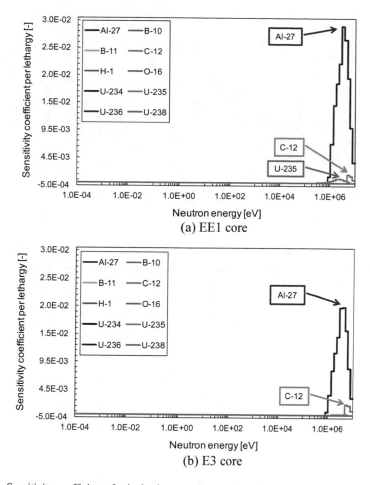

Fig. 8.5 Sensitivity coefficients for inelastic scattering reactions in excess reactivity (Ref. [1])

dominant over the sum of all contributions, including the capture and fission reactions of ^{235}U; a large contribution [19] was attributable to the sensitivity coefficients of ^{235}U capture and fission reactions. For the worth of C1 control rod in E3 core, the total uncertainty was 164.1 pcm, although the reaction-wise contribution was slight in the boron isotopes, as shown in Table 8.7, demonstrating the same tendency of excess reactivity as in EE1 core shown in Table 8.6. Finally, the uncertainty of k_{eff} in excess reactivity and control rod worth (C1, C2 and C3 rods) was summarized around 150 pcm in the EE1 and E3 cores.

Furthermore, the effect of decreasing uncertainty induced by nuclear data on calculated k_{eff} was investigated by the cross-section adjustment method [20, 21], with the use of uncertainty of k_{eff} values. The result of this investigation demonstrated a great improvement from around 150 to 3 pcm induced by nuclear data of JENDL-4.0 in all excess reactivity and control rod worth in EE1 and E3 cores. Moreover, by

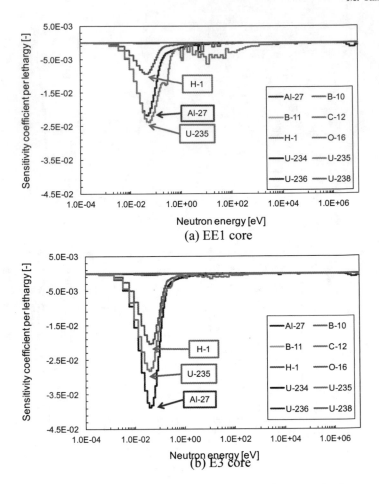

Fig. 8.6 Sensitivity coefficients for capture reactions in excess reactivity (Ref. [1])

applying the cross-section adjustment method to the uncertainty analyses, the C/E value of all excess reactivity and control rod worth in both EE1 and E3 cores reached around a unit.

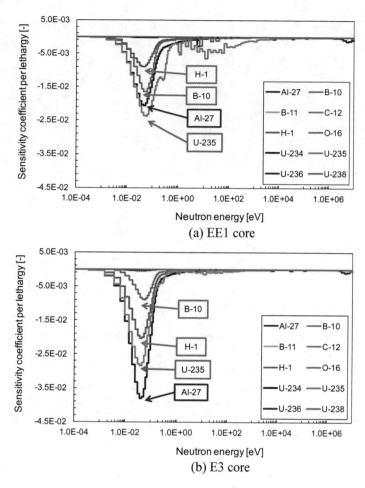

Fig. 8.7 Sensitivity coefficients for capture reactions in C1 rod worth (Ref. [1])

Table 8.6 Reaction-wise contribution [pcm] to excess reactivity induced by covariance data of JENDL-4.0 in EE1 core (Ref. [1])

Isotopes	Reactions					
	Capture	Elastic	Inelastic	Fission	$(n, 2n)$	Total
^{10}B	0.0	0.0	–	–	–	0.0
^{11}B	0.0	0.0	–	–	–	0.0
^{16}O	0.0	0.0	0.0	–	0.0	0.0
^{235}U	111.7	5.3	12.9	44.7	0.7	121.1
^{238}U	2.8	0.2	0.9	0.1	0.1	2.9
					Total	135.8

Table 8.7 Reaction-wise contribution [pcm] to the worth of C1 control rod induced by covariance data of JENDL-4.0 in E3 core (Ref. [1])

Isotopes	Reactions					
	Capture	Elastic	Inelastic	Fission	$(n, 2n)$	Total
^{10}B	9.0	0.0	–	–	–	9.0
^{11}B	0.0	4.4	–	–	–	4.4
^{16}O	0.0	1.2	0.0	–	0.0	1.2
^{235}U	113.0	1.1	2.8	81.8	0.2	139.5
^{238}U	0.7	0.1	0.2	0.0	0.0	0.8
					Total	164.1

8.3 Benchmarks

8.3.1 Experimental Analyses

8.3.1.1 Reactivity Measurements

Before quantifying the uncertainty in criticality, the validity of the standard modeling of the core configuration was verified by the comparison of the excess reactivity and the control rod worth between the calculation and the experiment, without consideration of variation in the position and material property. The calculation of the reactivity was performed through two eigenvalue calculations with MCNP6.1 and KENO-VI module of SCALE6.2 code system [22] together with ENDF/B-VII.1. The calculated reactivity was obtained as follows:

$$\rho_{excess}^{cal} = \frac{1}{k_{eff}^{critical}} - \frac{1}{k_{eff}^{clean}}, \tag{8.18}$$

$$\rho_{rod}^{cal} = \frac{1}{k_{eff}^{rod}} - \frac{1}{k_{eff}^{critical}}, \tag{8.19}$$

where ρ_{excess}^{cal} and ρ_{rod}^{cal} are the calculated excess reactivity and control rod worth, respectively, $k_{eff}^{critical}$ the k_{eff} value at the critical state, k_{eff}^{clean} the k_{eff} value at the withdrawal of all control rods and k_{eff}^{rod} the k_{eff} value at the insertion of control rod C1, C2, or C3 in the critical state.

8.3.1.2 Numerical Simulations

Through the development of the methodology of the adjoint flux [23, 24], sensitivity and uncertainty analyses were easily conducted with the use of the Monte Carlo method applying to rigorous modeling without homogenization [25]. Sensitivity and uncertainty analyses of k_{eff} induced by nuclear data were performed by TSUNAMI-3D module of SCALE6.2 together with ENDF/B-VII.1 and its 56-group covariance library (56groupcov7.1) for standard modeling of EE1 and E3 core configurations. Here, the adjoint flux was obtained by the CLUTCH method [26].

Variation in the manufacturing tolerance of HEU plates is provided by statistically processing the measured results (enrichment and thickness) of all plates at KUCA and with specification sheet in the fabrication for length of each side, as shown in Table 8.1. On the basis of the variation in HEU plates, the uncertainty analyses for the enrichment and tolerance of length of both sides were conducted with KPERT option in MCNP6.1 together with ENDF/B-VII.1. Since the variation is too small to explicitly calculate the impact (difference in k_{eff} values), a pseudo variation was considered 10%, instead of the actual variation shown in Table 8.1, in the perturbation for the enrichment and length of sides of the HEU plates. Here, while varying the length of the sides of HEU plates, their mass was maintained by retaining the original number of atoms. The pseudo uncertainty of k_{eff} attributed to varying the enrichment and length was converted to the actual one (Table 8.1) on the basis of the standard method provided in the guideline by OECD/NEA [27] as follows:

$$\Delta k_{eff} = \sqrt{\sum_{i=1}^{N} \frac{1}{n_i} \left(\frac{\Delta k_{eff,i}}{\Delta q_i} \cdot \frac{\Delta q_i}{\Delta x_i} \right)^2}, \qquad (8.20)$$

where, Δk_{eff} is the difference between k_{eff} values before and after variation Δq_i (the intensity of the pseudo variation), Δx_i the real uncertainty shown in Table 8.1, n_i the number of HEU plates involved in the variation, finally, i and N the region of perturbation and the total number of regions, respectively. In the calculation for the variation in the enrichment and the length of HEU plates, the core was divided into fuel rod regions in the x-y plane (25 divisions in EE1 core; 21 divisions in E3 core) and into three regions in z axis ($N = 75$ and 63 in EE1 and E3 cores, respectively). The uncertainty of k_{eff} by varying the thickness of HEU plates was evaluated by setting all HEU plates 3.1% decreasing in thickness according to Ref. [14].

The uncertainty attributed to the reproducibility of control rod position was approximately evaluated by eigenvalue calculations (Eq. (8.20)) with MCNP6.1, and by varying control rod position (x-y directions) in the casing for 5.15 mm radially (Δq_i in Eq. (8.20)) and by 8 segments circumferentially. Then, the actual uncertainty of control rod position in the x-y plane is limited to a radius of 1.15 mm (Δx_i in Eq. (8.20)). In the case of one (two) control rod(s) inserted, N in Eq. (8.20) becomes 8 (64). The uncertainty was finally deduced by subtracting the k_{eff} value obtained at the reference position from that obtained by varying the position.

8.3.2 Uncertainty

8.3.2.1 Eigenvalue Bias

The eigenvalue calculations with the use of MCNP6.1 and SCALE code system were performed for a total of 1E + 08 histories (1E + 03 active cycles of 1E + 05 each); the statistical error was less than 10 pcm. Through the comparison of the results of k_{eff} between MCNP6.1 and SCALE6.2/KENO-VI, the difference was found about 100 pcm in EE1 core and about 200 pcm in E3 core, as shown in Tables 8.8 and 8.9, respectively. Moreover, eigenvalue bias was dependent on the neutron spectrum of the core: about 350 pcm and about 600 pcm in EE1 and E3 cores, respectively. Interestingly, the uncertainty induced by nuclear data showed about 950 pcm in the k_{eff} evaluation by the SCALE code. Also, the uncertainty was found almost same value regardless of the control rod position, indicating that the neutronic characteristics are the same even at insertion and withdrawal of control rods.

To validate the modeling of core configuration and eigenvalue calculations, the calculated results of excess reactivity and control rod worth compared with the measured ones in EE1 and E3 cores, as shown in Tables 8.10 and 8.11, respectively. Although eigenvalue bias was smaller than the uncertainty induced by nuclear

Table 8.8 Comparison of k_{eff} values between experiments and calculations (MCNP6.1 and SCALE6.2) in EE1 core (Ref. [2])

Case	Core	Inserted control rod (position)	Effective multiplication factor k_{eff}		
			Experiment	MCNP6.1	SCALE6.2
I-1	EE1	–	1.00202 ± 0.00006	1.00534 ± 0.00009	1.00574 ± 0.00009 (947.7 ± 0.3 pcm)*
I-2	EE1	C1 (critical position)	1.00000	1.00348 ± 0.00009	1.00395 ± 0.00010 (948.3 ± 0.3 pcm)
I-3	EE1	C3 (critical position)	1.00000	1.00374 ± 0.00009	1.00383 ± 0.00010 (949.0 ± 0.3 pcm)
I-4	EE1	C3 (critical position) C1 (full inserted)	0.99199 ± 0.00001	0.99651 ± 0.00009	0.99533 ± 0.00010 (956.5 ± 0.3 pcm)
I-5	EE1	C3 (critical position) C2 (full inserted)	0.99861 ± 0.00002	1.00359 ± 0.00009	1.00242 ± 0.00010 (949.7 ± 0.3 pcm)
I-6	EE1	C1 (critical position) C3 (full inserted)	0.99501 ± 0.00008	1.00002 ± 0.00009	0.99899 ± 0.00010 (953.7 ± 0.3 pcm)

*Uncertainty by SCALEW6.2/TSUNAMI-3D

Table 8.9 Comparison of k_{eff} values between experiments and calculations (MCNP6.1 and SCALE6.2) in E3 core (Ref. [2])

Case	Core	Inserted control rod (position)	Effective multiplication factor k_{eff}		
			Experiment	MCNP6.1	SCALE6.2
II-1	E3	–	1.00253 ± 0.00002	1.00823 ± 0.00008	1.01045 ± 0.00009 $(911.7 \pm 0.3$ pcm$)*$
II-2	E3	C1 (critical position)	1.00000	1.00578 ± 0.00008	1.00811 ± 0.00009 $(912.6 \pm 0.3$ pcm$)$
II-3	E3	C2 (critical position)	1.00000	1.00614 ± 0.00008	1.00820 ± 0.00009 $(913.4 \pm 0.3$ pcm$)$
II-4	E3	C2 (critical position) C1 (full inserted)	0.99477 ± 0.00006	1.00275 ± 0.00008	1.00312 ± 0.00009 $(918.6 \pm 0.3$ pcm$)$
II-5	E3	C1 (critical position) C2 (full inserted)	0.99592 ± 0.00004	1.00212 ± 0.00008	1.00396 ± 0.00009 $(918.0 \pm 0.3$ pcm$)$
II-6	E3	C1 (critical position) C3 (full inserted)	0.99687 ± 0.00004	1.00074 ± 0.00008	1.00495 ± 0.00010 $(916.3 \pm 0.3$ pcm$)$

*Uncertainty by SCALEW6.2/TSUNAMI-3D

Table 8.10 Comparison between measured and calculated reactivity at critical experiments in EE1 core (Ref. [2])

Reactivity	Experiment [pcm]	MCNP6.1 [pcm]	SCALE6.2/KENO-VI [pcm]
Excess	202 ± 6	184 ± 13 (0.91 ± 0.07)	189 ± 19 (0.93 ± 010)
C1 rod	807 ± 11	876 ± 13 (1.09 ± 0.02)	862 ± 14 (1.07 ± 0.02)
C2 rod	139 ± 3	151 ± 13 (1.09 ± 0.09)	152 ± 14 (1.09 ± 0.10)
C3 rod	502 ± 8	525 ± 13 (1.05 ± 0.03)	494 ± 14 (0.98 ± 0.03)

(): C/E (calculation/experiment) value

data, the comparison revealed the relative difference of 10% in C/E (calculation/experiment) values, indicating that the calculations of EE1 and E3 cores were valid and pertinently modeled for experimental analyses.

Table 8.11 Comparison between measured and calculated reactivity at critical experiments in E3 core (Ref. [2])

Reactivity	Experiment[pcm]	MCNP6.1 [pcm]	SCALE-6.2/KENO-VI [pcm]
Excess	252 ± 2	242 ± 11 (0.96 ± 0.04)	229 ± 13 (0.91 ± 0.05)
C1 rod	526 ± 6	536 ± 11 (1.02 ± 0.02)	502 ± 13 (0.95 ± 0.03)
C2 rod	410 ± 4	377 ± 11 (0.92 ± 0.03)	410 ± 13 (1.00 ± 0.03)
C3 rod	314 ± 4	315 ± 11 (1.04 ± 0.04)	312 ± 13 (0.99 ± 0.04)

(): C/E (calculation/experiment) value

8.3.2.2   Core Components

The impact of ^{235}U was observed about a significant 900 pcm in both EE1 and E3 cores, as shown in Tables 8.12 and 8.13, respectively. Especially, in the EE1 core, the uncertainty was mostly composed of the values of χ, ν and (n, γ) (capture cross section) reactions of ^{235}U that are related to the infinite multiplication factor. In fact, sensitivity profiles of the χ value of ^{235}U shown in Fig. 8.8 were large with a large number of standard deviations. A marked difference was observed in the $\bar{\nu}$ value of ^{235}U shown in Fig. 8.9 between EE1 and E3 cores at thermal neutron and resonance regions in sensitivity profiles; also, core spectrum dependence on sensitivity was effectively canceled by constant standard deviation at these regions.

The uncertainty of capture reactions of polyethylene with the consideration of the thermal scattering law $S(\alpha, \beta)$ (termed $*^1$H in Tables 8.12 and 8.13) was effective in E3 core more than scattering ($*^1$H (n, n)) reactions caused by the soft spectrum core with a large number of moderators as indicated by sensitivity profiles in Figs. 8.10 and 8.11.

Furthermore, the core spectrum was found to be dependent on the uncertainty of aluminum (Al) in the HEU plate (U-Al alloy) and aluminum in the core component (Al sheath) at KUCA indicated by "^{27}Al" and "$*^{27}$Al," respectively, as shown in Figs. 8.12 and 8.13. In the EE1 core, elastic scattering and inelastic scattering (^{27}Al (n, n')) reactions in HEU plates affected the uncertainty because of the large sensitivity at the fast neutron region as shown in Figs. 8.14 and 8.15, respectively. Also, at the hard spectrum core, elastic scattering reactions of the Al sheath indicated higher value than that of ^{27}Al in HEU plates. Here, uncertainty that is not varied by control rod insertion should be emphasized, because the impact of nuclear reactions related to boron are very small and not shown in Tables 8.12 and 8.13. Accordingly, the calculated results of excess reactivity and control rod worth were considered in good agreement with the measured ones within a 10% difference because the uncertainty of boron is not induced by control rod insertions.

Table 8.12 Breakdown of the uncertainty [pcm] of k_{eff} in nuclides and nuclear reactions in EE1 core (Ref. [2])

Component	Nuclides reaction versus reaction		Case					
			I-1	I-2	I-3	I-4	I-5	I-6
HEU	^{235}U χ	^{235}U χ	770.86 $\pm$ 0.27	772.51 $\pm$ 0.27	772.28 $\pm$ 0.27	779.51 $\pm$ 0.28	773.71 $\pm$ 0.27	777.73 $\pm$ 0.27
	^{235}U $\bar{\nu}$	^{235}U $\bar{\nu}$	360.18 $\pm$ 0.01	360.12 $\pm$ 0.01	360.13 $\pm$ 0.01	359.77 $\pm$ 0.01	360.06 $\pm$ 0.01	359.88 $\pm$ 0.01
	^{235}U (n, γ)	^{235}U (n, γ)	216.32 $\pm$ 0.01	216.33 $\pm$ 0.01	216.36 $\pm$ 0.01	216.34 $\pm$ 0.01	216.34 $\pm$ 0.01	216.34 $\pm$ 0.01
	^{235}U (n, f)	^{235}U (n, γ)	103.38 $\pm$ 0.01	103.38 $\pm$ 0.01	103.41 $\pm$ 0.01	103.39 $\pm$ 0.01	103.38 $\pm$ 0.01	103.38 $\pm$ 0.01
	^{235}U (n, f)	^{235}U (n, f)	69.71 $\pm$ 0.01	69.77 $\pm$ 0.01	69.80 $\pm$ 0.01	69.95 $\pm$ 0.01	69.80 $\pm$ 0.01	69.88 $\pm$ 0.01
	^{27}Al (n, n)	^{27}Al (n, n)	147.30 $\pm$ 0.07	144.61 $\pm$ 0.07	146.57 $\pm$ 0.07	154.01 $\pm$ 0.08	143.87 $\pm$ 0.07	149.45 $\pm$ 0.08
	^{27}Al (n, n')	^{27}Al (n, n')	142.17 $\pm$ 0.03	142.97 $\pm$ 0.03	143.74 $\pm$ 0.03	146.16 $\pm$ 0.03	144.26 $\pm$ 0.03	144.64 $\pm$ 0.03
	^{27}Al (n, γ)	^{27}Al (n, γ)	43.41 $\pm$ 0.01	43.45 $\pm$ 0.01	43.48 $\pm$ 0.01	43.63 $\pm$ 0.01	43.50 $\pm$ 0.01	43.56 $\pm$ 0.01
Polyethylene	*^{1}H (n, n)	*^{1}H (n, n)	192.70 $\pm$ 0.02	191.94 $\pm$ 0.02	192.73 $\pm$ 0.02	196.16 $\pm$ 0.02	192.90 $\pm$ 0.02	193.34 $\pm$ 0.02
	*^{1}H (n, γ)	*^{1}H (n, γ)	120.60 $\pm$ 0.01	119.22 $\pm$ 0.01	118.78 $\pm$ 0.01	113.26 $\pm$ 0.01	117.20 $\pm$ 0.01	115.26 $\pm$ 0.01
	C (n, n)	C (n, n)	61.06 $\pm$ 0.01	60.62 $\pm$ 0.01	60.81 $\pm$ 0.01	61.24 $\pm$ 0.01	60.90 $\pm$ 0.01	60.97 $\pm$ 0.01
	C (n, n')	C (n, n')	24.88 $\pm$ 0.01	25.41 $\pm$ 0.01	24.66 $\pm$ 0.01	24.53 $\pm$ 0.01	24.94 $\pm$ 0.01	24.66 $\pm$ 0.01
	C (n, n)	C (n, n)	-24.07 $\pm$ 0.01	-24.37 $\pm$ 0.01	-23.94 $\pm$ 0.01	-23.88 $\pm$ 0.01	-24.10 $\pm$ 0.01	-23.95 $\pm$ 0.01
Aluminum sheath	*^{27}Al (n, n)	Al (n, n)	93.22 $\pm$ 0.03	92.56 $\pm$ 0.03	95.85 $\pm$ 0.03	96.56 $\pm$ 0.03	95.91 $\pm$ 0.03	96.67 $\pm$ 0.03
	*^{27}Al (n, n')	Al (n, n')	49.57 $\pm$ 0.01	49.80 $\pm$ 0.01	49.80 $\pm$ 0.01	50.28 $\pm$ 0.01	50.28 $\pm$ 0.01	49.92 $\pm$ 0.01
Control rod	^{10}B (n, γ)	^{10}B (n, γ)	9.5E-08	1.0E-03	9.5E-04	5.42E-03	1.59E-03	3.52E-03

*With consideration of thermal scattering law $S(\alpha, \beta)$

Table 8.13 Breakdown of the uncertainty [pcm] of k_{eff} in nuclides and nuclear reactions in E3 core (Ref. [2])

Component	Nuclides reaction versus reaction		Case					
			II-1	II-2	II-3	II-4	II-5	II-6
HEU	^{235}U χ	^{235}U χ	726.35 ± 0.25	729.64 ± 0.25	731.48 ± 0.25	735.58 ± 0.25	735.63 ± 0.25	735.89 ± 0.25
	^{235}U $\bar{\nu}$	^{235}U $\bar{\nu}$	379.52 ± 0.01	379.50 ± 0.01	379.49 ± 0.01	379.46 ± 0.01	379.47 ± 0.01	379.47 ± 0.01
	^{235}U (n, γ)	^{235}U (n, γ)	182.95 ± 0.01	182.85 ± 0.01	182.86 ± 0.01	182.70 ± 0.01	182.77 ± 0.01	182.75 ± 0.01
	^{235}U (n, f)	^{235}U (n, f)	126.42 ± 0.01	126.44 ± 0.01	126.44 ± 0.01	126.46 ± 0.01	126.45 ± 0.01	126.46 ± 0.01
	^{235}U (n, f)	^{235}U (n, γ)	109.66 ± 0.01	109.76 ± 0.01	109.74 ± 0.01	109.87 ± 0.01	109.82 ± 0.01	109.84 ± 0.01
	^{27}Al (n, n)	^{27}Al (n, n)	40.01 ± 0.01	39.38 ± 0.01	40.38 ± 0.01	40.25 ± 0.01	41.52 ± 0.01	41.07 ± 0.01
	^{27}Al (n, n')	^{27}Al (n, n')	37.25 ± 0.01	37.68 ± 0.01	38.11 ± 0.01	37.84 ± 0.01	37.66 ± 0.01	37.64 ± 0.01
Polyethylene	*^{1}H (n, γ)	*^{1}H (n, γ)	217.00 ± 0.01	215.12 ± 0.01	214.89 ± 0.01	211.03 ± 0.01	211.60 ± 0.01	212.23 ± 0.01
	*^{1}H (n, n)	*^{1}H (n, n)	188.58 ± 0.03	188.17 ± 0.03	187.72 ± 0.03	191.17 ± 0.03	190.96 ± 0.03	190.40 ± 0.03
	C (n, n)	C (n, n)	57.98 ± 0.01	57.70 ± 0.01	57.90 ± 0.01	58.40 ± 0.01	58.44 ± 0.01	58.42 ± 0.01
	C (n, n')	C (n, n')	31.96 ± 0.01	32.25 ± 0.01	32.29 ± 0.01	31.79 ± 0.01	32.25 ± 0.01	32.37 ± 0.01
	C (n, n)	C (n, n')	-28.83 ± 0.01	-28.94 ± 0.01	-29.01 ± 0.01	-28.70 ± 0.01	-28.98 ± 0.01	-28.94 ± 0.01
Aluminum sheath	*^{27}Al (n, n)	*^{27}Al (n, n)	68.36 ± 0.02	69.46 ± 0.02	68.11 ± 0.02	68.72 ± 0.02	70.06 ± 0.02	69.49 ± 0.02
	*^{27}Al (n, n')	*^{27}Al (n, n')	37.69 ± 0.01	37.50 ± 0.01	37.67 ± 0.01	37.61 ± 0.01	37.42 ± 0.01	37.60 ± 0.01
	*^{27}Al (n, γ)	*^{27}Al (n, γ)	30.81 ± 0.01	30.43 ± 0.01	30.39 ± 0.01	29.52 ± 0.01	29.66 ± 0.01	29.82 ± 0.01
Control rod	^{10}B (n, γ)	^{10}B (n, γ)	1.1E-07	1.1E-03	1.0E-03	3.5E-03	2.9E-03	2.5E-03

*With consideration of thermal scattering law $S(\alpha, \beta)$

8.3.2.3 Tolerance of HEU Plate

For the variation in manufacturing tolerance of HEU plates shown in Table 8.1, uncertainty was evaluated with MCNP6.1 together with ENDF/B-VII.1 with the same number of histories as was reactivity evaluation, as shown in Tables 8.14 and 8.15. Based on the evaluation of the uncertainty for HEU plates, the uncertainty of the

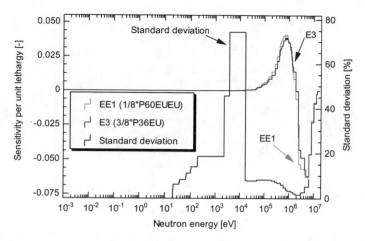

Fig. 8.8 Sensitivity profile and standard deviation covariance data for χ value of ^{235}U (Ref. [2])

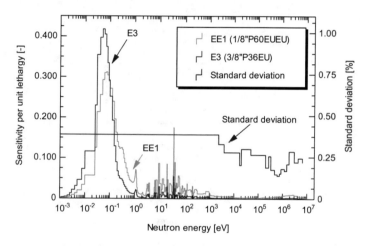

Fig. 8.9 Sensitivity profile and standard deviation covariance data for $\bar{\upsilon}$ value of ^{235}U (Ref. [2])

enrichment (4–14 pcm), length of sides (5 pcm), and thickness (10 pcm) was small regardless of the position of the control rods. Nonetheless, the difference of neutron spectra demonstrated its dependence on the uncertainty between EE1 and E3 cores, emphasizing especially the variation of total number of fuel plates: 3000 and 756 plates in EE1 and E3 cores, respectively. Additionally, the uncertainty attributed to variation in manufacturing tolerance was smaller than that induced by nuclear data, indicating that uncertainty induced by nuclear data is reasonable for the accuracy of criticality at KUCA.

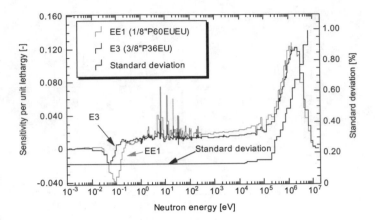

Fig. 8.10 Sensitivity profile and standard deviation covariance data for H (n, n) reactions (polyethylene) (Ref. [2])

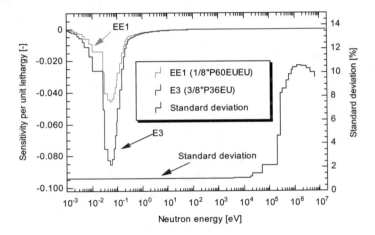

Fig. 8.11 Sensitivity profile and standard deviation in covariance data for H (n, γ) reactions (polyethylene) (Ref. [2])

8.3.2.4 Reproducibility of Control Rod Position

The uncertainty induced by the reproducibility of control rod positions, as shown in Tables 8.14 and 8.15, indicated about 3 pcm even at the withdrawal (no insertion) of control rods in Cases I-1 and II-1, showing an index of accuracy induced by the approximation in Eq. (8.20). By considering the index from the control rod insertion pattern, the uncertainty attributed to varying control rod position was evaluated about 8 pcm depending slightly on the insertion pattern. Here, the evaluated uncertainty was nearly the same as the experimental uncertainty in the reactivity shown in Tables 8.10

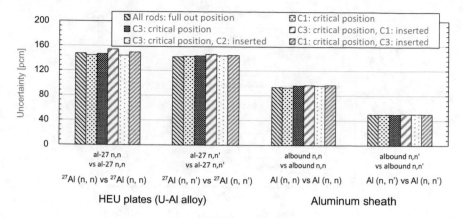

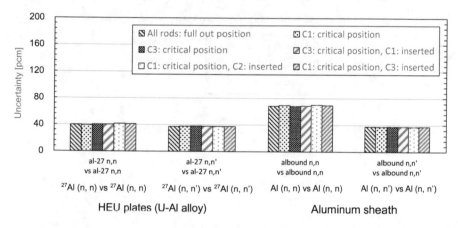

Fig. 8.12 Uncertainty of aluminum in HEU (U-Al alloy) plates, aluminum sheath, and control rod guide tube induced by nuclear data in EE1 core (Ref. [2])

Fig. 8.13 Uncertainty of aluminum in HEU (U-Al alloy) plates, aluminum sheath, and control rod guide tube induced by nuclear data in E3 core (Ref. [2])

and 8.11 (2 pcm through 11 pcm), demonstrating that the measured reactivity was slightly varied by the position of control rods.

8.4   Conclusion

Sensitivity and uncertainty analyses were conducted with the combined use of experimental results (excess reactivity and control rod worth) carried out at KUCA and numerical simulations by the MCNP6.1 calculations, the SRAC2006 and MARBLE code systems. The experimental value was compared with the calculated one by

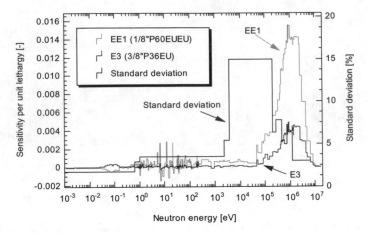

Fig. 8.14 Sensitivity profile and standard deviation in covariance data for ^{27}Al (n, γ) reactions (HEU plate) (Ref. [2])

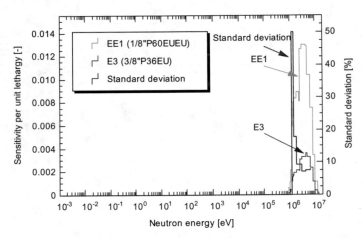

Fig. 8.15 Sensitivity profile and standard deviation in covariance data for ^{27}Al (n, n') reactions (HEU plate) (Ref. [2])

Table 8.14 Uncertainty [pcm] of k_{eff} attributed to manufacturing variation in HEU plates in Table 8.1 and reproducibility of control rods for EE1 core (Ref. [2])

Case	Enrichment of ^{235}U	Vertical length	Horizontal length	Thickness	Reproducibility of control rod
I-1	4.25 ± 0.09	4.19 ± 0.38	3.68 ± 0.43	9.44 ± 0.23	3.23 ± 3.55
I-2	4.21 ± 0.09	3.80 ± 0.40	4.10 ± 0.41	9.60 ± 0.23	2.22 ± 3.77
I-3	4.08 ± 0.09	4.51 ± 0.34	3.59 ± 0.40	9.68 ± 0.23	7.22 ± 3.57
I-4	4.21 ± 0.09	4.59 ± 0.39	4.40 ± 0.38	9.68 ± 0.23	5.54 ± 3.56
I-5	4.14 ± 0.09	3.69 ± 0.37	4.22 ± 0.41	10.08 ± 0.23	4.65 ± 3.56
I-6	4.13 ± 0.09	4.15 ± 0.38	4.23 ± 0.41	9.80 ± 0.23	4.28 ± 3.56

Table 8.15 Uncertainty [pcm] of k_{eff} attributed to manufacturing variation in HEU plates in Table 8.1 and reproducibility of control rods for E3 core (Ref. [2])

Case	Enrichment of ^{235}U	Vertical length	Horizontal length	Thickness	Reproducibility of control rod
II-1	13.20 ± 0.11	3.00 ± 0.25	3.09 ± 0.28	8.40 ± 0.41	2.47 ± 3.16
II-2	13.22 ± 0.11	3.15 ± 0.31	3.41 ± 0.35	7.71 ± 0.41	2.90 ± 3.17
II-3	13.27 ± 0.11	3.00 ± 0.2 9	3.19 ± 0.30	8.11 ± 0.41	3.56 ± 317
II-4	13.36 ± 0.11	3.10 ± 0.29.	2.60 ± 0.31	8.04 ± 0.41	6.78 ± 3.14
II-5	13.50 ± 0.11	2.90 ± 0.32	3.03 ± 0.31	7.42 ± 0.41	3.99 ± 3.16
II-6	13.30 ± 0.11	3.08 ± 0.28	2.89 ± 0.32	7.93 ± 0.41	4.10 ± 3.15

the deterministic approach with the covariance data of JENDL-4.0. Sensitivity and uncertainty analyses demonstrated that the impact of ^{27}Al and ^{235}U was remarkably large in the KUCA A cores, respectively. Moreover, the numerical results revealed the quantitative evaluation (about 150 pcm) of uncertainty induced by the JENDL-4.0 data library in the A cores. Also, these results indicated that further investigation is needed of the numerical analyses of uncertainty of ^{27}Al composed mainly of core components in the A cores, with the use of ^{27}Al covariance data, in order to assess the effect of the uncertainty of ^{27}Al cross sections on reactivity.

To ensure the accuracy of criticality by experimental analyses at KUCA, the modeling of core configuration was examined through the comparison of excess reactivity and control rod worth between the calculation and the experiment in hard and soft spectrum cores. Furthermore, uncertainty was evaluated for manufacturing tolerance in HEU plates, for reproducibility of the control rod position, and for nuclear data. In the validation estimation of calculated k_{eff} values in the modeling of reference core materials and core configurations with MCNP6.1 and SCALE6.2/KENO-VI, the bias in calculated k_{eff} values showed that the difference in the spectrum was about 350 pcm in EE1 core and about 600 pcm in E3 core. Moreover, uncertainty of k_{eff} induced by nuclear data indicated about 950 pcm for k_{eff} evaluation with a slight variation in control rod position and core spectrum. In the breakdown of the uncertainty of k_{eff} induced by nuclear data, the impact of ^{235}U was significantly dominant in over 90% for both cores. The sensitivities of Al in HEU plates and in Al sheaths were marked in fast neutron and resonance regions, leading to large uncertainty about 100 pcm and 40 pcm in EE1 and E3 cores, respectively. Also, uncertainty was evaluated about 10 and 8 pcm in the tolerance of HEU plates and the reproducibility of control rod positions, respectively.

References

1. Pyeon CH, Yamanaka M, Ito M et al (2018) Uncertainty quantification of criticality in solid-moderated and -reflected cores at Kyoto University Critical Assembly. J Nucl Sci Technol 55:812

2. Yamanaka M, Pyeon CH (2019) Benchmarks of criticality in solid-moderated and solid-reflected core in at Kyoto University Critical Assembly. Nucl Sci Eng 193:404
3. Fukushima M, Kitamura Y, Kugo T et al (2016) Benchmark models for criticalities of FCA-IX assemblies with systematically changed neutron spectra. J Nucl Sci Technol 53:406
4. Goorley JT, James MR, Booth TE et al (2013) Initial MCNP6 release overview—MCNP6 version 1.0. LA-UR-13-22934
5. Shibata K, Iwamoto O, Nakagawa T et al (2011) JENDL-4.0: a new library for nuclear science and technology. J Nucl Sci Technol 48:1
6. Chadwick MB, Oblozinsky P, Herman M et al (2006) ENDF/V-II.0: next generation evaluated nuclear data library for nuclear science and technology. Nucl Data Sheet 107:2931
7. Okumura K, Kugo T, Kaneko K et al (2007) SRAC2006: a comprehensive neutronic calculation code system. JAERI-Data/Code 2007-004
8. Yokoyama K, Hazama T, Numata K et al (2014) Development of comprehensive and versatile framework for reactor analysis, MARBLE. Ann Nucl Energy 66:51
9. Fowler TB, Vondy DR (1969) Nuclear reactor core analysis code: Citation. ORNL-TM-2496, rev. 2
10. Hara A, Takeda T, Kikuchi Y (1984) SAGEP: two-dimensional sensitivity analysis code based on generalized perturbation theory. JAERI-M 84-027
11. Hazama T, Chiba G, Sugino K et al (2006) Development of fine and ultra-fine group cell calculation code SLAROM-UF for fast reactor analysis. J Nucl Sci Technol 43:908
12. Pyeon CH, Fujimoto A, Sugawara T et al (2016) Validation of Pb nuclear data by Monte Carlo analyses of sample reactivity experiments at Kyoto University Critical Assembly. J Nucl Sci Technol 53:602
13. Usachev LN (1964) Perturbation theory for the breeding ratio and for other number ratios pertaining to various reactor processes. J Nucl Energy 18:571
14. Gandini A (1967) A generalized perturbation method for bi-linear functionals of the real and adjoint neutron fluxes. J Nucl Energy 21:755
15. Cecchini GP, Salvatores M (1971) Advances in the generalized perturbation theory. Nucl Sci Eng 46:304
16. Kobayashi K (1996) Reactor Physics. Corona Publishing Co. Ltd. Tokyo, Japan. [in Japanese]
17. Cacuci DG (2004) On the neutron kinetics and control of accelerator-driven systems. Nucl Sci Eng 148:55
18. Broadhead BL, Rearden BT, Hopper CM et al (2004) Sensitivity- and uncertainty-based criticality safety validation techniques. Nucl Sci Eng 146:340
19. Pyeon CH, Fujimoto A, Sugawara T et al (2017) Sensitivity and uncertainty analyses of lead sample reactivity experiments at Kyoto University Critical Assembly. Nucl Sci Eng 185:460
20. Dragt JB, Dekker JM, Grupperlaar H et al (1977) Methods of adjustment and error evaluation of neutron capture cross sections; application to fission product nuclides. Nucl Sci Eng 62:117
21. Cacuci DG, Inoescu-bujor M (2010) Best-estimate model calibration and prediction through experimental data assimilation: I. mathematical framework Nucl Sci Eng 165:18
22. Rearden BT, Jessee MA (2016) SCALE code system. ORNL/TM-2005/39 version 6.2.1
23. Nauchi Y, Kameyama T (2005) Proposal of direct calculation of kinetic parameters β_{eff} and Λ based on continuous energy Monte Carlo method. J Nucl Sci Technol 42:503
24. Meulekamp RK, van der Marck SC (2006) Calculating the effective delayed neutron fraction with Monte Carlo. Nucl Sci Eng 152:142
25. Rearden BT, Williams ML, Jessee MA et al (2011) Sensitivity and uncertainty analysis capabilities and data in SCALE. Nucl Technol 174:236
26. Perfetti CM, Rearden BT (2016) Development of a generalized perturbation theory method for sensitivity analysis using continuous-energy Monte Carlo methods. Nucl Sci Eng 182:354
27. OECD/NEA (2008) ICSBEP guide to the expression of uncertainties. https://www.oecd-nea.org/science/wpncs/icsbep/documents/UncGuide.pdf, Accessed 30 June 2020

Appendix A1: Experimental Benchmarks on ADS at Kyoto University Critical Assembly

A1.1 Experimental Settings of ADS Benchmarks

A1.1.1 Core Components

See (Figs. A1.1, A1.2, A1.3, A1.4, A1.5, A1.6, A1.7, A1.8, A1.9, A1.10, A1.11, A1.12 and A1.13).

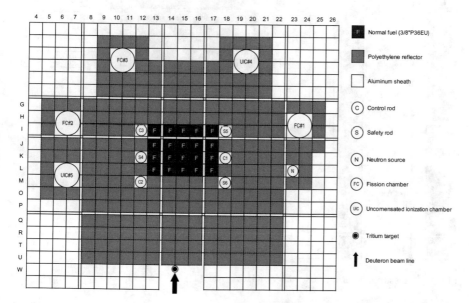

Fig. A1.1 Top view of KUCA core configuration (Ref. [1])

© The Editor(s) (if applicable) and The Author(s) 2021
C. H. Pyeon (ed.), *Accelerator-Driven System at Kyoto University Critical Assembly*,
https://doi.org/10.1007/978-981-16-0344-0

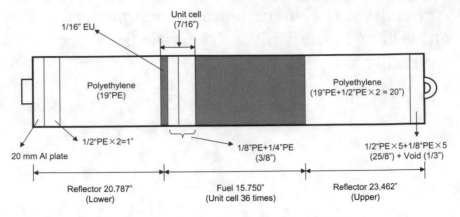

Fig. A1.2 Fall sideway view of "F" (3/8″P36EU) fuel assembly (Ref. [1])

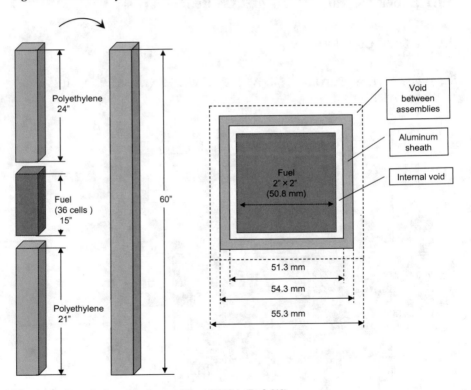

Fig. A1.3 Description of fuel assembly at KUCA (Ref. [1])

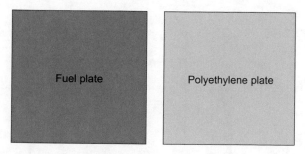

Fig. A1.4 Description of fuel and polyethylene plates (Ref. [1])

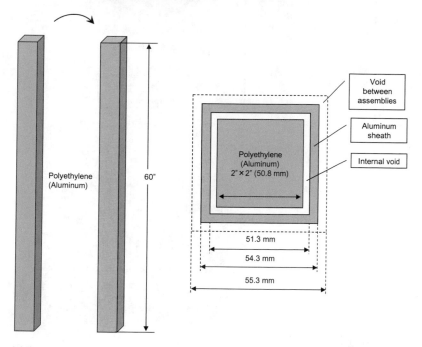

Fig. A1.5 Description of polyethylene (Aluminum) reflector at KUCA (Ref. [1])

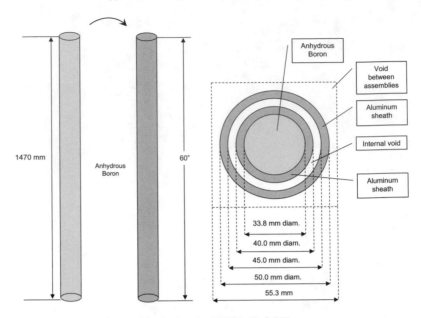

Fig. A1.6 Description of control (safety) rod at KUCA (Ref. [1])

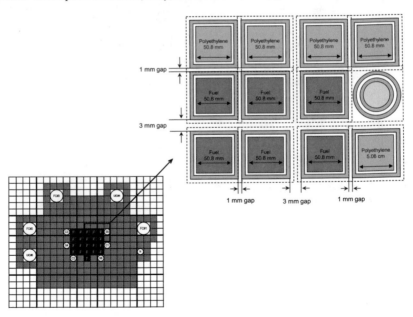

Fig. A1.7 Description of fuel assembly, polyethylene reflector and control rod (Ref. [1])

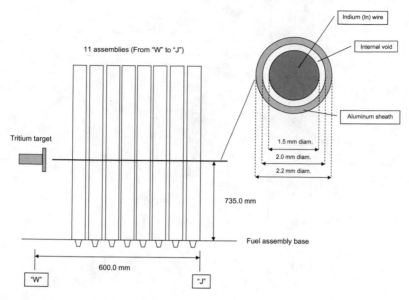

Fig. A1.8 Setting of Indium (In) wire (Ref. [2])

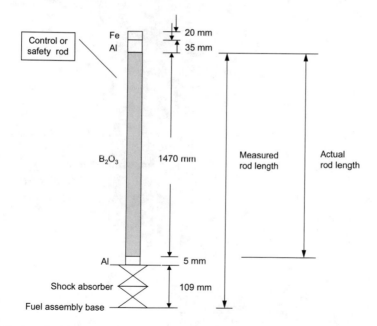

Fig. A1.9 Actual position of control (safety) rod. (Actual position = Measured position − 114 mm) (Ref. [1])

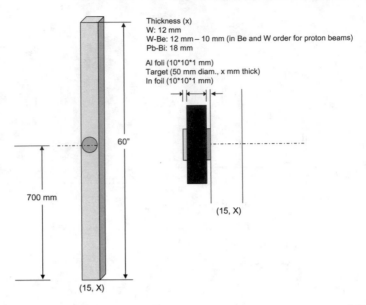

Fig. A1.10 Attachment of target, Al and In foils at location of polyethylene rod (Ref. [3])

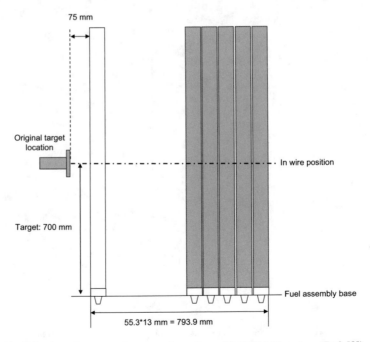

Fig. A1.11 Side view of target and core configuration with 100 MeV protons (Ref. [3])

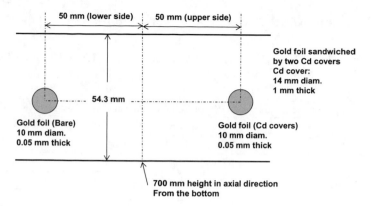

Fig. A1.12 Fall sideway view of HEU fuel rod (Ref. [3])

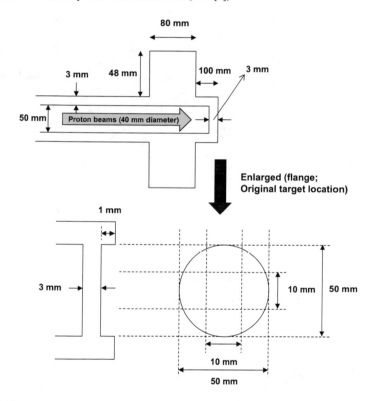

Fig. A1.13 Target configuration of location of original target (Ref. [3])

A1.1.2 Atomic Number Density of Core Elements

See (Tables A1.1, A1.2, A1.3, A1.4, A1.5, A1.6, A1.7, A1.8, A1.9, A1.10, A1.11 and A1.12)

Table A1.1 Highly-enriched uranium (HEU or EU) fuel plate made of U-Al alloy (Ref. [1])

Isotope	Atomic density [$\times 10^{24}$ cm^{-3}]
	1/16″EU plate [1.5875 mm]
^{234}U	1.13659E-05
^{235}U	1.50682E-03
^{236}U	4.82971E-06
^{238}U	9.25879E-05
^{27}Al	5.56436E-02

Table A1.2 Polyethylene reflector (PE) (Ref. [1])

Isotope	Atomic density [$\times 10^{24}$ cm^{-3}]			
	1/8″PE plate [3.086 mm]	1/4″PE plate [6.300 mm]	1/2″PE plate [12.500 mm]	19″PE rod [483.085 mm]
^{1}H	8.02167E-02	8.08711E-02	8.06560E-02	8.00083E-02
^{6}C	4.01084E-02	4.04356E-02	4.03280E-02	4.00042E-02

Table A1.3 Control and safety rods (Ref. [1])

Isotope	Atomic density [$\times 10^{24}$ cm^{-3}]
^{10}B	3.87448E-03
^{11}B	1.68447E-02
^{16}O	3.10787E-02

Table A1.4 Aluminum sheath for the core element and 1/16″Al plate (Ref. [1])

Isotope	Atomic density [$\times 10^{24}$ cm^{-3}]
	1/16″Al plate [1.5875 mm]
^{27}Al	6.00385E-02

Table A1.5 ^{115}In, ^{56}Fe, ^{27}Al, ^{93}Nb, ^{197}Au and Cd foils (Ref. [1])

Foil	Isotope	Abundance (%)	Purity (%)	Atomic density [$\times 10^{24}$ cm^{-3}]
^{115}In	^{113}In	4.29	99.99	1.64406E-03
	^{115}In	95.71	99.99	3.66790E-02
^{56}Fe	^{54}Fe	5.845	99.5	4.93395E-03
	^{56}Fe	91.754	99.5	7.74524E-02
	^{57}Fe	2.119	99.5	1.78871E-03
	^{58}Fe	0.282	99.5	2.38045E-04

(continued)

Table A1.5 (continued)

Foil	Isotope	Abundance (%)	Purity (%)	Atomic density [$\times 10^{24}$ cm^{-3}]
^{27}Al	^{27}Al	100	99.5	5.99156E-02
^{93}Nb	^{93}Nb	100	99.9	5.54750E-02
^{197}Au	^{197}Au	100	99.95	5.90193E-02
Cd	^{106}Cd	1.25	99.99	5.39648E-04
	^{108}Cd	0.89	99.99	3.91477E-04
	^{110}Cd	12.51	99.99	5.59564E-03
	^{111}Cd	12.81	99.99	5.78677E-02
	^{112}Cd	24.13	99.99	1.10072E-02
	^{113}Cd	12.22	99.99	5.62419E-03
	^{114}Cd	28.72	99.99	1.33398E-02
	^{116}Cd	7.47	99.99	3.53884E-03

Table A1.6 W, Be and Pb–Bi targets (Ref. [3])

Target	Isotope	Abundance (%)	Atomic density [$\times 10^{24}$ cm^{-3}]
W (50 mm diam. and 12.0 mm thick)	^{180}W	0.12	8.04702E-05
	^{182}W	26.50	1.64613E-02
	^{183}W	14.31	9.00328E-03
	^{184}W	30.64	1.93829E-02
	^{186}W	28.43	1.81807E-02
Be (50 mm diam. and 10.0 mm thick)	^{9}Be	100	1.23487E-01
Pb-Bi (44.5/55.5) (50 mm diam. and 18.0 mm thick) and 1/8″Pb-Bi plate [3.426 mm]	^{204}Pb	1.4	1.87461E-04
	^{206}Pb	24.1	3.25860E-03
	^{207}Pb	22.1	3.00266E-03
	^{208}Pb	52.4	7.15378E-03
	^{209}Bi	100	1.67670E-02

Table A1.7 Beam tube component (SUS304) shown in Fig. A1.3 (Ref. [3])

Isotope	Atomic density [$\times 10^{24}$ cm^{-3}]
^{54}Fe	3.55712E-03
^{56}Fe	5.58391E-02
^{57}Fe	1.28957E-03
^{58}Fe	1.71618E-04
^{50}Cr	7.51530E-04
^{52}Cr	1.44925E-02

(continued)

Table A1.7 (continued)

Isotope	Atomic density [$\times 10^{24}$ cm^{-3}]
^{53}Cr	1.64333E-03
^{54}Cr	4.09060E-04
^{58}Ni	5.10587E-03
^{60}Ni	1.96674E-03
^{61}Ni	8.54932E-05
^{62}Ni	2.72597E-04
^{64}Ni	6.94130E-05

Table A1.8 Polyethylene moderator "p" (Ref. [3])

Isotope	Atomic density [$\times 10^{24}$ cm^{-3}]	
	1/8″p plate [3.158 mm]	10″p rod [254.00 mm]
H	7.77938E-02	7.97990E-02
C	3.95860E-02	4.08960E-02

Table A1.9 Coating materials (Face; 1st and Base 2nd layers) over Pb-Bi plate (Ref. [3])

Nuclide	Isotope	Atomic density [$\times 10^{24}$ cm^{-3}] (Face; 1st layer)	Atomic density [$\times 10^{24}$ cm^{-3}] (Base; 2nd layer)
H	^{1}H	2.83301E-03	3.78991E-03
	^{2}H	4.25015E-07	5.64072E-07
C	–	2.27058E-03	4.03671E-03
O	^{16}O	2.06885E-03	4.58782E-04
	^{17}O	7.88039E-07	1.74753E-07
	^{18}O	4.14757E-06	9.19753E-07
Ti	^{46}Ti	2.50941E-05	5.36896E-05
	^{47}Ti	2.28983E-05	4.89918E-05
	^{48}Ti	2.31493E-04	4.95287E-04
	^{49}Ti	1.72522E-05	3.69116E-05
	^{50}Ti	1.69385E-05	3.62405E-05
Si	^{28}Si	–	1.86243E-05
	^{29}Si	–	9.43026E-07
	^{30}Si	–	6.25992E-07
S	^{32}S	4.16818E-04	–
	^{33}S	3.28997E-06	–
	^{34}S	1.84677E-05	–
	^{35}S	8.77326E-08	–

(continued)

Table A1.9 (continued)

Nuclide	Isotope	Atomic density [$\times 10^{24}$ cm^{-3}] (Face; 1st layer)	Atomic density [$\times 10^{24}$ cm^{-3}] (Base; 2nd layer)
Ba	^{132}Ba	4.43050E-07	–
	^{134}Ba	1.06025E-05	–
	^{135}Ba	2.89167E-05	–
	^{136}Ba	3.44526E-05	–
	^{137}Ba	4.92619E-05	–
	^{138}Ba	3.14521E-04	–

Table A1.10 1/8″Pb plate (Refs. [4-5])

Isotope	Abundance (%)	Atomic density [$\times 10^{24}$ cm^{-3}] [3.011 mm]
^{204}Pb	1.4	4.668193E-04
^{206}Pb	24.1	8.035960E-03
^{207}Pb	22.1	7.369076E-03
^{208}Pb	52.4	1.747238E-02

Table A1.11 1/8″Bi plate (Ref. [4])

Isotope	Abundance (%)	Atomic density [$\times 10^{24}$ cm^{-3}] [3.002 mm]
^{209}Bi	100	2.65467E-02

Table A1.12 Core elements of 2″Gr, 1/8″Th, 1/8″NU, 1/2″Gr and 1/2″Be (Refs. [2–5])

Elements	Isotope	Atomic density [$\times 10^{24}$ cm^{-3}]
2″Gr block [50.80 mm]	C	8.64182E-02
1/8″Th (thorium) fuel plate [3.175 mm]	^{232}Th	2.86384E-02
1/8″NU (natural uranium) fuel plate [1.05 mm × 3]	^{235}U	3.25792E-04
	^{238}U	4.48577E-02
1/2″Ge (graphite) plate [12.7 mm]	C	8.60664E-02
1/2″Be (beryllium) plate [12.5 mm]	Be	1.22932E-01

References

1. Pyeon CH (2012) Experimental benchmarks for accelerator-driven system (ADS) at Kyoto University Critical Assembly. KURRI-TR-444
2. Pyeon CH (2015) Experimental benchmarks on thorium-loaded accelerator-driven system (ADS) at Kyoto University Critical Assembly. KURRI-TR(CD)-48
3. Pyeon CH (2017) Experimental benchmarks of neutronics on solid Pb-Bi in accelerator-driven system with 100 MeV protons at Kyoto University Critical Assembly. KURRI-TR-447
4. Pyeon CH and Yamanaka M (2018) Experimental benchmarks of neutron characteristics on uranium-lead zoned core in accelerator-driven system at Kyoto University Critical Assembly. KURNS-EKR-001
5. Pyeon CH (2020) Experimental benchmarks of medium-fast spectrum core (EE1 core) in accelerator-driven system at Kyoto University Critical Assembly. KURNS-EKR-007

Appendix A2: ^{235}U-Fueled and Pb–Bi–Zoned ADS Core

A2.1 Pb–Bi Target

A2.1.1 Core Configurations

See (Figs. A2.1, A2.2, A2.3 and A2.4).

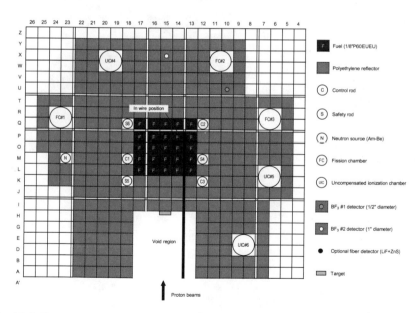

Fig. A2.1 Top view of core configuration of ADS with 100 MeV protons (in case of the In wire reaction rate distribution) (Ref. [1])

© The Editor(s) (if applicable) and The Author(s) 2021
C. H. Pyeon (ed.), *Accelerator-Driven System at Kyoto University Critical Assembly*,
https://doi.org/10.1007/978-981-16-0344-0

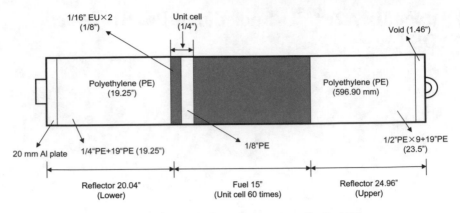

Fig. A2.2 Fall sideway view of "F" (1/8″P60EUEU) fuel assembly (Ref. [1])

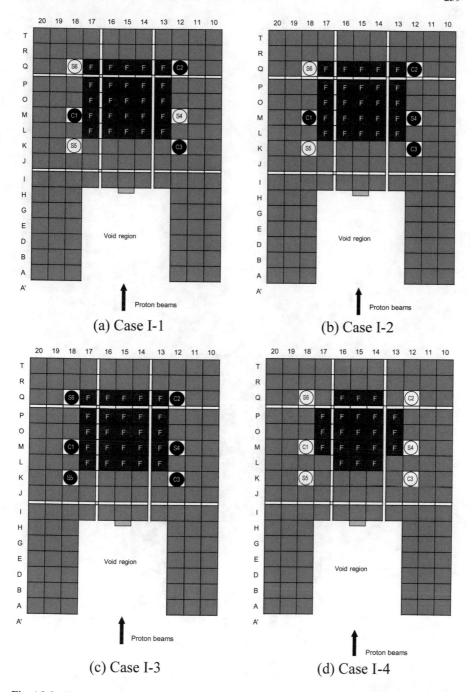

Fig. A2.3 Core configurations of ADS with 100 MeV protons (Black: Full-in; White: Full-out) (in cases of subcriticality measurement by the PNS and the Noise methods) (Ref. [1])

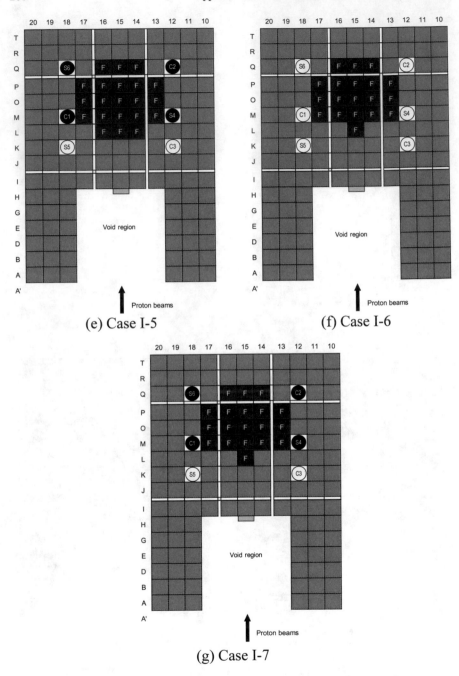

(e) Case I-5

(f) Case I-6

(g) Case I-7

Fig. A2.3 (continued)

Fig. A2.4 Schema of an optical fiber detection system (Ref. [1])

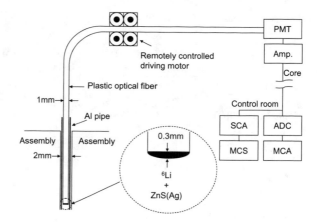

A2.1.2 Results of Experiments

A2.1.2.1 Reaction Rate Distribution

See (Tables A2.1, A2.2, A2.3, A2.4, A2.5 and A2.6, Figs. A2.5, A2.6, A2.7 and A2.8).

Table A2.1 Positions of control and safety rods at the critical state (Ref. [1])

Rod	Rod position (mm)
C1	676.92
C2	1201.16
A3	1201.43
S4	1200.00
S5	1200.00
S6	1200.00
Excess reactivity [pcm]	180 ± 1

Table A2.2 Control rod worth (Ref. [1])

Rod	Rod worth [pcm]
C1 (S4)	805 ± 4
C2 (S6)	596 ± 4
A3 (S5)	139 ± 4

Table A2.3 Measured reaction rates [s^{-1} cm^{-3}] and Cd ratio (Ref. [1])

	W target	W–Be target	Pb–Bi target
Au foil (Bare)	(3.017 ± 0.734) E+06	(3.330 ± 0.026) E+06	(2.070 ± 0.158) E+06
Au foil (Cd)	(2.240 ± 0.546) E+06	(2.406 ± 0.020) E+06	(1.497 ± 0.116) E+06
^{115}In(n, n')^{115m}In	(2.176 ± 0.078) E+05	(1.777 ± 0.015) E+05	(1.578 ± 0.051) E+05
^{27}Al(n, n+3p)^{24}Na	(1.036 ± 0.062) E+06	(1.185 ± 0.080) E+06	(0.928 ± 0.062) E+06
Cd ratio	1.35 ± 0.27	1.39 ± 0.02	1.38 ± 0.28

Table A2.4 Core condition of indium reaction rate distribution in subcritical state (Ref. [1])

Case	Number of fuel rods	Rod insertion
Case I-3 (refer to Table A2.7)	25	C1, C2, A3, S4, S5, S6

Table A2.5 Specification of measurement of reaction rate distribution (Ref. [1])

Reaction	Location	Foil/Wire
^{115}In(n, n')^{115m}In	Target (refer to Fig. A1.10)	10 × 10 × 1 mm
^{115}In(n, γ)^{116}In	Core (refer to Fig. A1.11)	1 mm diameter, 800 mm long

Table A2.6 Dimensions of targets (W, W–Be and Pb–Bi) (Ref. [1])

Target	Diameter (mm)	Thickness (mm)
W	50	12.0
W–Be	50	W: 12.0; Be: 10.0
Pb–Bi	50	Pb–Bi: 18.0

Fig. A2.5 Comparison between neutron spectra with the use of W, W–Be and Pb–Bi targets (Refs. [1–2])

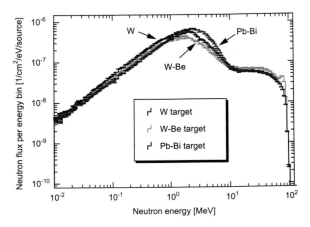

Fig. A2.6 Comparison between neutron spectra with the use of W and W–Be targets (Refs. [1–2])

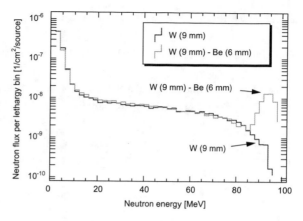

Fig. A2.7 Proportionality of cross sections of ^{115}In capture and ^{235}U fission reactions (Refs. [1–2])

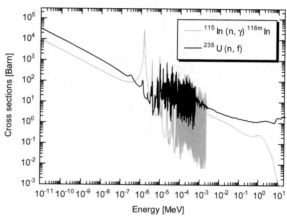

Fig. A2.8 Comparison between normalized reaction rates obtained in ADS experiments (Refs. [1–2])

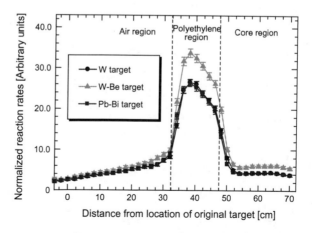

A2.1.2.2 PNS and Feynman-α Methods

See (Tables A2.7, A2.8, A2.9, A2.10, A2.11 and A2.12 and Figs. A2.9 and A2.10).

Table A2.7 List of core condition in all the cores shown in Fig. A2.2 (Ref. [1])

Case	# of fuel rods	Rod insertion	100 MeV protons	
			PNS	Noise
Case I-1	25	C1, C2, A3	Available	Available
Case I-2	25	C1, C2, A3, S4	Available	Available
Case I-3	25	C1, C2, A3, S4, S5, S6	Available	Available
Case I-4	21	All six rods withdrawn	Available	Available
Case I-5	21	C1, C2, S4, S6	Available	Available

PNS method: Pulsed-neutron source method; Noise method: Feynman-α method

Table A2.8 Beam characteristics of 100 MeV protons (Ref. [1])

Case	100 MeV protons	
	Repetition (Hz)	Width (ns)
Case I-1	20	100
Case I-2	20	100
Case I-3	20	100
Case I-4	20	100
Case I-5	20	100

Proton beam intensity was 1 nA; Spot size of proton beams: 40 mm.

Table A2.9 Measured prompt neutron decay constants α [s^{-1}] deduced by least-squared fitting in the PNS method (Ref. [1])

Neutron decay constants α [s^{-1}]			
Target	BF$_3$ #1 in (10, U)	BF$_3$ #2 in (15, X)	Optical fiber
Case I-1			
W	737.1 ± 56.0	739.6 ± 9.5	706.4 ± 83.3
W–Be	732.5 ± 5.2	747.8 ± 5.5	782.0 ± 5.1
Pb–Bi	737.7 ± 10.9	731.9 ± 10.6	756.5 ± 12.5
Case I-2			
W	1059.9 ± 10.0	1075.0 ± 5.8	1215.2 ± 12.5
W–Be	1062.7 ± 10.0	1085.6 ± 4.0	1101.9 ± 5.3
Pb–Bi	1070.1 ± 21.5	1072.0 ± 20.0	1080.8 ± 8.3
Case I-3			
W	1364.2 ± 8.4	1372.5 ± 9.9	1379.6 ± 68.4
W–Be	1377.2 ± 6.5	1383.2 ± 6.8	1402.3 ± 10.3
Pb–Bi	1338.0 ± 3.7	1358.3 ± 4.0	1381.5 ± 7.1

(continued)

Table A2.9 (continued)

Neutron decay constants α [s^{-1}]			
Target	BF$_3$ #1 in (10, U)	BF$_3$ #2 in (15, X)	Optical fiber
Case I-4			
W	1052.0 ± 7.5	1033.3 ± 8.8	1204.5 ± 16.6
W–Be	1033.2 ± 3.1	1027.7 ± 3.5	1073.8 ± 4.1
Pb–Bi	1006.3 ± 5.6	1004.1 ± 6.1	1083.2 ± 5.3
Case I-5			
W	1844.6 ± 21.3	1845.9 ± 17.7	1922.5 ± 33.3
W–Be	1770.5 ± 3.7	1697.0 ± 4.6	1909.1 ± 3.9
Pb–Bi	1785.2 ± 9.4	1742.1 ± 8.7	1895.5 ± 11.1

Table A2.10 Measured subcriticality [pcm] deduced by the extrapolated area ratio method (Ref. [1])

Subcriticality ρ [pcm] (by Area ratio method)			
Target	BF$_3$ #1 in (10, U)	BF$_3$ #2 in (15, X)	Optical fiber
Case I-1 (Reference: ρ_{exp} = 1360 pcm by experiment)			
W	1132 ± 17	1393 ± 20	1888 ± 27
W–Be	1145 ± 18	1458 ± 21	1473 ± 21
Pb–Bi	1166 ± 18	1387 ± 20	1390 ± 20
Case I-2 (Reference: ρ_{exp} = 2165 pcm by experiment)			
W	1877 ± 27	2363 ± 35	3002 ± 44
W–Be	1879 ± 28	2463 ± 36	2229 ± 33
Pb–Bi	1926 ± 29	2356 ± 35	2252 ± 33
Case I-3 (Reference: ρ_{exp} = 2900 pcm by experiment)			
W	2398 ± 35	3156 ± 46	3822 ± 58
W–Be	2502 ± 39	3405 ± 51	$2939 + 45$
Pb–Bi	2624 ± 39	3238 ± 48	2977 ± 45
Case I-4 (Reference: ρ_{MCNPX} = 2773 pcm by MCNP)			
W	2682 ± 39	3291 ± 48	3672 ± 55
W–Be	2688 ± 41	3483 ± 52	2910 ± 43
Pb–Bi	2738 ± 41	3302 ± 49	2922 ± 44
Case I-5 (Reference: ρ_{MCNPX} = 4902 pcm by MCNP)			
W	4944 ± 78	6386 ± 101	6228 ± 99
W–Be	5004 ± 83	7050 ± 114	4929 ± 77
Pb–Bi	4903 ± 77	6356 ± 92	4912 ± 74

(β_{eff} = 807 pcm; Λ = 3.050E-05 s by MCNP6.1 with JENDL-4.0)

Table A2.11 Measured prompt neutron decay constants α [s^{-1}] deduced by least-squared fitting in the Feynman-α method (Ref. [1])

Neutron decay constants α [s^{-1}]			
Target	BF$_3$ #1 in (10, U)	BF$_3$ #2 in (15, X)	Optical fiber
Case I-1			
W	632.5 ± 7.1	665.4 ± 12.5	636.6 ± 10.8
W–Be	641.8 ± 21.2	683.0 ± 13.5	705.6 ± 11.8
Pb–Bi	616.2 ± 30.4	650.7 ± 26.8	714.2 ± 19.4
Case I-2			
W	878.7 ± 22.2	869.1 ± 28.2	1089.8 ± 60.2
W–Be	874.7 ± 12.6	912.8 ± 23.7	1073.4 ± 60.4
Pb–Bi	885.0 ± 25.3	898.5 ± 36.3	970.2 ± 62.6
Case I-3			
W	1086.4 ± 21.6	1103.1 ± 26.8	1407.5 ± 57.1
W–Be	1094.3 ± 11.8	1159.6 ± 16.4	1296.0 ± 18.6
Pb–Bi	1124.8 ± 13.9	1132.8 ± 23.4	1308.5 ± 19.5
Case I-4			
W	889.0 ± 11.6	941.8 ± 20.5	1072.8 ± 23.8
W–Be	880.1 ± 12.8	975.5 ± 35.0	999.7 ± 11.4
Pb–Bi	900.0 ± 17.5	931.1 ± 10.5	1021.5 ± 23.3
Case I-5			
W	1487.9 ± 48.9	1451.6 ± 35.4	1965.8 ± 91.7
W–Be	1456.5 ± 18.8	1575.3 ± 28.5	1746.6 ± 23.2
Pb–Bi	1390.4 ± 95.4	1431.6 ± 96.1	1682.6 ± 142.9

Table A2.12 Measured subcriticality [pcm] deduced by the Feynman-α method (Ref. [1])

Subcriticality ρ [pcm] (by Feynman-α method)			
Target	BF$_3$ #1 in (10, U)	BF$_3$ #2 in (15, X)	Optical fiber
Case I-1 (Reference: ρ_{exp} = 1360 pcm)			
W	1122 ± 25	1515 ± 40	1406 ± 35
W–Be	1151 ± 66	1581 ± 43	1667 ± 38
Pb–Bi	1073 ± 94	1459 ± 83	1699 ± 61
Case I-2 (Reference: ρ_{exp} = 2165 pcm)			
W	1873 ± 69	2285 ± 87	3119 ± 184
W–Be	1861 ± 41	2450 ± 74	3057 ± 185
Pb–Bi	1892 ± 78	2396± 112	2667 ± 191
Case I-3 (Reference: ρ_{exp} = 2900 pcm)			
W	2507 ± 67	3169 ± 83	4320 ± 175
W–Be	2531 ± 39	3383 ± 52	3898 ± 59

(continued)

Table A2.12 (continued)

Subcriticality ρ [pcm] (by Feynman-α method)			
Target	BF$_3$ #1 in (10, U)	BF$_3$ #2 in (15, X)	Optical fiber
Pb–Bi	2624 ± 45	3282 ± 73	3946 ± 61
Case I-4 (Reference: ρ$_{MCNPX}$ = 2773 pcm)			
W	1905 ± 38	2560 ± 64	3055 ± 74
W–Be	1877 ± 41	2884 ± 108	2778 ± 37
Pb–Bi	1938 ± 55	2519 ± 35	2861 ± 72
Case I-5 (Reference: ρ$_{MCNPX}$ = 4902 pcm)			
W	3731 ± 150	4486 ± 109	6430 ± 280
W–Be	3634 ± 60	4954 ± 86	5601 ± 73
Pb–Bi	3434 ± 291	4411 ± 294	5360 ± 436

(β_{eff} = 807 pcm; Λ = 3.050E-05 s by MCNP6.1 with JENDL-4.0)

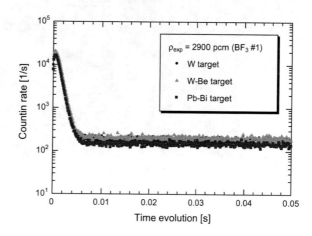

Fig. A2.9 Experimental results of time evolution on prompt and delayed neutron behaviors at position (10, U) of BF$_3$ detector #1 in Case I-3 (Refs. [1–2])

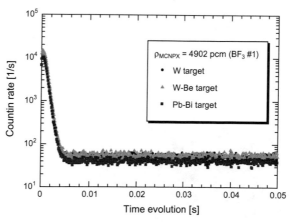

Fig. A2.10 Experimental results of time evolution on prompt and delayed neutron behaviors at position (10, U) of BF$_3$ detector #1 in Case I-5 (Refs. [1–2])

A2.2 Subcriticality Measurements

A2.2.1 Core Configurations

See (Figs. A2.11, A2.12, A2.13, A2.14, A2.15 and A2.16).

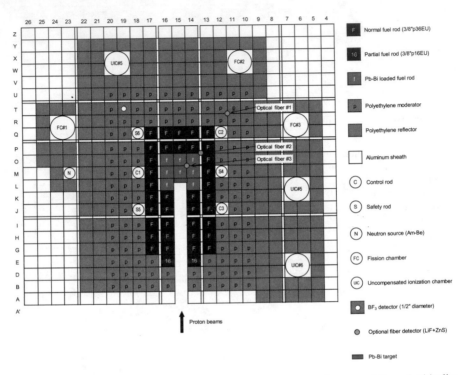

Fig. A2.11 Core configuration of ADS with 100 MeV protons (in case of the subcriticality measurements) (Refs. [1, 3])

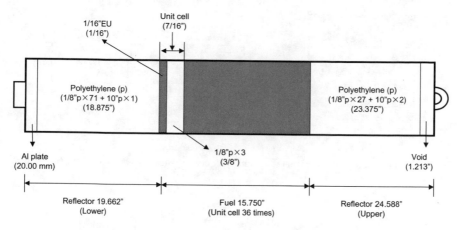

Fig. A2.12 Fall sideway view of "F" (3/8″p36EU) fuel assembly in Fig. A2.11 (Ref. [1])

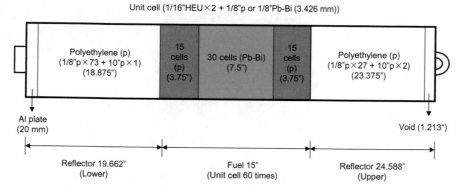

Fig. A2.13 Fall sideway view of fuel assembly "f" shown in Fig. A2.11. (f: 1/8″p15EUEU < 1/8″PbBi30EUEU >1/8″p15EUEU) (Refs. [1, 3])

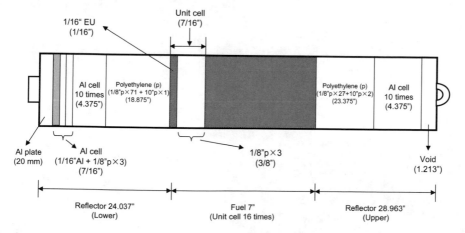

Fig. A2.14 Fall sideway view of fuel assembly "16" (3/8″p16EU) shown in Fig. A2.11 (Refs. [1, 3])

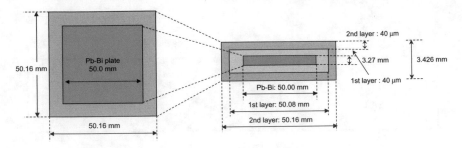

Fig. A2.15 Description of Pb–Bi plate covering over coating materials shown in Fig. A2.11 (Ref. [1])

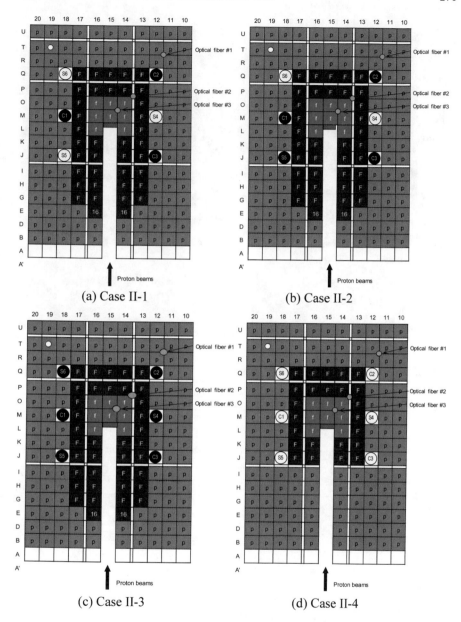

Fig. A2.16 Core configurations of ADS with 100 MeV protons (Black: Full-in; White: Full-out) (in cases of subcriticality measurement by the PNS and the Feynman-α methods) (Refs. [1, 3])

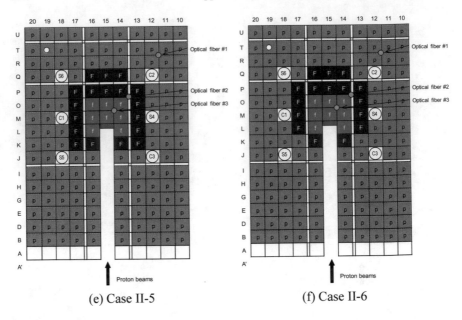

(e) Case II-5   (f) Case II-6

Fig. A2.16 (continued)

A2.2.2 Results of Experiments

See (Tables A2.13, A2.14, A2.15, A2.16 and A2.17, Fig. A2.17)

Table A2.13 List of core condition in all the cores shown in Fig. A2.16 (Ref. [1])

Case	# of fuel rods	Rod insertion	100 MeV protons	
			PNS	Noise
Case II-1	46	C1, C2, A3	Available	Available
Case II-2	46	C1, C2, A3, S5	Available	Available
Case II-3	46	C1, C2, A3, S4, S5, S6	Available	Available
Case II-4	32	All six rods withdrawn	Available	Available
Case II-5	26	All six rods withdrawn	Available	Available
Case II-6	24	All six rods withdrawn	Available	Available

PNS method: Pulsed-neutron source method; Noise method: Feynman-α method

Table A2.14 Positions of control and safety rods at critical state in Fig. A2.11 (Ref. [1])

Rod	Rod position (mm)
C1	1200.00
C2	525.39
A3	1200.00
S4	1200.00
S5	1200.00
S6	1200.00
Excess reactivity [pcm]	149 ± 3

Table A2.15 Control rod worth (Ref. [1])

Rod	Rod worth [pcm]
C1 (S4)	549 ± 3
C2 (S6)	194 ± 1
A3 (S5)	483 ± 2

($\beta_{eff} = 781 \pm 4$ [pcm] by MCNP6.1 with ENDF/B-VII.0)

Table A2.16 Effective multiplication factor in Cases II-1 through II-6 (Ref. [1])

Case	k_{eff} (MCNP6.1 with ENDF/B-VII.0)
Critical state	1.00378 ± 0.0004
Case II-1	0.99219 ± 0.0004
Case II-2	0.98723 ± 0.0004
Case II-3	0.97950 ± 0.0004
Case II-4	0.95784 ± 0.0004
Case II-5	0.91355 ± 0.0004
Case II-6	0.90006 ± 0.0004

Table A2.17 Beam characteristics of 100 MeV protons (Ref. [1])

Case	100 MeV protons	
	Repetition (Hz)	Width (ns)
Case II-1	20	50
Case II-2	20	50
Case II-3	20	50
Case II-4	20	50
Case II-5	20	50
Case II-6	20	50

Proton beam intensity was 1 nA; Spot size of proton beams: 40 mm.

Fig. A2.17 Calibration curve of C2 (as well as in C1 and A3) control rod (Ref. [1])

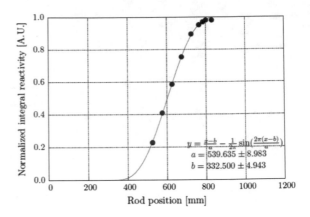

Rod position [mm]

$$y = \frac{x-b}{a} - \frac{1}{2\pi}\sin\left(\frac{2\pi(x-b)}{a}\right)$$
$$a = 539.635 \pm 8.983$$
$$b = 332.500 \pm 4.943$$

A2.2.3 PNS and Feynman-α Methods

See (Tables A2.18, A2.19, A2.20 and A2.21).

Table A2.18 Measured prompt neutron decay constants α [s^{-1}] deduced by least-squared fitting in the PNS method (Ref. [1])

Neutron decay constants α [s^{-1}]			
Case	Fiber #1	Fiber #2	Fiber #3
Case II-1	398.0 ± 2.5	326.8 ± 2.7	494.5 ± 10.2
Case II-2	506.7 ± 3.0	452.6 ± 2.8	685.4 ± 10.8
Case II-3	672.6 ± 3.5	631.9 ± 1.8	1019.7 ± 8.3
Case II-4	982.5 ± 4.4	971.3 ± 3.1	1378.0 ± 18.4
Case II-5	1665.4 ± 8.7	1680.7 ± 5.1	1827.7 ± 22.8
Case II-6	1910.9 ± 7.4	1930.7 ± 3.3	2061.0 ± 37.8

Table A2.19 Measured subcriticality [pcm] deduced by the extrapolated area ratio method (Ref. [1])

			Subcriticality ρ [pcm] (PNS method)		
Case	β_{eff} [pcm]	Reference [pcm]	Fiber #1	Fiber #2	Fiber #3
Case II-1	785 ± 4	1160 ± 5	1224 ± 10	1153 ± 8	2108 ± 59
Case II-2	785 ± 4	1684 ± 6	1880 ± 15	1604 ± 11	4490 ± 109
Case II-3	783 ± 5	2483 ± 6	2824 ± 22	2250 ± 13	10347 ± 116
Case II-4	788 ± 5	4812 ± 6	5656 ± 62	4177 ± 31	22939 ± 1279
Case II-5	816 ± 5	9895 ± 6	12819 ± 137	8187 ± 54	14985 ± 1136
Case II-6	806 ± 5	11556 ± 6	16312 ± 160	9738 ± 64	19109 ± 2563

(β_{eff} : MCNP6.1 with JENDL-4.0)
Note that subcriticalities in Cases II-1, II-2 and II-3 were deduced by the excess reactivity and the control rod worth in Tables A2.14 and A2.15. In Cases II-4, II-5 and II-6, the subcriticality was deduced with the use of MCNP calculations.

Table A2.20 Measured prompt neutron decay constants α [s^{-1}] deduced by least-squared fitting in the Feynman-α method (Ref. [1])

Neutron decay constants α [s^{-1}]			
Case	Fiber #1	Fiber #2	Fiber #3
Case II-1	400.6 ± 9.2	367.1 ± 5.8	554.0 ± 16.8
Case II-2	498.1 ± 4.6	463.5 ± 4.0	700.0 ± 6.7
Case II-3	654.5 ± 2.4	620.0 ± 1.8	877.3 ± 4.8
Case II-4	814.6 ± 5.8	822.0 ± 5.3	1029.0 ± 11.7
Case II-5	1365.1 ± 9.1	1400.0 ± 7.4	1669.1 ± 16.9
Case II-6	1556.7 ± 12.8	1636.1 ± 9.9	1917.4 ± 21.9

Table A2.21 Measured subcriticality [pcm] deduced by the Feynman-α (Noise) method (Ref. [1])

Case	β_{eff} [pcm]	Reference [pcm]	Subcriticality ρ [pcm] (by Noise method)		
			Fiber #1	Fiber #2	Fiber #3
Case II-1	785 ± 4	1160 ± 5	1491 ± 57	1283 ± 36	2445 ± 105
Case II-2	785 ± 4	1684 ± 5	2024 ± 28	1814 ± 24	3250 ± 41
Case II-3	783 ± 5	2483 ± 6	2881 ± 15	2677 ± 11	4203 ± 29
Case II-4	788 ± 5	4812 ± 6	4336 ± 38	4384 ± 35	5740 ± 77
Case II-5	816 ± 5	9895 ± 6	8087 ± 61	8316 ± 50	10110 ± 113
Case II-6	806 ± 5	11556 ± 6	9633 ± 88	10175 ± 68	12096 ± 150

(β_{eff} : MCNP6.1 with JENDL-4.0)

A2.3 Reaction Rates

A2.3.1 Core Configurations

See (Figs. A2.18 and A2.19).

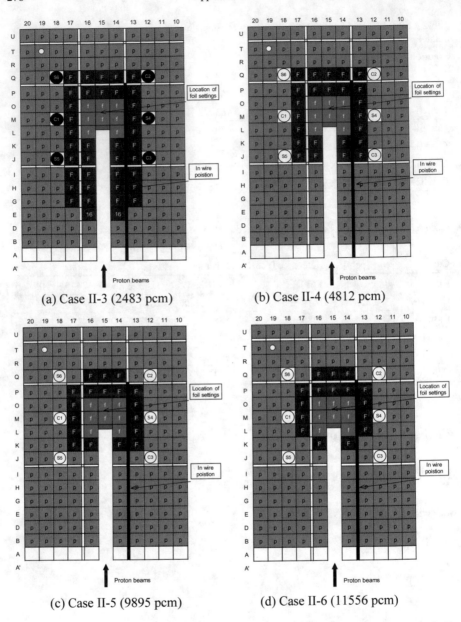

Fig. A2.18 Top view of ^{235}U-fueled and Pb-Bi-zoned ADS core with 100 MeV protons (Refs. [1, 3])

Fig. A2.19 Foil
arrangement at a gap
between Pb-Bi target and
fuel rod in position (15, M)
shown in Fig. A2.18 (Refs.
[1, 3])

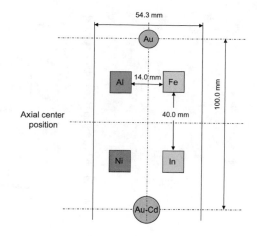

A2.3.2 Reaction Rate Distributions

See (Fig. A2.20).

Fig. A2.20 Comparison
between measured and
calculated ^{115}In$(n, \gamma)^{116m}$In
reaction rate distributions
along (14-13, P-A) region
shown in Fig. A2.18 (Refs.
[1, 3])

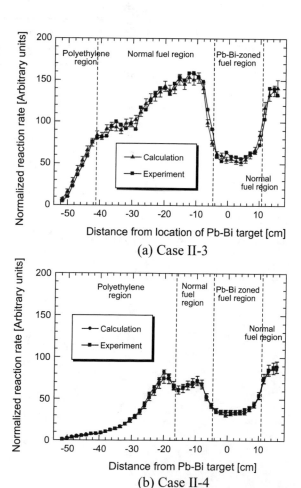

Fig. A2.20 (continued)

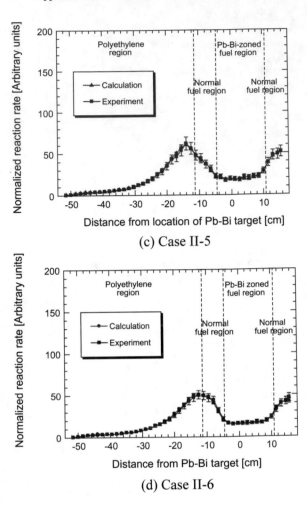

(c) Case II-5

(d) Case II-6

A2.3.3 Reaction Rates of Activation Foils

See (Tables A2.22 and A2.23).

Table A2.22 Main characteristics and description of activation foils (Ref. [1])

Reaction	Dimension	Threshold (MeV)	$T_{1/2}$	γ-ray energy (keV)	Emission rate (%)
^{197}Au$(n, \gamma)^{198}$Au (bare and Cd*)	8 mm diam. 0.05 mm thick	–	2.697 d	411.9	95.51
Cd plate	10 mm diam. 1 mm thick	–	–	–	–
^{115}In$(n, \gamma)^{116m}$In (wire)	1 mm diam. 680 mm long	–	54.12 m	1097.3 1293.54	55.7 85
^{115}In$(n, n')^{115m}$In (foil)	10 × 10 × 1 mm	0.4	4.486 h	336.2	45.08
^{58}Ni$(n, p)^{58}$Co	10 × 10 × 1 mm	0.9	70.82 d	810.8	99.4
^{56}Fe$(n, p)^{56}$Mn	10 × 10 × 1 mm	5.0	2.578 h	846.8 1810.7	98.9 27.2
^{27}Al$(n, \alpha)^{24}$Na	10 × 10 × 1 mm	5.6	14.96 h	1368.6	100

Cd*: Au foil (Cd covered) was sandwiched between two Cd plates (10 mm diam. and 1 mm thick).

Table A2.23 Measured reaction rates of activation foils in Cases II-3 through II-6 (Ref. [1])

Reaction	Measured reaction rate [s^{-1} cm^{-3}]			
	Case II-3	Case II-4	Case II-5	Case II-6
^{197}Au$(n, \gamma)^{198}$Au (bare)	(8.88 ± 0.02) E+06	(4.88 ± 0.04) E+06	(3.51 ± 0.08) E+06	(2.53 ± 0.04) E+06
^{197}Au$(n, \gamma)^{198}$Au (Cd)	(7.84 ± 0.09) E+06	(4.46 ± 0.04) E+6	(3.11 ± 0.04) E+6	(2.30 ± 0.03) E+06
^{115}In$(n, n')^{115m}$In	(8.60 ± 0.13) E+04	(4.27 ± 0.22) E+04	(4.27 ± 0.03) E+04	(2.86 ± 0.06) E+04
^{58}Ni$(n, p)^{58}$Co	(4.90 ± 0.08) E+04	(3.18 ± 0.03) E+04	(3.23 ± 0.16) E+04	(2.00 ± 0.10) E+04
^{56}Fe$(n, p)^{56}$Mn	(1.82 ± 0.07) E+03	(1.26 ± 0.02) E+03	(1.55 ± 0.03) E+03	(1.39 ± 0.02) E+03
^{27}Al$(n, \alpha)^{24}$Na	(1.54 ± 0.05) E+03	(1.11 ± 0.02) E+03	(1.62 ± 0.03) E+03	(1.10 ± 0.01) E+03

References

1. Pyeon CH (2017) Experimental benchmarks of neutronics on solid Pb–Bi in accelerator-driven system with 100 MeV protons at Kyoto University Critical Assembly. KURRI-TR-447
2. Pyeon CH, Nakano H, Yamanaka M et al (2015) Neutron characteristics of solid targets in accelerator-driven system with 100 MeV protons at Kyoto University Critical Assembly. Nucl Technol 192: 181
3. Pyeon CH, Vu TM, Yamanaka M et al (2018) Reaction rate analyses of accelerator-driven system experiments with 100 MeV protons at Kyoto University Critical Assembly. J Nucl Sci Technol 55: 190

Appendix A3: ^{235}U-Fueled and Pb-Zoned ADS Core

A3.1 Core Configurations

See (Figs. A3.1, A3.2, A3.3, A3.4 and A3.5).

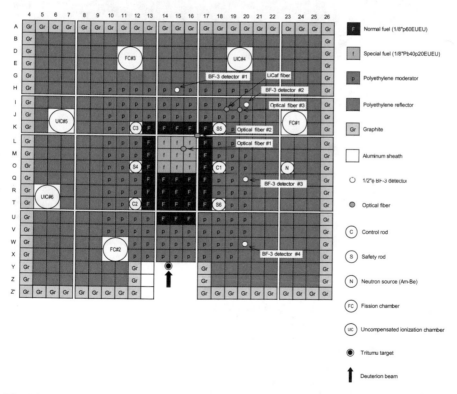

Fig. A3.1 Top view of U-Pb zoned core configuration with 14 MeV neutrons (Ref. [1])

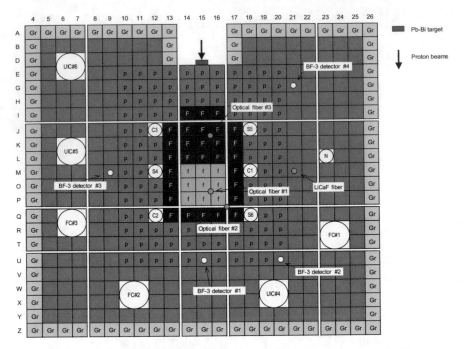

Fig. A3.2 Top view of U-Pb zoned core configuration with 100 MeV protons (Ref. [1])

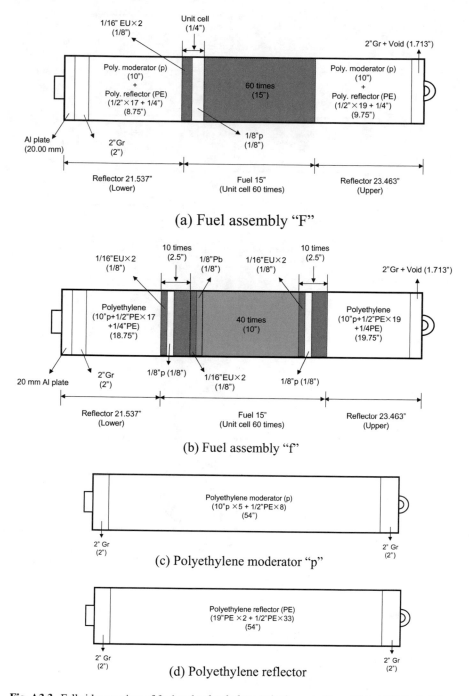

(a) Fuel assembly "F"

(b) Fuel assembly "f"

(c) Polyethylene moderator "p"

(d) Polyethylene reflector

Fig. A3.3 Fall sideway view of fuel and polyethylene rods shown in Figs. A3.1, A3.2 (Ref. [1])

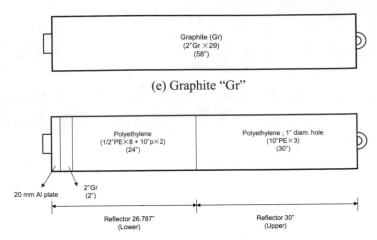

(e) Graphite "Gr"

(f) Polyethylene with 1" diameter hole

Fig. A3.3 (continued)

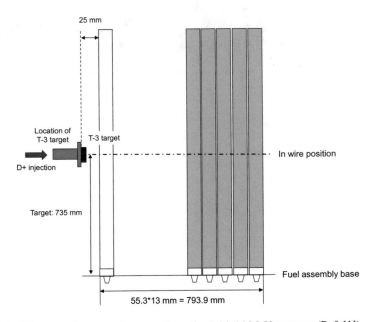

Fig. A3.4 Side view of target and core configuration with 14 MeV neutrons (Ref. [1])

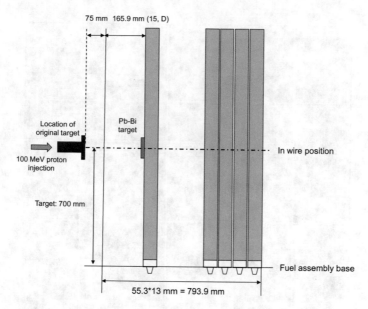

Fig. A3.5 Side view of target and core configuration with 100 MeV protons (Ref. [1])

A3.1.1 ADS with 14 MeV Neutrons

See (Fig. A3.6).

A3.1.2 ADS with 100 MeV Protons

See (Fig. A3.7).

A3.2 Kinetics Parameters

A3.2.1 ADS with 14 MeV Neutrons

A3.2.1.1 Core Condition at Critical State

See (Tables A3.1, A3.2 and A3.3).

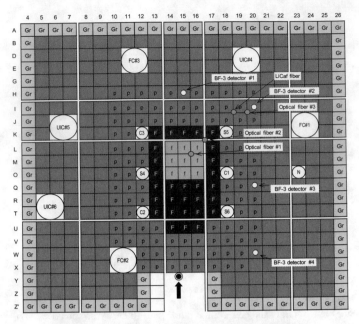

(a) Case D1 (Reference core; 4560 HEU plates)

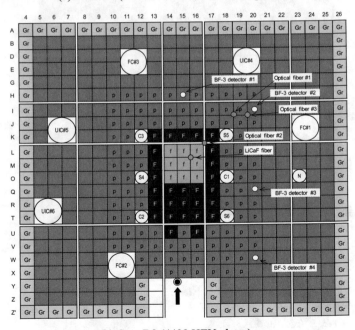

(b) Case D2 (4400 HEU plates)

Fig. A3.6 Subcritical core configurations of ADS with 14 MeV neutrons (Refs. [1–2]). *Note* Change of location of detectors (comparing Case D1 with Case D2, D3, A4, D5 and D6), Optical fiber #1: (15-16, L-M) → (18-19, I-J), LiCaF fiber: (18-19, I-J) → (15-16, L-M)

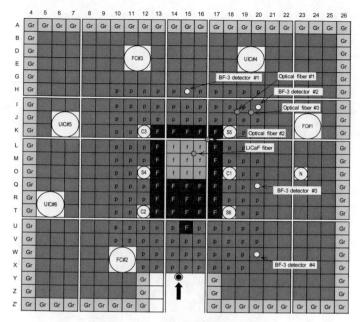

(c) Case D3 (4320 HEU plates)

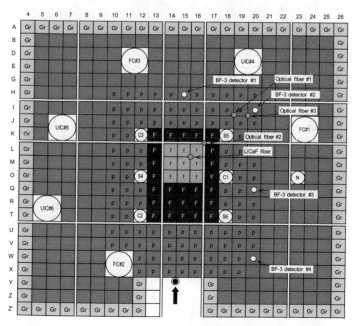

(d) Case D4 (4200 HEU plates)

Fig. A3.6 (continued)

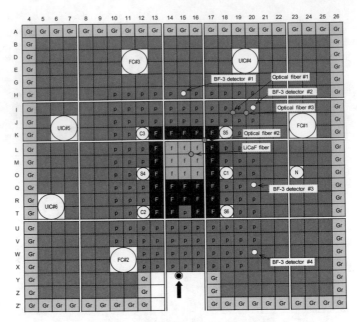

(e) Case D5 (4080 HEU plates)

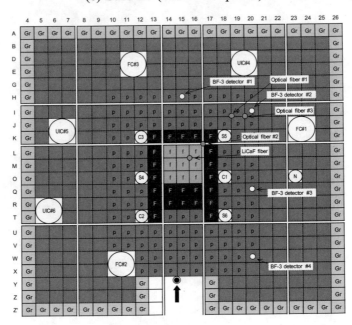

(f) Case D6 (3840 HEU plates)

Fig. A3.6 (continued)

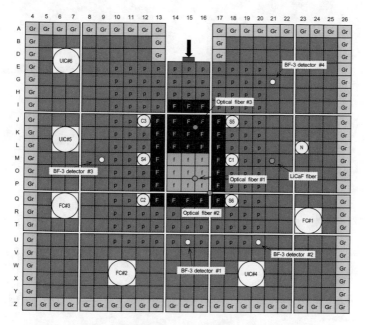

(a) Case F1 (Reference core; 4560 HEU plates)

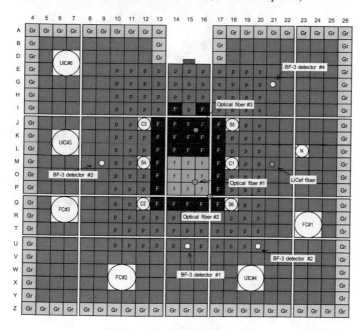

(b) Case F2 (4440 HEU plates)

Fig. A3.7 Subcritical core configurations of ADS with 100 MeV protons (Pb–Bi target) (Refs. [1–2])

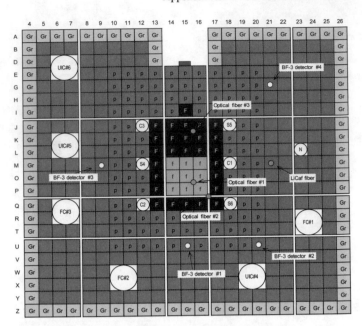

(c) Case F3 (Number of fuel plates: 4320)

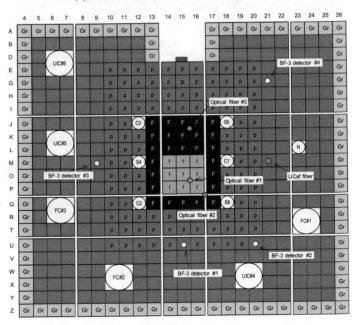

(d) Case F4 (4200 HEU plates)

Fig. A3.7 (continued)

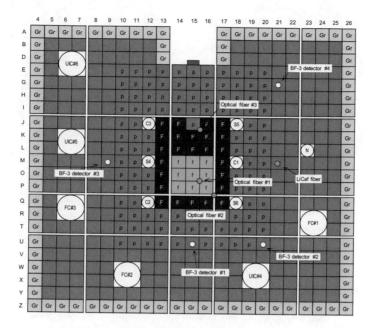

(e) Case F5 (4080 HEU plates)

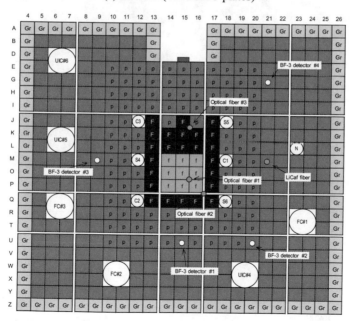

(f) Case F6 (3960 HEU plates)

Fig. A3.7 (continued)

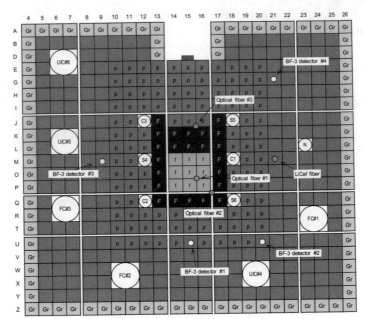

(g) Case F7 (3840 HEU plates)

Fig. A3.7 (continued)

Table A3.1 Control rod positions at critical state in reference core (4560 HEU plates; Case D1) (Ref. [1])

Rod	Rod position (mm)
C1	766.56
C2	1200.00
A3	1200.00
S4	1200.00
S5	1200.00
S6	1200.00
Excess reactivity [pcm]	80 ± 2

Table A3.2 Control rod (reactivity) worth (Ref. [1])

Rod	Rod worth [pcm]
C1 (S4)	902 ± 27
C2 (S6)	696 ± 21
A3 (S5)	232 ± 7

Table A3.3 Kinetic parameters by MCNP6.1 with JENDL-4.0

β_{eff}	853 ± 3 [pcm]
Λ	$(3.24 \pm 0.03) \times 10^{-5}$ (s)

A3.2.1.2 Case D1 (4560 HEU Plates)

See (Tables A3.4, A3.5, A3.6, A3.7, A3.8, A3.9 and A3.10).

Table A3.4 Core condition in Case D1-1 to Case D1-5 (4560 HEU plates) (Ref. [1])

Case	C1	C2	A3	S4	S5	S6	k_{eff}
D1-1	0.00	1200.00	1200.00	1200.00	1200.00	1200.00	0.99178
D1-2	0.00	1200.00	1200.00	1200.00	1200.00	1200.00	0.99178
D1-3	0.00	1200.00	1200.00	1200.00	1200.00	1200.00	0.99178
D1-4	0.00	1200.00	0.00	1200.00	1200.00	1200.00	0.98947
D1-5	0.00	0.00	0.00	1200.00	1200.00	1200.00	0.98225

Table A3.5 Beam characteristics of 14 MeV neutrons in Cases D1-1 to D1-5 (Ref. [1])

Case	Frequency [Hz]	Width (μs)	Current (mA)
D1-1	20	80	0.20
D1-2	50	80	0.20
D1-3	100	80	0.20
D1-4	20	80	0.20
D1-5	20	80	0.20

Table A3.6 Measured results of α [s^{-1}] and ρ [$] in Case D1-1 (Ref. [1])

Detector	α [s^{-1}]		ρ [$]	
	Feynman-α	α-fitting	Area (Sjostrand)	Area (Gozani)
BF-3 #1 (Ch#1)	422.96 ± 12.93	426.81 ± 28.34	0.846 ± 0.033	0.857 ± 0.120
BF-3 #2 (Ch#2)	497.90 ± 14.80	493.62 ± 31.92	0.868 ± 0.015	1.069 ± 0.119
BF-3 #3 (Ch#3)	486.88 ± 11.37	494.12 ± 21.71	0.902 ± 0.009	1.064 ± 0.079
BF-3 #4 (Ch#4)	562.18 ± 18.08	448.21 ± 53.07	0.915 ± 0.004	0.857 ± 0.141
Fiber #1 (Ch#5)	–	–	–	–
Fiber #2 (Ch#6)	–	402.58 ± 57.99	0.913 ± 0.001	0.744 ± 0.134
Fiber #3 (Ch#7)	–	–	–	–

Table A3.7 Measured results of α [s^{-1}] and ρ [$] in Case D1-2 (Ref. [1])

| Detector | α [s^{-1}] | | ρ [$] | |
	Feynman-α	α-fitting	Area (Sjostrand)	Area (Gozani)
BF-3 #1 (Ch#1)	458.03 ± 15.40	476.24 ± 6.51	0.853 ± 0.020	1.018 ± 0.030
BF-3 #2 (Ch#2)	494.00 ± 14.20	484.22 ± 6.17	0.887 ± 0.010	1.070 ± 0.016
BF-3 #3 (Ch#3)	481.88 ± 8.95	484.81 ± 4.34	0.850 ± 0.006	0.970 ± 0.009
BF-3 #4 (Ch#4)	835.59 ± 29.96	511.58 ± 11.18	0.776 ± 0.003	0.808 ± 0.011
Fiber #1 (Ch#5)	–	–	–	–
Fiber #2 (Ch#6)	501.50 ± 146.80	384.68 ± 53.93	0.776 ± 0.003	0.704 ± 0.042
Fiber #3 (Ch#7)	–	–	–	–

Table A3.8 Measured results of α [s^{-1}] and ρ [$] in Case D1-3 (Ref. [1])

| Detector | α [s^{-1}] | | ρ [$] | |
	Feynman-α	α-fitting	Area (Sjostrand)	Area (Gozani)
BF-3 #1 (Ch#1)	434.36 ± 123.56	521.50 ± 26.53	0.739 ± 0.014	0.918 ± 0.034
BF-3 #2 (Ch#2)	643.90 ± 99.97	511.42 ± 26.28	0.769 ± 0.007	0.939 ± 0.029
BF-3 #3 (Ch#3)	502.81 ± 92.37	544.23 ± 48.33	0.763 ± 0.004	0.918 ± 0.049
BF-3 #4 (Ch#4)	1639.61 ± 131.63	677.76 ± 359.00	0.698 ± 0.002	0.861 ±0.340
Fiber #1 (Ch#5)	–	–	–	–
Fiber #2 (Ch#6)	–	508.13 ± 88.37	0.696 ± 0.001	0.713 ± 0.069
LiCaF (Ch#7)	538.30 ± 16.81	–	0.604 ± 0.002	0.632 ± 0.012

Table A3.9 Measured results of α [s^{-1}] and ρ [$] in Case D1-4 (Ref. [1])

| Detector | α [s^{-1}] | | ρ [$] | |
	Feynman-α	α-fitting	Area (Sjostrand)	Area (Gozani)
BF-3 #1 (Ch#1)	565.29 ± 4.61	568.41 ± 9.41	1.118 ± 0.012	1.400 ± 0.024
BF-3 #2 (Ch#2)	589.68 ± 5.40	559.85 ± 9.51	1.130 ± 0.006	1.403 ± 0.018
BF-3 #3 (Ch#3)	555.32 ± 2.11	570.25 ± 9.56	1.158 ± 0.013	1.306 ± 0.024
BF-3 #4 (Ch#4)	–	514.20 ± 21.79	1.212 ± 0.005	1.076 ± 0.027
Fiber #1 (Ch#5)	–	–	–	–
Fiber #2 (Ch#6)	490.46 ± 41.67	521.64 ± 35.67	1.049 ± 0.001	1.245 ± 0.049
LiCaF (Ch#7)	1023.42±277.51	–	1.022 ± 0.001	2.107 ± 0.643

Table A3.10 Measured results of α [s^{-1}] and ρ [$] in Case D1-5 (Ref. [1])

Detector	α [s^{-1}]		ρ [$]	
	Feynman-α	α-fitting	Area (Sjostrand)	Area (Gozani)
BF-3 #1 (Ch#1)	498.46 ± 7.46	480.22 ± 5.36	0.899 ± 0.018	1.090 ± 0.028
BF-3 #2 (Ch#2)	489.43 ± 6.55	483.02 ± 5.09	0.887 ± 0.008	1.068 ± 0.014
BF-3 #3 (Ch#3)	488.59 ± 3.46	483.93 ± 3.76	0.890 ± 0.005	1.024 ± 0.009
BF-3 #4 (Ch#4)	–	497.04 ± 9.23	0.891 ± 0.002	0.951 ± 0.010
Fiber #1 (Ch#5)	–	–	–	–
Fiber #2 (Ch#6)	541.11 ± 127.67	709.57 ± 185.61	0.872 ± 0.001	1.181 ± 0.241
LiCaF (Ch#7)	–	–	–	–

A3.2.1.3 Case D2 (4400 HEU Plates)

See (Table A3.11, A3.12, A3.13, A3.14 and A3.15).

Table A3.11 Core condition in Cases D2-1 to D2-3 (4400 HEU plates) (Ref. [1])

Case	C1	C2	A3	S4	S5	S6	k_{eff}
D2-1	1200.00	1200.00	1200.00	1200.00	1200.00	1200.00	0.99328
D2-2	1200.00	1200.00	1200.00	1200.00	1200.00	1200.00	0.99328
D2-3	1200.00	1200.00	1200.00	1200.00	1200.00	1200.00	0.99328

Table A3.12 Beam characteristics of 14 MeV neutrons in Cases D2-1 to D2-3 (Ref. [1])

Case	Frequency (Hz)	Width (μs)	Current (mA)
D2-1	20	75	0.10
D2-2	50	80	0.20
D2-3	100	80	0.15

Table A3.13 Measured results of α [s^{-1}] and ρ [$] in Case D2-1 (Ref. [1])

Detector	α [s^{-1}]		ρ [$]	
	Feynman-α	α-fitting	Area (Sjostrand)	Area (Gozani)
BF-3 #1 (Ch#1)	376.67 ± 8.46	370.33 ± 4.94	0.565 ± 0.010	0.652 ± 0.014
BF-3 #2 (Ch#2)	389.96 ± 7.21	370.56 ± 4.21	0.578 ± 0.004	0.665 ± 0.007
BF-3 #3 (Ch#3)	370.68 ± 2.88	369.87 ± 2.84	0.587 ± 0.003	0.437 ± 0.002
BF-3 #4 (Ch#4)	501.71 ± 11.71	376.97 ± 8.13	0.612 ± 0.001	0.647 ± 0.006
Fiber #1 (Ch#5)	–	–	0.585 ± 0.001	–
Fiber #2 (Ch#6)	703.56 ± 71.90	–	0.570 ± 0.001	0.861 ± 0.068
Fiber #3 (Ch#7)	–	–	–	–

Table A3.14 Measured results of α [s^{-1}] and ρ [$] in Case D2-2 (Ref. [1])

| Detector | α [s^{-1}] | | ρ [$] | |
	Feynman-α	α-fitting	Are (Sjostrand)	Area (Gozani)
BF-3 #1 (Ch#1)	370.30 ± 15.26	378.35 ± 5.69	0.589 ± 0.006	0.691 ± 0.009
BF-3 #2 (Ch#2)	431.86 ± 12.01	376.62 ± 4.97	0.582 ± 0.003	0.676 ± 0.005
BF-3 #3 (Ch#3)	367.38 ± 7.31	369.18 ± 3.58	0.595 ± 0.002	0.666 ± 0.003
BF-3 #4 (Ch#4)	–	365.64 ± 12.33	0.637 ± 0.001	0.689 ± 0.009
Fiber #1 (Ch#5)	–	–	0.630 ± 0.001	0.455 ± 0.001
Fiber #2 (Ch#6)	–	386.76 ± 22.42	0.629 ± 0.001	0.696 ± 0.017
Fiber #3 (Ch#7)	–	–	–	–

Table A3.15 Measured results of α [s^{-1}] and ρ [$] in Case D2-3 (Ref. [1])

| Detector | α [s^{-1}] | | ρ [$] | |
	Feynman-α	α-fitting	Area (Sjostrand)	Area (Gozani)
BF-3 #1 (Ch#1)	249.68 ± 154.15	367.36 ± 9.62	0.484 ± 0.009	0.550 ± 0.014
BF-3 #2 (Ch#2)	789.84 ± 57.23	359.50 ± 8.38	0.479 ± 0.004	0.536 ± 0.007
BF-3 #3 (Ch#3)	384.44 ± 37.91	372.34 ± 4.49	0.466 ± 0.003	0.506 ± 0.004
BF-3 #4 (Ch#4)	–	398.44 ± 15.90	0.507 ± 0.001	0.558 ± 0.010
Fiber #1 (Ch#5)	–	–	0.504 ± 0.001	0.358 ± 0.0001
Fiber #2 (Ch#6)	–	384.18 ± 29.34	0.503 ± 0.001	0.546 ± 0.018
Fiber #3 (Ch#7)	–	–	–	–

A3.2.1.4 Case D3 (4320 HEU Plates)

See (Tables A3.16, A3.17 and A3.18).

Table A3.16 Core condition in Case D3-1 (4320 HEU plates) (Ref. [1])

Case	C1	C2	A3	S4	S5	S6	k_{eff}
D3-1	1200.00	1200.00	1200.00	1200.00	1200.00	1200.00	0.98004

Table A3.17 Beam characteristics of 14 MeV neutrons in Case D3-1 (Ref. [1])

Case	Frequency (Hz)	Width (μs)	Current (mA)
D3-1	20	80	0.20

Table A3.18 Measured results of α [s^{-1}] and ρ [$] in Case D3-1 (Ref. [1])

Detector	α [s^{-1}]		ρ [$]	
	Feynman-α	α-fitting	Area (Sjostrand)	Area (Gozani)
BF-3 #1 (Ch#1)	669.23 ± 5.43	678.46 ± 4.75	1.911 ± 0.024	2.589 ± 0.041
BF-3 #2 (Ch#2)	693.11 ± 4.60	681.85 ± 4.28	1.880 ± 0.011	2.610 ± 0.022
BF-3 #3 (Ch#3)	684.25 ± 1.93	686.16 ± 2.50	1.972 ± 0.007	2.663 ± 0.013
BF-3 #4 (Ch#4)	1296.59 ± 35.04	715.47 ± 9.24	2.244 ± 0.002	2.757 ± 0.028
Fiber #1 (Ch#5)	–	–	–	–
Fiber #2 (Ch#6)	703.56 ± 71.90	–	1.916 ± 0.001	2.323 ± 0.184
LiCaF (Ch#7)	–	–	–	–

A3.2.1.5 Case D4 (4200 HEU Plates)

See (Tables A3.19, A3.20 and A3.21).

Table A3.19 Core condition in Case A4-1 (4200 HEU plates) (Ref. [1])

Case	C1	C2	A3	S4	S5	S6	k_{eff}
A4-1	1200.00	1200.00	1200.00	1200.00	1200.00	1200.00	0.96603

Table A3.20 Beam characteristics of 14 MeV neutrons in Case A4-1 (Ref. [1])

Case	Frequency (Hz)	Width (μs)	Current [mA]
A4-1	10	80	0.20

Table A3.21 Measured results of α [s^{-1}] and ρ [$] in Case A4-1 (Ref. [1])

Detector	α [s^{-1}]		ρ [$]	
	Feynman-α	α-fitting	Area (Sjostrand)	Area (Gozani)
BF-3 #1 (Ch#1)	878.01 ± 26.55	969.40 ± 33.08	2.687 ± 0.116	3.618 ± 0.242
BF-3 #2 (Ch#2)	979.74 ± 27.61	970.87 ± 26.68	2.785 ± 0.055	3.789 ± 0.147
BF-3 #3 (Ch#3)	997.15 ± 13.85	1045.50 ± 14.28	2.843 ± 0.034	3.793 ± 0.084
BF-3 #4 (Ch#4)	2345.71 ± 86.12	1265.82 ± 92.59	2.924 ± 0.019	3.929 ± 0.402
Fiber #1 (Ch#5)	–	–	–	–
Fiber #2 (Ch#6)	–	–	–	–
LiCaF (Ch#7)	–	–	–	–

A3.2.1.6 Case D5 (4080 HEU Plates)

See (Tables A3.22, A3.23, A3.24, A3.25 and A3.26).

Table A3.22 Core condition in Cases D5-1 to Case D5-3 (4080 HEU plates) (Ref. [1])

Case	C1	C2	A3	S4	S5	S6	k_{eff}
D5-1	1200.00	1200.00	1200.00	1200.00	1200.00	1200.00	0.95560
D5-2	1200.00	1200.00	1200.00	1200.00	1200.00	1200.00	0.95560
D5-3	1200.00	1200.00	1200.00	1200.00	1200.00	1200.00	0.95560

Table A3.23 Beam characteristics of 14 MeV neutrons in Cases D5-1 to D5-3 (Ref. [1])

Case	Frequency (Hz)	Width (μs)	Current (mA)
D5-1	20	90	0.20
D5-2	50	90	0.20
D5-3	100	90	0.20

Table A3.24 Measured results of α [s^{-1}] and ρ [$] in Case D5-1 (Ref. [1])

| Detector | α [s^{-1}] | | ρ [$] | |
	Feynman-α	α-fitting	Area (Sjostrand)	Area (Gozani)
BF-3 #1 (Ch#1)	1143.72 ± 17.67	1150.34 ± 15.00	3.641 ± 0.084	5.436 ± 0.178
BF-3 #2 (Ch#2)	1203.76 ± 15.21	1148.70 ± 13.09	3.642 ± 0.038	5.329 ± 0.103
BF-3 #3 (Ch#3)	1188.02 ± 6.15	1186.95 ± 9.38	3.608 ± 0.023	4.849 ± 0.064
BF-3 #4 (Ch#4)	2957.04 ± 86.78	1291.39 ± 30.21	2.618 ± 0.012	2.543 ± 0.087
Fiber #1 (Ch#5)	–	–	1.948 ± 0.006	0.457 ± 0.003
Fiber #2 (Ch#6)	–	–	1.900 ± 0.002	0.449 ± 0.001
Fiber #3 (Ch#7)	–	–	–	–

Table A3.25 Measured results of α [s^{-1}s] and ρ [$] in Case D5-2 (Ref. [1])

| Detector | α [s^{-1}] | | ρ [$] | |
	Feynman-α	α-fitting	Area (Sjostrand)	Area (Gozani)
BF-3 #1 (Ch#1)	1093.68 ± 23.47	1167.71 ± 14.40	3.957 ± 0.093	6.047 ± 0.192
BF-3 #2 (Ch#2)	1193.46 ± 17.93	1162.44 ± 12.44	4.073 ± 0.044	6.133 ± 0.114
BF-3 #3 (Ch#3)	1138.36 ± 8.72	1185.21 ± 9.52	3.596 ± 0.024	4.733 ± 0.063
BF-3 #4 (Ch#4)	–	1337.28 ± 29.46	2.558 ± 0.011	2.303 ± 0.077
Fiber #1 (Ch#5)	–	–	2.260 ± 0.005	–
Fiber #2 (Ch#6)	823.99 ± 179.48	–	2.187 ± 0.003	0.451 ± 0.001
Fiber #3 (Ch#7)	–	–	–	–

Table A3.26 Measured results of α [s^{-1}] and ρ [$] in Case D5-3 (Ref. [1])

| Detector | α [s^{-1}] | | ρ [$] | |
	Feynman-α	α-fitting	Area (Sjostrand)	Area (Gozani)
BF-3 #1 (Ch#1)	959.79 ± 48.45	1160.10 ± 17.13	2.710 ± 0.123	3.766 ± 0.224
BF-3 #2 (Ch#2)	1215.05 ± 44.44	1166.86 ± 14.88	2.585 ± 0.053	3.460 ± 0.108
BF-3 #3 (Ch#3)	1105.17 ± 16.49	1174.34 ± 10.96	1.493 ± 0.020	1.086 ± 0.040
BF-3 #4 (Ch#4)	3583.95±186.81	1352.44 ± 42.55	0.333 ± 0.007	–
Fiber #1 (Ch#5)	–	–	0.324 ± 0.001	–
Fiber #2 (Ch#6)	–	1157.06 ± 80.98	0.328 ± 0.001	–
Fiber #3 (Ch#7)	–	–	–	–

A3.2.1.7 Case D6 (3840 HEU Plates)

See (Table A3.27, A3.28, A3.29, A3.30 and A3.31).

Table A3.27 Core condition in Cases D6-1 to Case D6-3 (3840 HEU plates) (Ref. [1])

Case	C1	C2	A3	S4	S5	S6	k_{eff}
D6-1	1200.00	1200.00	1200.00	1200.00	1200.00	1200.00	0.93000
D6-2	1200.00	1200.00	1200.00	1200.00	1200.00	1200.00	0.93000
D6-3	1200.00	1200.00	1200.00	1200.00	1200.00	1200.00	0.93000

Table A3.28 Beam characteristics of 14 MeV neutrons in Cases D6-1 to D6-3 (Ref. [1])

Case	Frequency (Hz)	Width (μs)	Current (mA)
D6-1	20	90	0.20
D6-2	50	90	0.20
D6-3	100	80	0.15

Table A3.29 Measured results of α [s^{-1}] and ρ [$] in Case D6-1 (Ref. [1])

| Detector | α [s^{-1}] | | ρ [$] | |
	Feynman-α	α-fitting	Area (Sjostrand)	Area (Gozani)
BF-3 #1 (Ch#1)	1493.13 ± 25.33	1604.09 ± 19.04	4.797 ± 0.122	8.069 ± 0.315
BF-3 #2 (Ch#2)	1635.64 ± 42.06	1734.05 ± 63.78	2.622 ± 0.033	5.053 ± 0.336
BF-3 #3 (Ch#3)	1641.15 ± 8.91	1686.35 ± 13.28	3.182 ± 0.024	3.992 ± 0.082
BF-3 #4 (Ch#4)	4166.57 ± 77.36	2118.77 ± 64.65	2.431 ± 0.012	–
Fiber #1 (Ch#5)	–	–	1.449 ± 0.006	–
Fiber #2 (Ch#6)	–	–	1.410 ± 0.001	–
Fiber #3 (Ch#7)	–	–	–	–

Table A3.30 Measured results of α [s^{-1}] and ρ [$] in Case D6-2 (Ref. [1])

| Detector | α [s^{-1}] | | ρ [$] | |
	Feynman-α	α-fitting	Area (Sjostrand)	Area (Gozani)
BF-3 #1 (Ch#1)	1671.52 ± 79.63	1672.67 ± 58.63	5.067 ± 0.289	9.059 ± 0.871
BF-3 #2 (Ch#2)	1790.58±101.01	1694.67 ± 43.44	4.847 ± 0.125	8.255 ± 0.483
BF-3 #3 (Ch#3)	1621.79 ± 48.81	1695.97 ± 29.62	4.070 ± 0.062	4.920 ± 0.203
BF-3 #4 (Ch#4)	4288.55±202.00	2125.10 ± 147.31	2.562 ± 0.027	0.015 ± 0.096
Fiber #1 (Ch#5)	–	–	2.105 ± 0.012	–
Fiber #2 (Ch#6)	–	–	2.042 ± 0.005	–
Fiber #3 (Ch#7)	–	–	–	–

Table A3.31 Measured results of α [s^{-1}] and ρ [$] in Case D6-3 (Ref. [1])

| Detector | α [s^{-1}] | | ρ [$] | |
	Feynman-α	α-fitting	Area (Sjostrand)	Area (Gozani)
BF-3 #1 (Ch#1)	1409.00 ± 46.46	1640.90 ± 18.09	4.329 ± 0.132	5.762 ± 0.251
BF-3 #2 (Ch#2)	1526.97±131.68	1666.31 ± 28.51	4.163 ± 0.058	5.448 ± 0.186
BF-3 #3 (Ch#3)	1640.72 ± 12.49	1711.10 ± 12.79	1.869 ± 0.019	0.047 ± 0.054
BF-3 #4 (Ch#4)	4356.49±139.02	2083.91 ± 46.16	0.596 ± 0.005	–
Fiber #1 (Ch#5)	–	–	0.557 ± 0.002	–
Fiber #2 (Ch#6)	–	–	0.553 ± 0.001	–
Fiber #3 (Ch#7)	–	–	–	–

A3.2.2 ADS with 100 MeV Protons

A3.2.2.1 Core Condition at Critical State

See (Tables A3.32, A3.33, A3.34 and A3.35).

Table A3.32 Control rod positions at the critical state in reference core (4560 HEU plates; Case F1) (Ref. [1])

Rod	Rod position [mm]
C1	786.14
C2	1200.00
A3	1200.00
S4	1200.00
S5	1200.00
S6	1200.00
Excess reactivity [pcm]	37 ± 1

Table A3.33 Control rod worth (Ref. [1])

Rod	Rod worth [pcm]
C1 (S4)	902 ± 27
C2 (S6)	696 ± 21
A3 (S5)	232 ± 7

Table A3.34 Kinetic parameters by MCNP6.1 with JENDL-4.0 (Ref. [1])

β_{eff}	853 ± 3 [pcm]
Λ	$(3.24 \pm 0.03) \times 10^{-5}$ (s)

Table A3.35 Proton beam characteristics obtained from FFAG accelerator (Ref. [1])

Energy [MeV]	Frequency (Hz)	Repetition (ns)	Current (nA)
100	20	100	0.05

A3.2.2.2 Case F1 (4560 HEU Plates)

See (Tables A3.36, A3.37, A3.38, A3.39, A3.40, A3.41 and A3.42).

Table A3.36 Core condition in Cases F1-1 to F1-6 (4560 HEU plates) (Ref. [1])

Case	C1	C2	A3	S4	S5	S6	k_{eff}
F1-1	0.00	0.00	0.00	1200.00	1200.00	1200.00	0.98208
F1-2	0.00	1200.00	1200.00	1200.00	1200.00	1200.00	0.99135
F1-3	0.00	0.00	1200.00	1200.00	1200.00	1200.00	0.98439
F1-4	0.00	0.00	1200.00	1200.00	1200.00	1200.00	0.98208
F1-5	0.00	0.00	1200.00	1200.00	1200.00	0.00	0.97744
F1-6	0.00	0.00	1200.00	0.00	1200.00	0.00	0.96842

Table A3.37 Measured results of α [s^{-1}] and ρ [\$] in Case F1-1 (Ref. [1])

Detector	α [s^{-1}]		ρ [\$]	
	Feynman-α	α-fitting	Area (Sjostrand)	Area (Gozani)
BF-3 #5 (Ch#1)	805.90 ± 7.13	775.65 ± 6.01	1.834 ± 0.004	1.529 ± 0.005
BF-3 #2 (Ch#2)	793.06 ± 7.12	791.55 ± 6.71	1.949 ± 0.018	1.921 ± 0.020
BF-3 #3 (Ch#3)	807.48 ± 3.88	803.10 ± 3.07	2.048 ± 0.012	1.952 ± 0.013
BF-3 #4 (Ch#4)	1145.43 ± 17.85	868.30 ± 11.13	1.801 ± 0.005	1.501 ± 0.008
BF-3 #1 (Ch#5)	771.43 ± 6.92	786.80 ± 6.76	1.964 ± 0.037	1.941 ± 0.040

Table A3.38 Measured results of α [s^{-1}] and ρ [$] in Case F1-2 (Ref. [1])

Detector	α [s^{-1}]		ρ [$]	
	Feynman-α	α-fitting	Area (Sjostrand)	Area (Gozani)
BF-3 #5 (Ch#1)	494.19 ± 2.79	479.32 ± 3.11	0.895 ± 0.001	0.798 ± 0.002
BF-3 #2 (Ch#2)	483.23 ± 3.19	481.73 ± 3.31	0.968 ± 0.006	0.953 ± 0.006
BF-3 #3 (Ch#3)	488.39 ± 1.90	484.30 ± 1.59	0.970 ± 0.004	0.931 ± 0.004
BF-3 #4 (Ch#4)	639.45 ± 8.43	516.73 ± 4.89	0.884 ± 0.001	0.783 ± 0.002
BF-3 #1 (Ch#5)	496.11 ± 2.97	478.76 ± 3.39	0.973 ± 0.011	0.963 ± 0.012

Table A3.39 Measured results of α [s^{-1}] and ρ [$] in Case F1-3 (Ref. [1])

Detector	α [s^{-1}]		ρ [$]	
	Feynman-α	α-fitting	Area (Sjostrand)	Area (Gozani)
BF-3 #5 (Ch#1)	541.28 ± 13.64	539.51 ± 10.25	1.310 ± 0.007	1.185 ± 0.008
BF-3 #2 (Ch#2)	544.49 ± 15.24	548.34 ± 10.76	1.194 ± 0.025	1.174 ± 0.028
BF-3 #3 (Ch#3)	555.53 ± 9.39	555.50 ± 4.92	1.210 ± 0.016	1.161 ± 0.017
BF-3 #4 (Ch#4)	704.34 ± 16.18	634.24 ± 15.92	1.320 ± 0.005	1.216 ± 0.009
BF-3 #1 (Ch#5)	542.98 ± 13.16	537.86 ± 10.37	1.209 ± 0.050	1.182 ± 0.055

Table A3.40 Measured results of α [s^{-1}] and ρ [$] in Case F1-4 (Ref. [1])

Detector	α [s^{-1}]		ρ [$]	
	Feynman-α	α-fitting	Area (Sjostrand)	Area (Gozani)
BF-3 #1 (Ch#1)	–	–	–	–
BF-3 #2 (Ch#2)	565.46 ± 6.56	548.85 ± 5.90	1.212 ± 0.024	1.199 ± 0.026
BF-3 #3 (Ch#3)	557.91 ± 4.08	557.84 ± 2.66	1.208 ± 0.009	1.160 ± 0.010
BF-3 #4 (Ch#4)	739.66 ± 10.29	645.39 ± 10.33	1.336 ± 0.003	1.214 ± 0.005
Fiber #1 (Ch#5)	–	–	–	–
Fiber #2 (Ch#6)	–	–	–	–
Fiber #3 (Ch#7)	541.36 ± 19.29	579.08 ± 12.96	1.314 ± 0.002	1.162 ± 0.006

Table A3.41 Measured results of α [s^{-1}] and ρ [$] in Case F1-5 (Ref. [1])

Detector	α [s^{-1}]		ρ [$]	
	Feynman-α	α-fitting	Area (Sjostrand)	Area (Gozani)
BF-3 #1 (Ch#1)	–	–	–	–
BF-3 #2 (Ch#2)	611.43 ± 5.24	612.90 ± 4.72	1.464 ± 0.025	1.421 ± 0.026
BF-3 #3 (Ch#3)	623.65 ± 2.41	621.46 ± 2.29	1.456 ± 0.008	1.397 ± 0.009
BF-3 #4 (Ch#4)	811.32 ± 11.35	704.50 ± 9.52	1.623 ± 0.003	1.458 ± 0.006
Fiber #1 (Ch#5)	–	–	–	–
Fiber #2 (Ch#6)	–	–	–	–
Fiber #3 (Ch#7)	658.73 ± 11.57	651.85 ± 8.92	1.613 ± 0.002	1.396 ± 0.005

Table A3.42 Measured results of α [s^{-1}] and ρ [$] in Case F1-6 (Ref. [1])

Detector	α [s^{-1}]		ρ [$]	
	Feynman-α	α-fitting	Area (Sjostrand)	Area (Gozani)
BF-3 #1 (Ch#1)	–	–	–	–
BF-3 #2 (Ch#2)	867.68 ± 17.35	894.89 ± 14.37	2.181 ± 0.087	2.111 ± 0.094
BF-3 #3 (Ch#3)	903.75 ± 10.95	911.20 ± 6.54	2.414 ± 0.038	2.266 ± 0.039
BF-3 #4 (Ch#4)	1351.25 ± 29.42	1041.78 ± 29.28	2.909 ± 0.016	2.441 ± 0.029
Fiber #1 (Ch#5)	–	–	–	–
Fiber #2 (Ch#6)	–	–	–	–
Fiber #3 (Ch#7)	1000.60 ± 49.77	908.43 ± 27.12	2.758 ± 0.012	2.196 ± 0.023

A3.2.2.3 Case F2 (4440 HEU Plates)

See (Tables A3.43 and A3.44).

Table A3.43 Core condition in Case F2-1 (4440 HEU plates) (Ref. [1])

Case	C1	C2	A3	S4	S5	S6	k_{eff}
F2-1	1200.00	1200.00	1200.00	1200.00	1200.00	1200.00	0.99328

Table A3.44 Measured results of α [s^{-1}] and ρ [$] in Case F2-1 (Ref. [1])

Detector	α [s^{-1}]		ρ [$]	
	Feynman-α	α-fitting	Area (Sjostrand)	Area (Gozani)
BF-3 #1 (Ch#1)	378.35 ± 4.00	378.90 ± 2.26	0.638 ± 0.003	0.639 ± 0.004
BF-3 #2 (Ch#2)	368.72 ± 4.36	375.31 ± 2.10	0.637 ± 0.002	0.630 ± 0.002
BF-3 #3 (Ch#3)	371.42 ± 3.98	377.47 ± 1.05	0.647 ± 0.002	0.626 ± 0.001
BF-3 #4 (Ch#4)	517.43 ± 10.05	377.55 ± 3.46	1.399 ± 0.006	0.631 ± 0.001
LiCaF (Ch#5)	371.82 ± 3.89	382.46 ± 1.72	0.644 ± 0.002	0.635 ± 0.001
Fiber #1 (Ch#6)	194.66 ± 220.45	–	–	0.594 ± 0.046
Fiber #2 (Ch#7)	472.97 ± 74.56	–	–	0.655 ± 0.017

A3.2.2.4 Case F3 (4320 HEU Plates)

See (Tables A3.45 and A3.46).

Table A3.45 Core condition in Case F3-1 (4320 HEU plates) (Ref. [1])

Case	C1	C2	A3	S4	S5	S6	k_{eff}
F3-1	1200.00	1200.00	1200.00	1200.00	1200.00	1200.00	0.98004

Table A3.46 Measured results of α [s^{-1}] and ρ [$] in Case F3-1 (Ref. [1])

| Detector | α [s^{-1}] | | ρ [$] | |
	Feynman-α	α-fitting	Area (Sjostrand)	Area (Gozani)
BF-3 #1 (Ch#1)	670.37 ± 4.72	697.62 ± 5.30	2.043 ± 0.012	2.092 ± 0.018
BF-3 #2 (Ch#2)	690.57 ± 9.73	691.99 ± 5.10	2.029 ± 0.010	2.063 ± 0.010
BF-3 #3 (Ch#3)	684.02 ± 4.10	699.73 ± 2.81	2.128 ± 0.007	2.047 ± 0.006
BF-3 #4 (Ch#4)	1255.70 ± 32.52	706.39 ± 8.95	8.132 ± 0.072	2.092 ± 0.007
LiCaF (Ch#5)	666.09 ± 4.93	698.59 ± 4.42	2.111 ± 0.010	2.096 ± 0.004
Fiber #3 (Ch#6)	725.66 ± 7.50	677.05 ± 12.28	3.502 ± 0.036	2.064 ± 0.009
Fiber #2 (Ch#7)	700.66 ± 115.19	–	–	2.075 ± 0.084

A3.2.2.5 Case F4 (4200 HEU Plates)

See (Tables A3.47, A3.48 and A3.49).

Table A3.47 Core condition in Cases F4-1 and F4-2 (4200 HEU plates) (Ref. [1])

Case	C1	C2	A3	S4	S5	S6	k_{eff}
F4-1	1200.00	1200.00	1200.00	1200.00	1200.00	1200.00	0.96603
F4-2	1200.00	1200.00	1200.00	1200.00	1200.00	1200.00	0.96603

Table A3.48 Measured results of α [s^{-1}] and ρ [$] in Case F4-1 (Ref. [1])

| Detector | α [s^{-1}] | | ρ [$] | |
	Feynman-α	α-fitting	Area (Sjostrand)	Area (Gozani)
BF-3 #1 (Ch#1)	981.43 ± 7.00	1047.39 ± 10.08	3.693 ± 0.024	3.756 ± 0.038
BF-3 #2 (Ch#2)	983.15 ± 8.05	1036.24 ± 10.26	3.661 ± 0.021	3.670 ± 0.022
BF-3 #3 (Ch#3)	1001.46 ± 4.15	1035.20 ± 5.49	3.921 ± 0.014	3.615 ± 0.013
BF-3 #4 (Ch#4)	2410.26 ± 60.24	1046.45 ± 19.88	27.699 ± 0.317	3.781 ± 0.026
LiCaF (Ch#5)	963.39 ± 5.85	1038.53 ± 8.12	3.789 ± 0.020	3.780 ± 0.012
Fiber #3 (Ch#6)	1059.11 ± 10.46	1019.33 ± 27.46	7.523 ± 0.086	3.699 ± 0.036
Fiber #2 (Ch#7)	1436.49±187.86	–	–	4.255 ± 0.280

Table A3.49 Measured results of α [s^{-1}] and ρ [$] in Case F4-2 (Ref. [1])

Detector	α [s^{-1}]		ρ [$]	
	Feynman-α	α-fitting	Area (Sjostrand)	Area (Gozani)
LiCaF (Ch#1)	950.13 ± 3.00	987.08 ± 3.65	3.816 ± 0.029	4.283 ± 0.037
BF-3 #2 (Ch#2)	984.69 ± 3.01	1014.65 ± 3.29	3.823 ± 0.014	4.323 ± 0.020
FC #2 (Ch#3)	944.57 ± 4.79	974.36 ± 6.49	3.973 ± 0.009	4.319 ± 0.023
FC #3 (Ch#4)	989.59 ± 3.61	997.24 ± 6.52	4.084 ± 0.005	4.440 ± 0.023
Fiber #1 (Ch#5)	–	–	–	–
Fiber #2 (Ch#6)	1125.74 ±52.42	1123.54 ± 45.39	3.772 ± 0.005	4.491 ± 0.153
Fiber #3 (Ch#7)	1412.79 ±68.42	–	–	–

A3.2.2.6 Case F5 (4080 HEU Plates)

See (Tables A3.50 and A3.51).

Table A3.50 Core condition in Case F5-1 (# of fuel plates: 4080) (Ref. [1])

Case	C1	C2	A3	S4	S5	S6	k_{eff}
F5-1	1200.00	1200.00	1200.00	1200.00	1200.00	1200.00	0.95560

Table A3.51 Measured results of α [s^{-1}] and ρ [$] in Case F5-1 (Ref. [1])

Detector	α [s^{-1}]		ρ [$]	
	Feynman-α	α-fitting	Area (Sjostrand)	Area (Gozani)
BF-3 #1 (Ch#1)	1100.61 ± 3.91	1142.68 ± 11.03	4.858 ± 0.137	5.441 ± 0.174
BF-3 #2 (Ch#2)	1110.85 ± 4.07	1143.44 ± 10.66	4.834 ± 0.067	5.332 ± 0.092
BF-3 #3 (Ch#3)	1114.54 ± 2.81	1167.77 ± 7.55	5.008 ± 0.049	5.388 ± 0.066
BF-3 #4 (Ch#4)	2632.74 ±26.57	2179.82 ±535.32	–	–
LiCaF (Ch#5)	1667.01 ±3.83	1131.01 ± 10.00	6.468 ± 0.034	6.676 ± 0.062
Fiber #2 (Ch#6)	1913.19 ±152.93	1494.85 ±130.19	6.346 ± 0.010	8.592 ± 0.839
Fiber #3 (Ch#7)	1299.97 ±87.56	1425.24 ±158.51	6.264 ± 0.008	8.043 ± 0.956

A3.2.2.7 Case F6 (3960 HEU Plates)

See (Tables A3.52 and A3.53).

Table A3.52 Core condition in Case F6-1 (3960 HEU plates) (Ref. [1])

Case	C1	C2	A3	S4	S5	S6	k_{eff}
F6-1	1200.00	1200.00	1200.00	1200.00	1200.00	1200.00	0.95047

Table A3.53 Measured results of α [s^{-1}] and ρ [$] in Case F6-1 (Ref. [1])

Detector	α [s^{-1}]		ρ [$]	
	Feynman-α	α-fitting	Area (Sjostrand)	Area (Gozani)
BF-3 #1 (Ch#1)	1147.81 ± 3.98	1214.68 ± 5.09	5.306 ± 0.061	6.025 ± 0.079
BF-3 #2 (Ch#2)	1170.63 ± 4.25	1227.65 ± 4.73	5.354 ± 0.030	6.023 ± 0.043
BF-3 #3 (Ch#3)	1178.31 ± 3.63	1236.21 ± 3.33	5.556 ± 0.022	5.973 ± 0.030
BF-3 #4 (Ch#4)	2810.88 ±24.95	1616.85 ±23.53	7.888 ± 0.010	8.108 ± 0.144
LiCaF (Ch#5)	1127.06 ± 4.03	–	–	–
Fiber #2 (Ch#6)	1198.27 ± 70.04	1301.00 ± 39.65	7.846 ± 0.004	6.366 ± 0.189
Fiber #3 (Ch#7)	1304.92 ± 40.08	–	–	–

A3.2.2.8 Case F7 (3840 HEU Plates)

See (Tables A3.54, A3.55 and A3.56).

Table A3.54 Core condition in Cases F7-1 and F7-2 (3840 HEU plates) (Ref. [1])

Case	C1	C2	A3	S4	S5	S6	k_{eff}
F7-1	1200.00	1200.00	1200.00	1200.00	1200.00	1200.00	0.92509
F7-2	1200.00	1200.00	1200.00	1200.00	1200.00	1200.00	0.92509

Table A3.55 Measured results of α [s^{-1}] and ρ [$] in Case F7-1 (Ref. [1])

Detector	α [s^{-1}]		ρ [$]	
	Feynman-α	α-fitting	Area (Sjostrand)	Area (Gozani)
BF-3 #1 (Ch#1)	1486.99 ± 6.33	1642.83 ± 9.39	8.718 ± 0.167	9.542 ± 0.209
BF-3 #2 (Ch#2)	1526.32 ± 5.96	1663.99 ± 9.55	8.364 ± 0.076	9.003 ± 0.111
BF-3 #3 (Ch#3)	1565.34 ± 4.34	1675.08 ± 7.43	8.932 ± 0.058	8.849 ± 0.080
BF-3 #4 (Ch#4)	3620.56 ± 17.16	2546.56 ±43.44	15.485 ± 0.029	17.614 ± 0.575
LiCaF (Ch#5)	1445.92 ± 5.88	1606.93 ± 8.73	14.215 ± 0.049	8.930 ± 0.068
Fiber #2 (Ch#6)	1972.45 ±169.96	–	–	–
Fiber #3 (Ch#7)	1723.10 ±113.58	–	–	–

Table A3.56 Measured results of α [s^{-1}] and ρ [$] in Case F7-2 (Ref. [1])

Detector	α [s^{-1}]		ρ [$]	
	Feynman-α	α-fitting	Area (Sjostrand)	Area (Gozani)
BF-3 #1 (Ch#1)	1502.75 ± 6.73	1615.9 ± 10.3	8.406 ± 0.182	8.958 ± 0.223
BF-3 #2 (Ch#2)	1522.17 ± 6.67	1659.6 ± 11.7	8.346 ± 0.088	8.974 ± 0.130
BF-3 #3 (Ch#3)	1480.47 ± 8.41	1589.6 ± 16.7	9.338 ± 0.057	9.285 ± 0.132
BF-3 #4 (Ch#4)	1561.98 ± 8.74	1529.3 ± 17.4	10.097 ± 0.035	9.256 ± 0.126
LiCaF (Ch#5)	1459.16 ± 6.75	1631.9 ± 9.7	9.699 ± 0.058	9.781 ± 0.095
Fiber #2 (Ch#6)	1582.17 ± 148.61	–	–	–
Fiber #3 (Ch#7)	1286.24 ± 105.37	–	–	–

A3.3 Reaction Rates

A3.3.1 Core Configurations

See (Fig. A3.8).

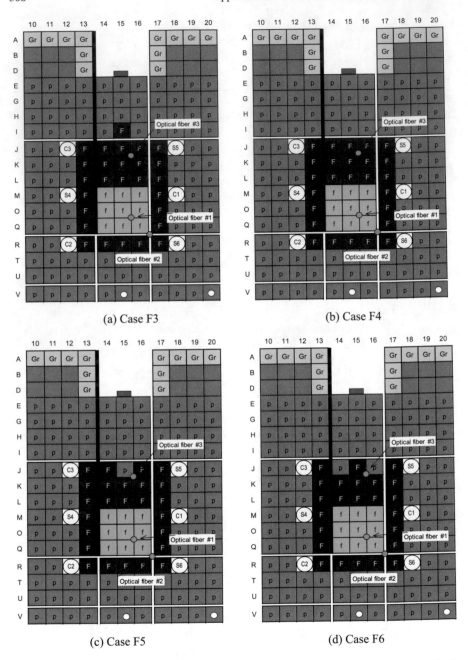

Fig. A3.8 Core configuration of reaction rate distributions in ADS with 100 MeV protons (Refs. [1, 3])

A3.3.2 Reaction Rate Distribution

See (Tables A3.57, A3.58 and A3.59, Fig. A3.9).

Table A3.57 Specification of measurement of reaction rate distribution in Fig. A3.8 (Ref. [1])

Reaction	Location	Foil/Wire
^{115}In$(n, n')^{115m}$In	Target (15, D; Fig. A3.8)	$10 \times 10 \times 1$ mm (Foil)
^{115}In$(n, \gamma)^{116}$In	Core (13-14, A-P; Fig. A3.8)	1 mm diameter, 800 mm long (Wire)

Table A3.58 Core condition in all the cores shown in Fig. A3.8 (Ref. [1])

Case	# of HEU plates	Positions of rods	k_{eff}
F3	4320	C1, C2, A3: 1200.00 [mm] S4, S5, S6: 1200.00 [mm]	0.98004
F4	4200	C1, C2, A3: 1200.00 [mm] S4, S5, S6: 1200.00 [mm]	0.96603
F5	4080	C1, C2, A3: 1200.00 [mm] S4, S5, S6: 1200.00 [mm]	0.95560
F6	3960	C1, C2, A3: 1200.00 [mm] S4, S5, S6: 1200.00 [mm]	0.95047

Table A3.59 Measured reaction rates of In foil at (15, D) in Cases F3 through F6 (Ref. [1])

	Measured reaction rate [s^{-1} cm^{-3}]			
Reaction	Case F3	Case F4	Case F5	Case F-6
^{115}In$(n, n')^{115m}$In	(1.970 ± 0.045)E+04	(1.320 ± 0.035)E+04	(2.385 ± 0.020)E+04	(3.482 ± 0.009)E+04

Fig. A3.9 Measured
(normalized) reaction rates
(^{115}In$(n,\gamma)^{116m}$In$/^{115}$In$(n,$
$n')^{115m}$In) (Refs. [1, 3])

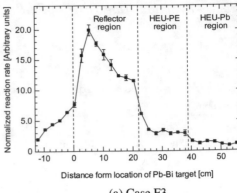

(a) Case F3

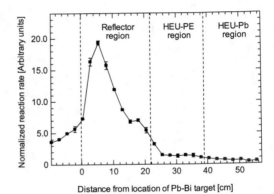

(b) Case F4

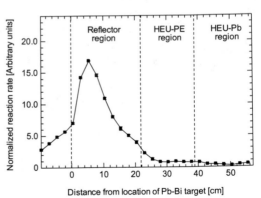

(c) Case F5

Fig. A3.9 (continued)

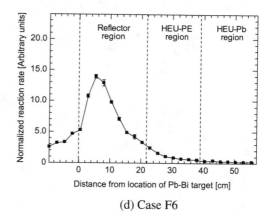

(d) Case F6

References

1. Pyeon CH and Yamanaka M (2018) Experimental benchmarks of neutron characteristics on uranium-lead zoned core in accelerator-driven system at Kyoto University Critical Assembly. KURNS-EKR-001
2. Yamanaka M, Pyeon CH, Endo T et al (2020) Experimental analyses of $\beta_{\text{eff}}/\Lambda$ in accelerator-driven system at Kyoto University Critical Assembly. J Nucl Sci Technol 57:205
3. Aizawa N, Yamanaka M, Iwasaki T et al (2019) Effect of neutron spectrum on subcritical multiplication factor in accelerator-driven system. Prog Nucl Energy 116:158

Appendix A4: ^{235}U-Fueled ADS Core in Medium-Fast Spectrum

A4.1 Core Configurations

A4.1.1 ADS with 14 MeV Neutrons

See (Figs. A4.1, A4.2 and A4.3).

© The Editor(s) (if applicable) and The Author(s) 2021
C. H. Pyeon (ed.), *Accelerator-Driven System at Kyoto University Critical Assembly*,
https://doi.org/10.1007/978-981-16-0344-0

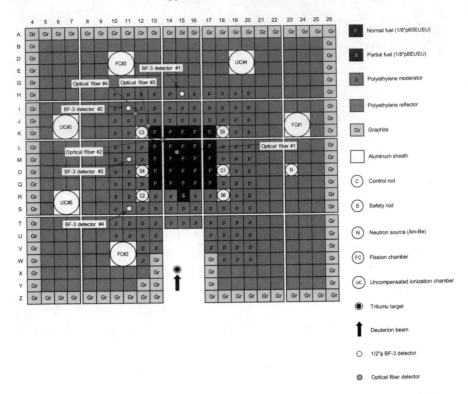

Fig. A4.1 Core configuration (EE1 core) of ADS with 14 MeV neutrons. (Case 1-D: Critical core; 3016 HEU fuel plates) (Refs. [1–2])

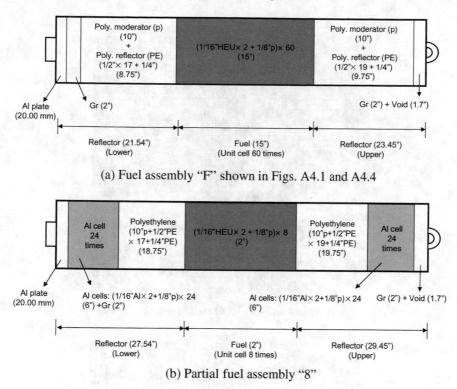

(a) Fuel assembly "F" shown in Figs. A4.1 and A4.4

(b) Partial fuel assembly "8"

Fig. A4.2 Fall sideway view of fuel rods shown in Fig. A4.1 (Ref. [1])

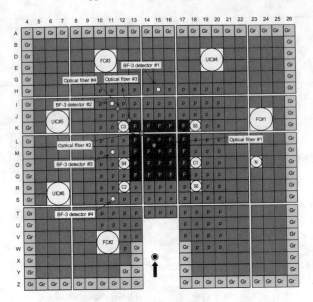

(a) Case 2-D (3000 HEU plates)

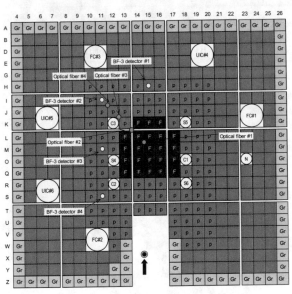

(b) Case 3-D (2760 HEU plates)

Fig. A4.3 Core configuration (EE1 core) of ADS with 14 MeV neutrons (Subcritical cores) (Ref. [1])

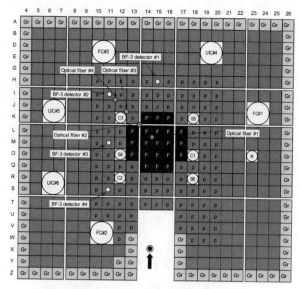

(c) Case 4-D (2520 HEU plates)

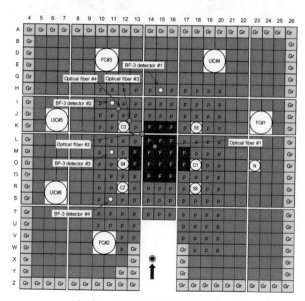

(d) Case 5-D (2280 HEU plates)

Fig. A4.3 (continued)

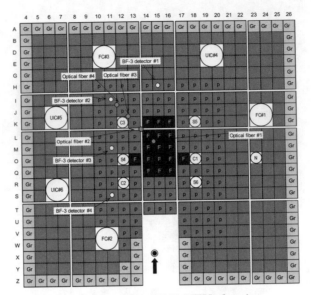

(e) Case 6-D (2040 HEU plates)

Fig. A4.3 (continued)

A4.1.2 ADS with 100 MeV Protons

See (Figs. A4.4, A4.5 and A4.6).

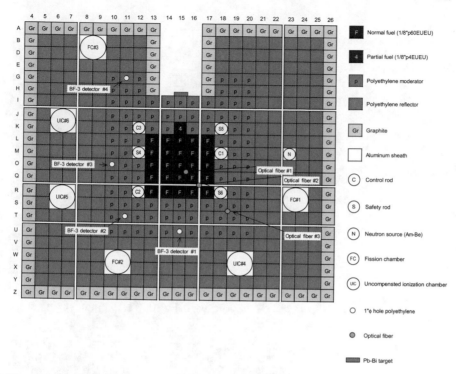

Fig. A4.4 Core configuration (EE1 core) of ADS with 100 MeV protons. (Case 1-F: Critical core; 3008 HEU plates) (Refs. [1–2])

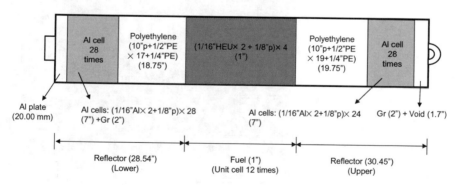

Fig. A4.5 Fall sideway view of fuel rod "4" shown in Fig. A4.4 (Ref. [1])

Fig. A4.6 Core configuration (EE1 core) of ADS with 100 MeV protons (Subcritical cores) (Ref. [1])

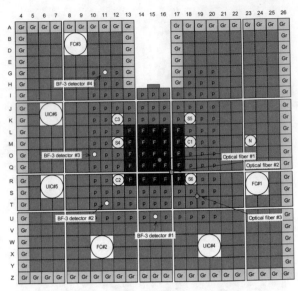

(a) Case 2-F (3000 HEU plates)

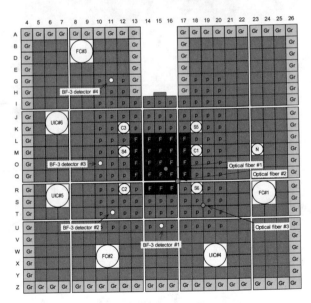

(b) Case 3-F (2760 HEU plates)

Fig. A4.6 (continued)

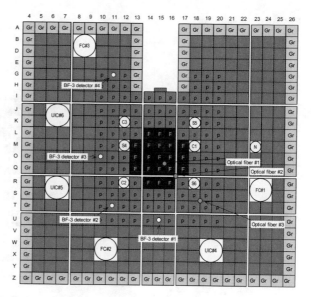

(c) Case 4-F (2520 HEU plates)

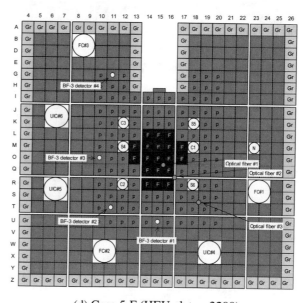

(d) Case 5-F (HEU plates: 2280)

Fig. A4.6 (continued)

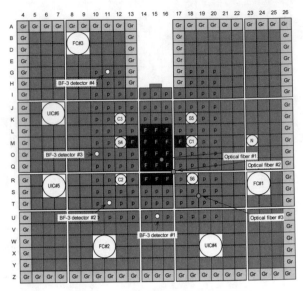

(e) Case 6-F (HEU plates: 2040)

A4.2 Results of Experiments

A4.2.1 Criticality and Control Rod Worth

See (Tables A4.1, A4.2, A4.3 and A4.4)

Table A4.1 Control rod positions at the critical state (Case 1-D) (Ref. [1])

Rod	Rod position [mm]
C1	1200.00
C2	1200.00
A3	630.01
S4	1200.00
S5	1200.00
S6	1200.00
Excess reactivity [pcm]	271.7 ± 1.0

Table A4.2 Control rod worth (Case 1-D) (Ref. [1])

Rod	Rod worth [pcm]
C1 (S4)	867.4 ± 3.9
C2 (S6)	142.8 ± 3.5
A3 (S5)	506.4 ± 4.3

Table A4.3 Control rod positions at the critical state (Case 1-F) (Ref. [1])

Rod	Rod position [mm]
C1	1200.00
C2	723.31
A3	1200.00
S4	1200.00
S5	1200.00
S6	1200.00
Excess reactivity [pcm]	102.5 ± 1.8

Table A4.4 Control rod worth (Case 1-F) (Ref. [1])

Rod	Rod worth [pcm]
C1 (S4)	866.7 ± 2.7
C2 (S6)	439.4 ± 3.9
A3 (S5)	148.6 ± 7.6

A4.2.2 PNS and Feynman-α Methods

See (Tables A4.5, A4.6, A4.7, A4.8, A4.9, A4.10 and A4.11).

Table A4.5 List of core condition in all the cores shown in Fig. A4.1 (Ref. [1])

Case	# of fuel plates	Rod insertion (down)	14 MeV neutrons		
			Repetition (Hz)	Width (μs)	Current (mA)
Case 1-D-1	3016	C1, C2, A3	100	97	0.12
Case 1-D-1T	3016	C1, C2, A3	100	97	0.12
Case 1-D-2	3016	C1, C2, A3	Am–Be		
Case 1-D-3	3016	C1, C2, A3	20	97	0.12
Case 1-D-4	3016	C1, C2, A3	20	97	0.12

Table A4.6 List of core condition in all the cores shown in Fig. A4.3a (Ref. [1])

Case	# of fuel plates	Rod	14 MeV neutrons		
			Repetition (Hz)	Width (μs)	Current (mA)
Case 2-D-1	3000	C1~S6 up	20	97	0.12
Case 2-D-2	3000	C1 down	20	97	0.12
Case 2-D-3	3000	C1 down	50	97	0.12

Table A4.7 List of core condition in all the cores shown in Fig. A4.3b (Ref. [1])

Case	# of fuel plates	Rod	14 MeV neutrons		
			Repetition (Hz)	Width (μs)	Current (mA)
Case 3-D-1	2760	C1~S6 up	20	97	0.12

Table A4.8 List of core condition in all the cores shown in Fig. A4.3c (Ref. [1])

Case	# of fuel plates	Rod (up)	14 MeV neutrons		
			Repetition (Hz)	Width (μs)	Current (mA)
Case 4-D-1	2520	C1~S6	20	85	0.08
Case 4-D-1T	2520	C1~S6	50	85	0.08
Case 4-D-2	2520	C1~S6	50	85	0.08
Case 4-D-3	2520	C1~S6	20	75	0.15
Case 4-D-4T	2520	C1~S6	50	85	0.15
Case 4-D-5	2520	C1~S6	50	85	0.15

Table A4.9 List of core condition in all the cores shown in Fig. A4.3d (Ref. [1])

Case	# of fuel plates	Rod (up)	14 MeV neutrons		
			Repetition (Hz)	Width (μs)	Current (mA)
Case 5-D-1	2280	C1~S6	20	97	0.08
Case 5-D-2	2280	C1~S6	100	97	0.08
Case 5-D-3T	2280	C1~S6	20	97	0.08
Case 5-D-3	2280	C1~S6	20	97	0.08
Case 5-D-4	2280	C1~S6	50	97	0.08

Table A4.10 List of core condition in all the cores shown in Fig. A4.3e (Ref. [1])

Case	# of fuel plates	Rod (up)	14 MeV neutrons		
			Repetition (Hz)	Width (μs)	Current (mA)
Case 6-D-1T	2040	C1~S6	20	93	0.10
Case 6-D-2	2040	C1~S6	100	97	0.10
Case 6-D-3	2040	C1~S6	50	97	0.08
Case 6-D-4T	2040	C1~S6	50	97	0.08
Case 6-D-5	2040	C1~S6	500	20	0.20

Table A4.11 List of core condition in all the cores shown in Fig. A4.4 (Ref. [1])

Case	# of fuel plates	Rod insertion (down)	100 MeV protons		
			Repetition (Hz)	Width (ns)	Current (nA)
Case 1-F-1	3008	C1, C2, A3	30	100	0.4
Case 1-F-2	3008	C1, C2, A3	30	100	0.4
Case 1-F-3	3008	C1, C2, A3, S4	30	100	0.4
Case 1-F-4	3008	C1~S6	30	100	0.4
Case 1-F-5	3008	C1~S6, CC	30	100	0.4

CC: Center core

Transient steps: All C1~S6 up (stable) → C1~A3 drop → S4 drop → S5 and S6 drop → Center core (CC) drop

No data on Case 2-F

See (Tables A4.12, A4.13 and A4.14).

Table A4.12 List of core condition in all the cores shown in Fig. A4.6c (Ref. [1])

Case	# of fuel plates	Rod insertion (down)	100 MeV protons		
			Repetition (Hz)	Width (ns)	Current (nA)
Case 3-F-1	2760	C1~S6 up	30	100	0.3
Case 3-F-2	2760	C1	30	100	0.3
Case 3-F-3	2760	C1, C2	30	100	0.3
Case 3-F-4	2760	C1, C2, A3	30	100	0.3
Case 3-F-5	2760	C1~S6, CC	30	100	0.3

CC: Center core

Transient steps: All C1~S6 up (stable) → C1 drop → C2 drop → A3 drop → CC drop

Table A4.13 List of core condition in all the cores shown in Fig. A4.6d (Ref. [1])

Case	# of fuel plates	Rod insertion (down)	100 MeV protons		
			Repetition (Hz)	Width (ns)	Current (nA)
Case 4-F-1	2520	C1~S6 up	30	100	0.3
Case 4-F-2	2520	C1	30	100	0.3
Case 4-F-3	2520	C1, C2	30	100	0.3
Case 4-F-4	2520	C1, C2, A3	30	100	0.3
Case 4-F-5	2520	C1~S6, CC	30	100	0.3

CC: Center core

Transient steps: All C1~S6 up (stable) → C1 drop → C2 drop → A3 drop → CC drop → CC drop

Table A4.14 List of core condition in all the cores shown in Fig. A4.6e (Ref. [1])

Case	# of fuel plates	Rod insertion (down)	100 MeV protons		
			Repetition (Hz)	Width (ns)	Current (nA)
Case 5-F-1	2280	C1~S6 up	30	100	0.3
Case 5-F-2	2280	C1	30	100	0.3
Case 5-F-3	2280	C1, C2	30	100	0.3
Case 5-F-4	2280	C1, C2, A3	30	100	0.3
Case 5-F-5	2280	C1~S6, CC	30	100	0.3

CC: Center core

Transient steps: All C1~S6 up (stable) → C1 drop → C2 drop → A3 drop → CC drop

A4.3 Kinetic Parameters

A4.3.1 ADS with 14 MeV Neutrons

See (Tables A4.15, A4.16, A4.17, A4.18, A4.19, A4.20, A4.21, A4.22, A4.23, A4.24, A4.25, A4.26, A4.27, A4.28, A4.29, A4.30, A4.31, A4.32, A4.33 and A4.34).

Table A4.15 Measured prompt neutron decay constants α [s^{-1}] deduced by PNS and Feynman-α methods and subcriticality [$] in dollar units (Ref. [1])

Case 1-D-1 (DT acc.) ($\rho_{pcm, exp}$ = 1245.0 ± 41.7 [pcm])

Detector	PNS method	Feynman-α method	Area $\rho_\$$[$]	Extended area $\rho_\$$[$]
BF-3 #1	717.63 ± 12.28	650.92 ± 55.23	1.50189 ± 0.02694	1.79660 ± 0.04535
BF-3 #2	710.94 ± 13.02	729.76 ± 60.58	1.56681 ± 0.02954	1.76655 ± 0.04770
BF-3 #3	709.12 ± 4.96	725.23 ± 50.43	1.54026 ± 0.01131	1.62362 ± 0.01756
BF-3 #4	739.00 ± 23.75	1226.41 ± 53.52	2.36997 ± 0.03507	1.50892 ± 0.05004
Fiber #1	713.88 ± 23.23	–	1.42923 ± 0.04583	1.35279 ± 0.06973
Fiber #2	733.90 ± 11.74	–	1.38885 ± 0.02078	1.47719 ± 0.03464
Fiber #3	711.48 ± 11.43	–	1.56890 ± 0.02685	1.75388 ± 0.04286

Table A4.16 Measured prompt neutron decay constants α [s^{-1}] deduced by PNS and Feynman-α methods (Ref. [1])

Case 1-D-2 (Am-Be) ($\rho_{pcm, exp}$ = 1245.0 ± 41.7 [pcm])

Detector	PNS method	Feynman-α method
BF-3 #1	–	745.76 ± 154.62
BF-3 #2	–	1051.355 ± 265.21
BF-3 #3	787.35 ± 68.45	604.28 ± 27.36
BF-3 #4	–	1057.10 ± 245.31
Fiber #1	–	865.09 ± 1492.43
Fiber #2	–	1179.31 ± 288.01
Fiber #3	–	645.96 ± 113.23

Table A4.17 Measured prompt neutron decay constants α [s^{-1}] deduced by PNS and Feynman-α methods (Ref. [1])

Case 1-D-3 (DT acc.) ($\rho_{pcm, exp} = 1245.0 \pm 41.7$ [pcm])

Detector	PNS method	Feynman-α method
BF-3 #1	–	–
BF-3 #2	–	–
BF-3 #3	715.18 ± 9.57	744.54 ± 8.70
Fiber #1	–	–
Fiber #2	724.51 ± 24.58	843.73 ± 139.74
Fiber #3	740.00 ± 26.794	977.05 ± 196.02

Table A4.18 Measured prompt neutron decay constants α [s^{-1}] deduced by PNS and Feynman-α methods (Ref. [1])

Case 1-D-4 (DT acc.) ($\rho_{pcm, exp} = 1245.0 \pm 41.7$ [pcm])

Detector	PNS method	Feynman-α method
BF-3 #1	–	–
BF-3 #2	–	–
BF-3 #3	713.43 ± 2.80	708.85 ± 3.28
Fiber #1	–	–
Fiber #2	715.59 ± 6.03	730.10 ± 5.92
Fiber #3	705.76 ± 6.19	728.07 ± 6.86

Table A4.19 Measured prompt neutron decay constants α [s^{-1}] deduced by PNS and Feynman-α methods, $\rho_\$$ in dollar units by the area ratio method, and β_{eff}/Λ [s^{-1}] by the α-fitting method (Ref. [1])

Detector	Case 2-D-1 (DT acc.) ($\rho_{pcm, MCNP} = 855.7 \pm 41.34$ [pcm])			
	α (PNS method)	α (Feynman-α method)	Extended area $\rho_\$$ [$]	β_{eff}/Λ [s^{-1}]
BF-3 #1	294.78 ± 29.27	319.16 ± 71.63	0.02283 ± 0.00703	288.2 ± 93.3
BF-3 #2	226.96 ± 23.31	363.98 ± 75.69	0.01576 ± 0.00537	223.4 ± 79.5
BF-3 #3	258.58 ± 12.34	270.42 ± 13.87	0.01938 ± 0.00270	253.7 ± 37.3
Fiber #1	248.38 ± 39.04	–	0.02066 ± 0.00559	243.4 ± 72.1
Fiber #2	229.95 ± 14.51	268.82 ± 20.68	0.01998 ± 0.00320	225.4 ± 38.8
Fiber #3	273.99 ± 26.24	310.39 ± 42.98	0.02613 ± 0.00559	267.0 ± 62.6

MCNP6.1 with ENDF/B-VII.0: $\beta_{eff} = 829.0$ [pcm], $\Lambda = 3.147E-05$ (s), $\beta_{eff}/\Lambda = 263.43$ [s^{-1}]

Table A4.20 Measured prompt neutron decay constants α [s^{-1}] deduced by PNS and Feynman-α methods and subcriticality [$] in dollar units (Ref. [1])

Case 2-D-2 (DT acc.) ($\rho_{pcm, exp} = 855.7 \pm 41.34$ [pcm])

Detector	PNS method	Feynman-α method	Area $\rho_\$$ [$]	Extended area $\rho_\$$ [$]
BF-3 #1	528.29 ± 9.04	511.67 ± 7.89	0.97699 ± 0.01798	1.12891 ± 0.02847
BF-3 #2	528.90 ± 8.16	525.47 ± 6.55	0.99300 ± 0.01691	1.13125 ± 0.02652
BF-3 #3	532.62 ± 3.41	524.58 ± 2.71	1.00291 ± 0.00762	1.03933 ± 0.01150
Fiber #1	551.62 ± 16.01	518.95 ± 15.28	1.05686 ± 0.03549	1.09990 ± 0.05433
Fiber #2	531.09 ± 6.56	535.64 ± 5.62	0.98490 ± 0.01442	1.03664 ± 0.02193
Fiber #4	538.75 ± 7.94	524.44 ± 7.70	1.00649 ± 0.01879	1.12482 ± 0.02905

Table A4.21 Measured prompt neutron decay constants α [s^{-1}] deduced by PNS and Feynman-α methods and subcriticality [$] in dollar units (Ref. [1])

Case 2-D-3 (DT acc.) ($\rho_{pcm, exp} = 855.7 \pm 41.34$ [pcm])

Detector	PNS method	Feynman-α method	Area $\rho_\$$ [$]	Extended area $\rho_\$$ [$]
BF-3 #1	520.68 ± 14.35	497.98 ± 28.83	0.99970 ± 0.02902	1.10363 ± 0.04398
BF-3 #2	540.67 ± 13.47	513.51 ± 29.28	1.04942 ± 0.02822	1.20361 ± 0.04376
BF-3 #3	539.67 ± 5.94	503.11 ± 19.54	0.99637 ± 0.01220	1.05783 ± 0.01841
Fiber #1	603.37 ± 31.75	644.03 ± 59.61	1.00039 ± 0.05132	1.06198 ± 0.08348
Fiber #2	556.54 ± 12.45	550.18 ± 24.54	0.98282 ± 0.02267	1.08083 ± 0.03559
Fiber #4	526.96 ± 13.58	507.13 ± 36.75	0.98791 ± 0.02879	1.09278 ± 0.04359

Table A4.22 Measured prompt neutron decay constants α [s^{-1}] deduced by PNS and Feynman-α methods and subcriticality [$] in dollar units (Ref. [1])

Case 3-D-1 (DT acc.) ($\rho_{pcm, MCNP} = 1512.9 \pm 5.7$ [pcm])

Detector	PNS method	Feynman-α method	Area $\rho_\$$ [$]	Extended area $\rho_\$$ [$]
BF-3 #1	667.29 ± 12.95	626.20 ± 12.85	1.65862 ± 0.04122	1.95981 ± 0.06664
BF-3 #2	655.80 ± 14.14	679.36 ± 15.34	1.72801 ± 0.04526	1.92831 ± 0.07057
BF-3 #3	672.77 ± 5.39	664.35 ± 6.05	1.72456 ± 0.01725	1.90308 ± 0.02698
Fiber #1	673.89 ± 23.09	724.98 ± 38.84	1.60421 ± 0.06756	1.60381 ± 0.10284
Fiber #2	667.31 ± 5.59	661.07 ± 6.15	1.66194 ± 0.01797	1.94418 ± 0.02886
Fiber #3	668.11 ± 11.27	658.91 ± 11.96	1.42546 ± 0.03085	1.65991 ± 0.05085

Table A4.23 Measured prompt neutron decay constants α [s^{-1}] deduced by PNS and Feynman-α methods and subcriticality [\$] in dollar units (Ref. [1])

Case 4-D-1 (DT acc.) ($\rho_{pcm, MCNP}$ = 3111.6 ± 5.8 [pcm])

Detector	PNS method	Feynman-α method	Area $\rho_\$$ [\$]	Extended area $\rho_\$$ [\$]
BF-3 #1	1056.02 ± 38.51	1113.92 ± 41.74	3.73361 ± 0.19036	4.78451 ± 0.34900
BF-3 #2	1119.68 ± 39.02	1064.70 ± 47.73	3.70243 ± 0.18643	5.17838 ± 0.37464
BF-3 #3	1076.31 ± 16.22	1078.63 ± 11.07	3.55740 ± 0.06986	4.26586 ± 0.12752
Fiber #1	1155.278 ± 157.26	1186.20 ± 144.92	–	–
Fiber #2	–	–		
Fiber #4	–	–		

Table A4.24 Measured prompt neutron decay constants α [s^{-1}] deduced by PNS and Feynman-α methods and subcriticality [\$] in dollar units (Ref. [1])

Case 4-D-2 (DT acc.) ($\rho_{pcm, MCNP}$ = 3111.6 ± 5.8 [pcm])

Detector	PNS method	Feynman-α method	Area $\rho_\$$ [\$]	Extended area $\rho_\$$ [\$]
BF-3 #1	1092.18 ± 19.15	999.68 ± 13.50	3.82788 ± 0.06968	5.00179 ± 0.14435
BF-3 #2	1082.50 ± 23.44	1013.01 ± 14.55	3.72918 ± 0.06925	4.51610 ± 0.14738
BF-3 #3	1076.16 ± 24.72	1024.44 ± 5.06	3.90435 ± 0.02899	4.31173 ± 0.11384
Fiber #1	1110.31 ± 56.56	986.93 ± 31.56	3.99660 ± 0.14054	4.59564 ± 0.32699
Fiber #2	1071.20 ± 17.88	993.30 ± 7.48	3.60796 ± 0.04005	4.24635 ± 0.09581
Fiber #4	1063.37 ± 22.56	–	3.85565 ± 0.07871	4.66042 ± 0.15531

Table A4.25 Measured prompt neutron decay constants α [s^{-1}] deduced by PNS and Feynman-α methods and subcriticality [\$] in dollar units (Ref. [1])

Case 4-D-3 (DT acc.) ($\rho_{pcm, MCNP}$ = 3111.6 ± 5.8 [pcm])

Detector	PNS method	Feynman-α method	Area $\rho_\$$ [\$]	Extended area $\rho_\$$ [\$]
BF-3 #1	1051.31 ± 10.10	1000.94 ± 6.79	3.60407 ± 0.05694	4.64020 ± 0.10113
BF-3 #2	1046.19 ± 10.80	1008.30 ± 6.44	3.77780 ± 0.06011	4.66594 ± 0.10417
BF-3 #3	1073.30 ± 5.42	1035.34 ± 2.38	3.76266 ± 0.02511	4.23217 ± 0.04311
Fiber #1	1069.91 ± 22.96	1049.00 ± 14.85	4.89473 ± 0.17728	5.05178 ± 0.25089
Fiber #2	1068.86 ± 6.59	1003.73 ± 3.41	3.61753 ± 0.03547	4.45911 ± 0.06205
Fiber #4	1065.15 ± 11.75	1000.72 ± 7.15	4.10600 ± 0.07930	5.12115 ± 0.13295

Table A4.26 Measured prompt neutron decay constants α [s^{-1}] deduced by PNS and Feynman-α methods and subcriticality [$] in dollar units (Ref. [1])

Case 4-D-5 (DT acc.) ($\rho_{pcm, MCNP} = 3111.6 \pm 5.8$ [pcm])

Detector	PNS method	Feynman-α method	Area $\rho_\$$ [$]	Extended area $\rho_\$$ [$]
BF-3 #1	1070.93 ± 14.87	981.40 ± 17.98	3.48096 ± 0.08413	4.87113 ± 0.15607
BF-3 #2	1041.43 ± 13.38	972.84 ± 16.43	3.79819 ± 0.09405	5.40366 ± 0.16873
BF-3 #3	1067.21 ± 7.71	995.28 ± 12.30	3.76012 ± 0.03970	4.45491 ± 0.06704
Fiber #1	1061.98 ± 27.72	1028.16 ± 34.83	3.77311 ± 0.16272	4.36965 ± 0.26057
Fiber #2	1059.13 ± 10.74	1003.92 ± 13.44	3.54025 ± 0.05283	4.27714 ± 0.09088
Fiber #4	1053.99 ± 17.86	970.70 ± 19.62	3.57903 ± 0.09639	4.81026 ± 0.17481

Table A4.27 Measured prompt neutron decay constants α [s^{-1}] deduced by PNS and Feynman-α methods and subcriticality [$] in dollar units (Ref. [1])

Case 5-D-1 (DT acc.) ($\rho_{pcm, MCNP} = 5466.8 \pm 5.8$ [pcm])

Detector	PNS method	Feynman-α method	Area $\rho_\$$ [$]	Extended area $\rho_\$$ [$]
BF-3 #1	1507.29 ± 45.15	1600.24 ± 58.30	13.12317 ± 1.02054	13.12317 ± 1.02054
BF-3 #2	–	–	–	–
BF-3 #3	1561.24 ± 16.71	1521.88 ± 11.71	8.04806 ± 0.22314	8.04806 ± 0.22314
Fiber #1	–	–	–	–
Fiber #2	1562.66 ± 25.08	1487.91 ± 22.39	6.74084 ± 0.20170	9.57870 ± 0.41788
Fiber #4	1567.86 ± 41.67	1597.84 ± 77.17	–	–

Table A4.28 Measured prompt neutron decay constants α [s^{-1}] deduced by PNS and Feynman-α methods and subcriticality [$] in dollar units (Ref. [1])

Case 5-D-2 (DT acc.) ($\rho_{pcm, MCNP} = 5466.8 \pm 5.8$ [pcm])

Detector	PNS method	Feynman-α method	Area $\rho_\$$ [$]	Extended area $\rho_\$$ [$]
BF-3 #1	1542.47 ± 32.10	1531.57 ± 94.15	6.69272 ± 0.26248	9.85874 ± 0.53993
BF-3 #2	1541.73 ± 37.74	1542.69 ± 104.74	7.09606 ± 0.31803	9.58566 ± 0.61002
BF-3 #3	1561.45 ± 12.48	1487.83 ± 19.03	7.12148 ± 0.11291	8.31547 ± 0.18855
Fiber #1	1559.87 ± 70.36	2076.12 ± 339.59	7.68274 ± 0.66021	7.74516 ± 0.96734
Fiber #2	1558.73 ± 18.72	1436.66 ± 39.58	6.44406 ± 0.15399	9.05447 ± 0.30220
Fiber #4	1569.88 ± 36.43	1308.68 ± 110.11	7.70664 ± 0.42332	10.96118 ± 0.77941

Table A4.29 Measured prompt neutron decay constants α [s^{-1}] deduced by PNS and Feynman-α methods and subcriticality [$] in dollar units (Ref. [1])

Case 5-D-3 (DT acc.) ($\rho_{pcm, MCNP} = 5466.8 \pm 5.8$ [pcm])

Detector	PNS method	Feynman-α method	Area $\rho_\$$ [$]	Extended area $\rho_\$$ [$]
BF-3 #1	1518.10 ± 19.77	1441.77 ± 14.03	8.60725 ± 0.25150	11.74861 ± 0.45170
BF-3 #2	1546.67 ± 20.78	1420.56 ± 15.18	8.18430 ± 0.24169	11.21470 ± 0.44724
BF-3 #3	1563.35 ± 9.26	1523.67 ± 3.69	6.77913 ± 0.06357	7.54184 ± 0.11621
Fiber #1	–	1623.23 ± 44.80	–	–
Fiber #2	1568.87 ± 13.17	1501.30 ± 6.54	7.03293 ± 0.11596	8.79780 ± 0.21192
Fiber #4	1504.68 ± 21.32	1381.16 ± 18.64	19.45768 ± 1.86598	29.33747 ± 2.93363

Table A4.30 Measured prompt neutron decay constants α [s^{-1}] deduced by PNS and Feynman-α methods and subcriticality [$] in dollar units (Ref. [1])

Case 5-D-4 (DT acc.) ($\rho_{pcm, MCNP} = 5466.8 \pm 5.8$ [pcm])

Detector	PNS method	Feynman-α method	Area $\rho_\$$ [$]	Extended area $\rho_\$$ [$]
BF-3 #1	1542.40 ± 26.77	1426.92 ± 26.99	7.48143 ± 0.27312	11.76062 ± 0.57855
BF-3 #2	1591.02 ±30.08	1367.56 ± 28.65	7.67026 ± 0.31138	12.84927 ± 0.70248
BF-3 #3	1551.31 ± 14.62	1513.58 ± 9.35	7.23510 ± 0.10362	9.32671 ± 0.21110
Fiber #1	1608.13 ± 81.13	1543.68 ± 77.62	8.70127 ± 0.71994	11.07538 ± 1.38422
Fiber #2	1557.72 ± 18.19	1493.76 ± 12.89	6.66444 ± 0.14187	10.45040 ± 0.32031
Fiber #4	1523.73 ± 40.46	1434.55 ± 37.42	9.42282 ± 0.55816	14.32678 ± 1.08688

Table A4.31 Measured prompt neutron decay constants α [s^{-1}] deduced by PNS and Feynman-α methods (Ref. [1])

Case 6-D-2 (DT acc.) ($\rho_{pcm, MCNP} = 8882.3 \pm 5.9$ [pcm])

Detector	PNS method	Feynman-α method
BF-3 #1	–	1780.35 ± 54.79
BF-3 #2	–	2213.13 ± 97.65
BF-3 #3	–	2022.98 ± 16.88
Fiber #1	–	4324.24 ± 1007.21
Fiber #2	–	1703.06 ± 32.53
Fiber #4	–	1926.80 ± 172.55

Table A4.32 Measured prompt neutron decay constants α [s^{-1}] deduced by PNS and Feynman-α methods and subcriticality [$] in dollar units (Ref. [1])

Case 6-D-3 (DT acc.) ($\rho_{pcm, MCNP}$ = 8882.3 ± 5.9 [pcm])

Detector	PNS method	Feynman-α method	Area $\rho_\$$ [$]	Extended area $\rho_\$$ [$]
BF-3 #1	2024.25 ± 61.49	1584.52 ± 128.30	14.92700 ± 1.82496	30.69158 ± 4.60981
BF-3 #2	2144.03 ± 84.83	2023.53 ± 174.81	15.44242 ± 2.04033	32.21389 ± 5.23709
BF-3 #3	2144.99 ± 34.08	1963.84 ± 32.34	12.80783 ± 0.55046	18.95650 ± 1.09303
Fiber #1	1998.46 ± 425.92	1829.06 ± 488.42	12.40719 ± 3.87931	5.56483 ± 2.31382
Fiber #2	1985.47 ± 44.19	1814.45 ± 82.31	12.33320 ± 0.93682	22.10653 ± 2.14379
Fiber #4	2064.13 ± 102.76	2480.72 ± 333.99	16.94444 ± 4.12772	32.17342 ± 8.66792

Table A4.33 Measured prompt neutron decay constants α [s^{-1}] deduced by PNS and Feynman-α methods (Ref. [1])

Case 6-D-5 (DT acc.) ($\rho_{pcm, MCNP}$ = 8882.3 ± 5.9 [pcm])

Detector	PNS method	Feynman-α method
BF-3 #1	2670.27 ± 125.23	1913.15 ± 270.08
BF-3 #2	3033.82 ± 242.75	3932.08 ± 634.72
BF-3 #3	3067.22 ± 118.67	2419.25 ± 83.22
Fiber #1	2560.30 ± 632.22	5538.75 ± 1156.03
Fiber #2	3148.90 ± 297.81	2914.43 ± 602.96
Fiber #4	2912.04 ± 453.49	5565.46 ± 3640.91

Table A4.34 Effective multiplication factor in Cases II-1 through II-6 (Ref. [1])

Case	k_{eff} (MCNP6.1 with JENDL-4.0)
Case 1-D (1-F)	–
Case 2-D (2-F); HEU 3000	0.99817 ± 0.00004
Case 3-D (3-F); HEU 2760	0.98400 ± 0.00004
Case 4-D (4-F); HEU 2520	0.96876 ± 0.00004
Case 5-D (5-F); HEU 2280	0.94715 ± 0.00004
Case 6-D (6-F); HEU 2040	0.91747 ± 0.00004

A4.3.2 ADS with 100 MeV Protons

See (Tables A4.35, A4.36, A4.37, A4.38, A4.39, A4.40, A4.41, A4.42, A4.43, A4.44, A4.45, A4.46, A4.47, A4.48, A4.49, A4.50, A4.51, A4.52, A4.53 and A4.54).

Table A4.35 Measured prompt neutron decay constants α [s^{-1}] deduced by PNS and Feynman-α methods and subcriticality [$] in dollar units (Ref. [1])

Case 1-F-1 (FFAG acc.) ($\rho_{pcm, exp} = 1352.2 \pm 38.82$ [pcm])

Detector	PNS method	Feynman-α method	Area $\rho_\$$ [$]	Extended area $\rho_\$$ [$]
BF-3 #1	689.89 ± 2.07	674.47 ± 1.20	1.41356 ± 0.00273	1.76629 ± 0.00565
BF-3 #2	752.28 ± 1.13	737.51 ± 1.25	1.67789 ± 0.00354	2.10922 ± 0.00611
BF-3 #3	666.36 ± 2.44	653.50 ± 1.25	1.39013 ± 0.00280	1.64844 ± 0.00588

Table A4.36 Measured prompt neutron decay constants α [s^{-1}] deduced by PNS and Feynman-α methods and subcriticality [$] in dollar units (Ref. [1])

Case 1-F-2 (FFAG acc.) ($\rho_{pcm, exp} = 1352.2 \pm 38.82$ [pcm])

Detector	PNS method	Feynman-α method	Area $\rho_\$$ [$]	Extended area $\rho_\$$ [$]
Fiber #1	768.70 ± 2.40	771.85 ± 1.71	1.71524 ± 0.00642	1.85089 ± 0.01061
Fiber #2	818.10 ± 2.61	805.01 ± 1.76	1.95777 ± 0.00726	2.20352 ± 0.01242
Fiber #3	784.56 ± 9.17	779.14 ± 2.04	2.00685 ± 0.00454	2.78333 ± 0.02660
BF-3 #1	704.67 ± 2.02	685.23 ± 1.74	1.46667 ± 0.00485	1.84440 ± 0.00859
BF-3 #2	752.37 ± 1.98	741.66 ± 2.07	1.67185 ± 0.00617	2.09335 ± 0.01068
BF-3 #3	688.21 ± 2.32	669.07 ± 1.62	1.49290 ± 0.00513	1.77403 ± 0.00899

Table A4.37 Measured prompt neutron decay constants α [s^{-1}] deduced by PNS and Feynman-α methods and subcriticality [$] in dollar units (Ref. [1])

Case 1-F-3 (FFAG acc.) ($\rho_{pcm, exp} = 2218.9 \pm 44.0$ [pcm])

Detector	PNS method	Feynman-α method	Area $\rho_\$$ [$]	Extended area $\rho_\$$ [$]
Fiber #1	1131.56 ± 14.49	1120.19 ± 9.39	2.76157 ± 0.04440	3.12409 ± 0.08529
Fiber #2	1179.28 ± 13.93	1157.56 ± 10.42	3.17980 ± 0.05207	3.68996 ± 0.09821
Fiber #3	1136.05 ± 13.75	1075.86 ± 7.17	3.12964 ± 0.03179	4.92435 ± 0.09021
BF-3 #1	1013.57 ± 4.94	966.32 ± 3.38	2.39188 ± 0.01619	3.38041 ± 0.03348
BF-3 #2	1065.40 ± 5.75	1027.56 ± 4.05	2.66907 ± 0.02091	3.67857 ± 0.04234
BF-3 #3	1026.57 ± 5.74	982.04 ± 3.60	2.55381 ± 0.01960	3.27812 ± 0.03800

Table A4.38 Measured prompt neutron decay constants α [s^{-1}] deduced by PNS and Feynman-α methods and subcriticality [$] in dollar units (Ref. [1])

Case 1-F-4 (FFAG acc.) ($\rho_{pcm, exp} = 2806.8 \pm 54.9$ [pcm])

Detector	PNS method	Feynman-α method	Area ρ$ [$]	Extended area ρ$ [$]
Fiber #1	1414.12 ± 9.70	1401.96 ± 5.90	3.53790 ± 0.03099	4.16379 ± 0.06793
Fiber #2	1474.29 ± 11.38	1439.92 ± 5.86	3.99097 ± 0.03643	4.72086 ± 0.08298
Fiber #3	1406.26 ± 13.80	1312.41 ± 4.50	3.88047 ± 0.02472	6.25959 ± 0.09960
BF-3 #1	1305.18 ± 6.34	1217.98 ± 4.13	3.06990 ± 0.02139	4.82552 ± 0.05332
BF-3 #2	1353.00 ± 7.60	1290.21 ± 5.22	3.49983 ± 0.02819	5.35192 ± 0.06915
BF-3 #3	1318.24 ± 7.39	1229.44 ± 4.19	3.31538 ± 0.02579	4.66721 ± 0.05952

Table A4.39 Measured prompt neutron decay constants α [s^{-1}] deduced by PNS and Feynman-α methods and subcriticality [$] in dollar units (Ref. [1])

Case 1-F-5 (FFAG acc.) ($\rho_{pcm, exp} = 13254.0 \pm 6.1$ [pcm])

Detector	PNS method	Feynman-α method	Area ρ$ [$]	Extended area ρ$ [$]
Fiber #1	2735.11 ± 70.62	2972.19 ± 36.03	14.70001 ± 0.47413	14.48612 ± 1.29327
Fiber #2	2548.37 ± 50.17	2375.37 ± 22.69	16.34603 ± 0.53004	22.42986 ± 1.45658
Fiber #3	2549.58 ± 67.04	2357.45 ± 16.72	18.24909 ± 0.50873	30.52158 ± 2.27163
BF-3 #1	2564.35 ± 22.04	2302.49 ± 9.52	11.86773 ± 0.16154	19.42186 ± 0.55781
BF-3 #2	2691.01 ± 45.25	2356.44 ± 21.71	15.75499 ± 0.47618	31.82251 ± 1.82153
BF-3 #3	2844.46 ± 42.48	2455.13 ± 18.55	16.94683 ± 0.45601	31.19015 ± 1.64630

Table A4.40 Measured prompt neutron decay constants α [s^{-1}] deduced by PNS and Feynman-α methods and subcriticality [$] in dollar units (Ref. [1])

Case 3-F-1 (FFAG acc.) ($\rho_{pcm, MCNP} = 1512.9 \pm 5.7$ [pcm])

Detector	PNS method	Feynman-α method	Area ρ$ [$]	Extended area ρ$ [$]
Fiber #1	705.96 ± 14.85	719.22 ± 1.29	1.82259 ± 0.00572	2.26361 ± 0.10200
Fiber #2	780.71 ± 27.33	900.39 ± 2.40	5.61627 ± 0.03135	10.24823 ± 0.75562
Fiber #3	800.02 ± 22.21	785.59 ± 2.43	5.76963 ± 0.02645	5.49433 ± 0.37515
BF-3 #1	620.25 ± 2.32	613.23 ± 1.34	1.43617 ± 0.00456	1.75078 ± 0.01056
BF-3 #2	689.60 ± 1.66	675.45 ± 1.39	1.77703 ± 0.00601	2.21869 ± 0.01243
BF-3 #3	579.38 ± 3.13	576.62 ± 1.44	1.32787 ± 0.00428	1.56141 ± 0.01104

Table A4.41 Measured prompt neutron decay constants α [s^{-1}] deduced by PNS and Feynman-α methods and subcriticality [$] in dollar units (Ref. [1])

Case 3-F-2 (FFAG acc.) ($\rho_{pcm, MCNP}$ = 2422.6 ± 5.7 [pcm])

Detector	PNS method	Feynman-α method	Area $\rho_\$$ [$]	Extended area $\rho_\$$ [$]
Fiber #1	1035.20 ± 3.73	1020.68 ± 2.66	2.84329 ± 0.01322	3.56877 ± 0.75391
Fiber #2	–	1303.54 ± 11.94	–	–
Fiber #3	1217.71 ± 10.58	1113.52 ± 3.77	9.07361 ± 0.07213	13.14405 ± 3.65436
BF-3 #1	965.21 ± 15.18	851.50 ± 2.79	2.31917 ± 0.01621	1.54458 ± 0.61125
BF-3 #2	985.52 ± 18.71	934.66 ± 3.39	2.84057 ± 0.02191	2.52729 ± 1.26345
BF-3 #3	968.37 ± 15.18	823.79 ± 2.83	2.19004 ± 0.01526	1.69662 ± 0.61436

Table A4.42 Measured prompt neutron decay constants α [s^{-1}] deduced by PNS and Feynman-α methods and subcriticality [$] in dollar units (Ref. [1])

Case 3-F-3 (FFAG acc.) ($\rho_{pcm, MCNP}$ = 2714.6 ± 5.7 [pcm])

Detector	PNS method	Feynman-α method	Area $\rho_\$$ [$]	Extended area $\rho_\$$ [$]
Fiber #1	1153.96 ± 5.91	1132.32 ± 3.60	3.16221 ± 0.02246	4.35512 ± 2.11887
Fiber #2	–	1397.59 ± 5.49	–	–
Fiber #3	1353.14 ± 11.60	1215.37 ± 4.95	10.10259 ± 0.12587	31.62532 ±11.28040
BF-3 #1	1090.15 ± 21.82	953.38 ± 3.41	2.59421 ± 0.01949	4.31016 ± 0.34241
BF-3 #2	1118.92 ± 30.49	1057.83 ± 4.58	3.21619 ± 0.03015	4.63895 ± 0.50563
BF-3 #3	1055.02 ± 18.26	904.09 ± 3.15	2.48810 ± 0.01875	3.98336 ± 0.27902

Table A4.43 Measured prompt neutron decay constants α [s^{-1}] deduced by PNS and Feynman-α methods and subcriticality [$] in dollar units (Ref. [1])

Case 3-F-4 (FFAG acc.) ($\rho_{pcm, MCNP}$ = 2884.1 ± 5.8 [pcm])

Detector	PNS method	Feynman-α method	Area $\rho_\$$ [$]	Extended area $\rho_\$$ [$]
Fiber #1	1222.67 ± 8.94	1207.66 ± 5.27	3.23913 ± 0.03420	3.89288 ± 0.06664
Fiber #2	–	1490.87 ± 8.31	–	–
Fiber #3	1454.23 ± 13.44	1285.38 ± 6.34	10.72912 ± 0.20364	20.01503 ± 0.48055
BF-3 #1	1125.12 ± 21.17	1012.46 ± 3.40	2.74824 ± 0.01767	4.25301 ± 0.32784
BF-3 #2	1199.67 ± 26.34	1118.91 ± 4.63	3.40634 ± 0.02744	5.30669 ± 0.51697
BF-3 #3	1139.12 ± 23.54	980.84 ± 3.04	2.63522 ± 0.01704	4.32898 ± 0.36156

Table A4.44 Measured prompt neutron decay constants α [s^{-1}] deduced by PNS and Feynman-α methods and subcriticality [$] in dollar units (Ref. [1])

Case 3-F-5 (FFAG acc.) ($\rho_{pcm, MCNP}$ = 14785.8 ± 6.1 [pcm])

Detector	PNS method	Feynman-α method	Area $\rho_\$$ [$]	Extended area $\rho_\$$ [$]
BF-3 #1	2232.48 ± 27.73	2019.60 ± 13.07	9.18032 ± 0.18320	17.95485 ± 0.66393
BF-3 #2	2410.92 ± 49.15	2164.72 ± 29.34	13.76546 ± 0.64883	30.58625 ± 2.22143
BF-3 #3	2475.27 ± 39.45	2139.70 ± 18.28	12.30041 ± 0.40731	28.35497 ± 1.56607

Table A4.45 Measured prompt neutron decay constants α [s^{-1}] deduced by PNS and Feynman-α methods and subcriticality [$] in dollar units (Ref. [1])

Case 4-F-1 (FFAG acc.) ($\rho_{pcm, MCNP} = 3111.6 \pm 5.8$ [pcm])

Detector	PNS method	Feynman-α method	Area $\rho_\$$ [$]	Extended area $\rho_\$$ [$]
Fiber #1	1125.38 ± 3.78	1108.71 ± 2.00	3.61854 ± 0.01663	4.82085 ± 0.03293
Fiber #2	1562.44 ± 8.03	1408.09 ± 3.66	11.70733 ± 0.11908	17.24558 ± 0.23977
Fiber #3	1375.56 ± 9.41	1228.52 ± 3.31	11.30293 ± 0.09279	20.07291 ± 0.26831
BF-3 #1	966.22 ± 6.00	918.48 ± 4.72	2.78717 ± 0.02319	3.77369 ± 0.04428
BF-3 #2	1074.85 ± 6.02	1026.24 ± 4.77	3.59930 ± 0.03212	5.05178 ± 0.06129
BF-3 #3	942.83 ± 6.41	887.37 ± 4.69	2.66206 ± 0.02245	3.49706 ± 0.04260

Table A4.46 Measured prompt neutron decay constants α [s^{-1}] deduced by PNS and Feynman-α methods and subcriticality [$] in dollar units (Ref. [1])

Case 4-F-2 (FFAG acc.) ($\rho_{pcm, MCNP} = 3906.2 \pm 5.8$ [pcm])

Detector	PNS method	Feynman-α method	Area $\rho_\$$ [$]	Extended area $\rho_\$$ [$]
Fiber #1	1391.13 ± 7.52	1373.04 ± 3.26	4.38045 ± 0.03693	5.64980 ± 0.07543
Fiber #2	1923.49 ± 17.25	1716.23 ± 6.16	14.51611 ± 0.28260	22.32025 ± 0.61495
Fiber #3	1702.94 ± 13.19	1472.83 ± 6.22	13.26441 ± 0.20715	23.86036 ± 0.50561
BF-3 #1	1207.87 ± 10.66	1126.86 ± 7.24	3.42911 ± 0.04326	5.05718 ± 0.09525
BF-3 #2	1330.03 ± 9.58	1243.59 ± 6.73	4.43292 ± 0.05867	6.80633 ± 0.12576
BF-3 #3	1174.22 ± 9.88	1065.23 ± 6.45	3.42483 ± 0.04214	4.81459 ± 0.08743

Table A4.47 Measured prompt neutron decay constants α [s^{-1}] deduced by PNS and Feynman-α methods and subcriticality [$] in dollar units (Ref. [1])

Case 4-F-3 (FFAG acc.) ($\rho_{pcm, MCNP} = 4216.6 \pm 5.8$ [pcm])

Detector	PNS method	Feynman-α method	Area $\rho_\$$ [$]	Extended area $\rho_\$$ [$]
Fiber #1	1510.24 ± 8.97	1496.27 ± 3.55	4.72259 ± 0.03967	5.79531 ± 0.08473
Fiber #2	2057.82 ± 19.03	1864.81 ± 7.14	15.98057 ± 0.31846	24.43068 ± 0.71408
Fiber #3	1869.31 ± 13.40	1603.28 ± 7.06	14.01938 ± 0.22056	26.06993 ± 0.56158
BF-3 #1	1289.31 ± 9.06	1187.38 ± 4.89	3.67468 ± 0.03501	5.55305 ± 0.08255
BF-3 #2	1446.93 ± 8.78	1349.98 ± 4.44	4.94667 ± 0.05298	7.56721 ± 0.11849
BF-3 #3	1297.96 ± 8.56	1163.16 ± 5.11	3.61958 ± 0.03220	5.43031 ± 0.07652

Table A4.48 Measured prompt neutron decay constants α [s^{-1}] deduced by PNS and Feynman-α methods and subcriticality [$] in dollar units (Ref. [1])

Case 4-F-4 (FFAG acc.) ($\rho_{pcm, MCNP} = 4257.9 \pm 5.8$ [pcm])

Detector	PNS method	Feynman-α method	Area $\rho_\$$ [$]	Extended area $\rho_\$$ [$]
Fiber #1	1556.12 ± 8.80	1522.92 ± 3.54	4.81807 ± 0.03981	6.79508 ± 0.09479
Fiber #2	2146.82 ± 19.38	1879.17 ± 3.54	16.81220 ± 0.32995	28.02081 ± 0.81516
Fiber #3	1848.02 ± 13.31	1601.44 ± 7.06	15.06196 ± 0.23488	34.22104 ± 0.71953
BF-3 #1	1287.36 ± 9.06	1188.78 ± 5.19	3.71551 ± 0.04354	5.31464 ± 0.09603
BF-3 #2	1446.88 ± 8.65	1375.82 ± 5.29	5.00395 ± 0.05329	7.66140 ± 0.11927
BF-3 #3	1325.30 ± 9.67	1182.60 ± 5.59	3.61555 ± 0.03216	5.48823 ± 0.08118

Table A4.49 Measured prompt neutron decay constants α [s^{-1}] deduced by PNS and Feynman-α methods and subcriticality [$] in dollar units (Ref. [1])

Case 4-F-5 (FFAG acc.) ($\rho_{pcm, MCNP} = 16947.3 \pm 6.2$ [pcm])

Detector	PNS method	Feynman-α method	Area $\rho_\$$ [$]	Extended area $\rho_\$$ [$]
Fiber #1	2562.00 ± 76.57	2848.42 ± 21.25	14.36405 ± 0.59092	14.36474 ± 0.59094
Fiber #2	–	2745.70 ± 27.44	–	–
Fiber #3	2911.88 ± 100.32	2686.22 ± 20.70	48.33320 ± 5.21077	80.00728 ± 12.21573
BF-3 #1	2277.16 ± 27.08	2022.44 ± 7.29	10.18877 ± 0.18311	15.18345 ± 0.54712
BF-3 #2	2580.40 ± 41.81	2313.12 ± 12.19	19.60222 ± 0.67824	35.40981 ± 2.04051
BF-3 #3	2643.02 ± 35.49	2212.25 ± 11.46	15.90766 ± 0.39806	30.82799 ± 1.43235

Table A4.50 Measured prompt neutron decay constants α [s^{-1}] deduced by PNS and Feynman-α methods and subcriticality [$] in dollar units (Ref. [1])

Case 5-F-1 (FFAG acc.) ($\rho_{pcm, MCNP} = 5466.8 \pm 5.9$ [pcm])

Detector	PNS method	Feynman-α method	Area $\rho_\$$ [$]	Extended area $\rho_\$$ [$]
BF-3 #1	1347.72 ± 7.55	1224.98 ± 3.24	4.2409 ± 0.0130	6.6210 ± 0.0557
BF-3 #2	1535.17 ± 3.86	1398.67 ± 4.98	6.4582 ± 0.0254	10.7421 ± 0.0638
BF-3 #3	1302.23 ± 9.671	1123.22 ± 3.32	3.6621 ± 0.0122	5.7880 ± 0.0607

Table A4.51 Measured prompt neutron decay constants α [s^{-1}] deduced by PNS and Feynman-α methods and subcriticality [\$] in dollar units (Ref. [1])

Case 5-F-2 (FFAG acc.) ($\rho_{pcm, MCNP} = 6283.4 \pm 5.9$ [pcm])

Detector	PNS method	Feynman-α method	Area $\rho_\$$ [\$]	Extended area $\rho_\$$ [\$]
BF-3 #1	1578.56 $\pm$ 8.75	1396.77 $\pm$ 4.37	4.9089 $\pm$ 0.0238	8.4536 $\pm$ 0.0893
BF-3 #2	1757.57 $\pm$ 6.66	1578.06 $\pm$ 9.59	7.4737 $\pm$ 0.0473	13.4353 $\pm$ 0.1328
BF-3 #3	1527.84 $\pm$ 10.921	1275.66 $\pm$ 4.17	4.3431 $\pm$ 0.0228	7.5644 $\pm$ 0.0959

Table A4.52 Measured prompt neutron decay constants α [s^{-1}] deduced by PNS and Feynman-α methods and subcriticality [\$] in dollar units (Ref. [1])

Case 5-F-3 (FFAG acc.) ($\rho_{pcm, MCNP} = 6402.4 \pm 5.9$ [pcm])

Detector	PNS method	Feynman-α method	Area $\rho_\$$ [\$]	Extended area $\rho_\$$ [\$]
BF-3 #1	1634.83 $\pm$ 10.43	1440.75 $\pm$ 4.85	4.9278 $\pm$ 0.0345	8.6750 $\pm$ 0.1175
BF-3 #2	1835.08 $\pm$ 9.22	1647.52 $\pm$ 6.36	7.6800 $\pm$ 0.0770	13.3188 $\pm$ 0.2001
BF-3 #3	1609.89 $\pm$ 11.47	1320.91 $\pm$ 4.98	4.3497 $\pm$ 0.0328	7.8715 $\pm$ 0.1169

Table A4.53 Measured prompt neutron decay constants α [s^{-1}] deduced by PNS and Feynman-α methods and subcriticality [\$] in dollar units (Ref. [1])

Case 5-F-4 (FFAG acc.) ($\rho_{pcm, MCNP} = 6449.0 \pm 5.9$ [pcm])

Detector	PNS method	Feynman-α method	Area $\rho_\$$ [\$]	Extended area $\rho_\$$ [\$]
BF-3 #1	1668.59 $\pm$ 10.09	1457.23 $\pm$ 4.78	5.0659 $\pm$ 0.0313	9.0312 $\pm$ 0.1142
BF-3 #2	1882.18 $\pm$ 9.29	1690.14 $\pm$ 7.51	7.8179 $\pm$ 0.0690	13.7674 $\pm$ 0.1928
BF-3 #3	1637.55 $\pm$ 11.21	1348.49 $\pm$ 5.03	4.4292 $\pm$ 0.0294	8.0330 $\pm$ 0.1122

Table A4.54 Measured prompt neutron decay constants α [s^{-1}] deduced by PNS and Feynman-α methods and subcriticality [\$] in dollar units (Ref. [1])

Case 5-F-5 (FFAG acc.) ($\rho_{pcm, MCNP} = 18537.6 \pm 6.2$ [pcm])

Detector	PNS method	Feynman-α method	Area $\rho_\$$ [\$]	Extended area $\rho_\$$ [\$]
BF-3 #1	2395.17 $\pm$ 16.79	1953.19 $\pm$ 6.48	8.7452 $\pm$ 0.0864	17.0252 $\pm$ 0.3619
BF-3 #2	2733.48 $\pm$ 33.44	2400.04 $\pm$ 17.18	22.7337 $\pm$ 0.6236	43.2796 $\pm$ 1.9802
BF-3 #3	2780.67 $\pm$ 26.47	2115.21 $\pm$ 9.68	13.4331 $\pm$ 0.2333	32.6171 $\pm$ 1.1065

A4.4 Reaction Rates

A4.4.1 Core Configurations

See (Figs. A4.7, A4.8 and A4.9).

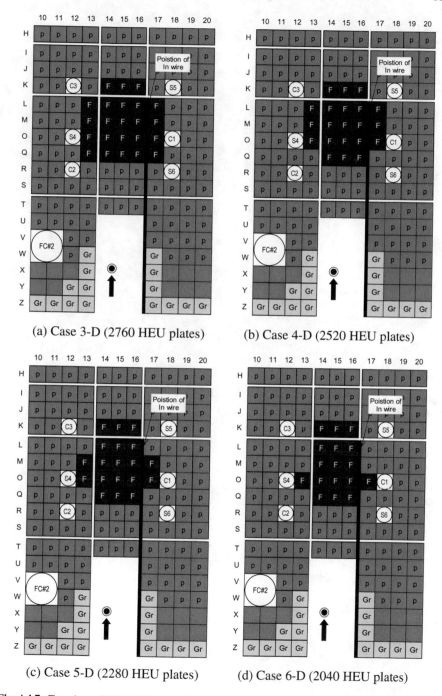

(a) Case 3-D (2760 HEU plates)

(b) Case 4-D (2520 HEU plates)

(c) Case 5-D (2280 HEU plates)

(d) Case 6-D (2040 HEU plates)

Fig. A4.7 Top view of EE1 ADS core with 14 MeV neutrons (Ref. [1])

Fig. A4.8 Foil arrangement at a gap between Pb-Bi target and fuel rod in position (15, M-O) shown in Fig. A4.7 (Ref. [1])

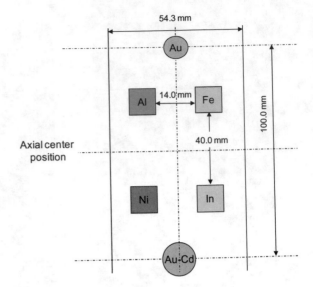

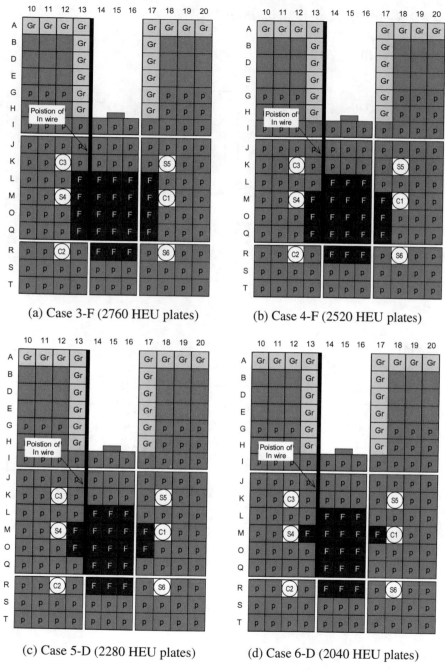

(a) Case 3-F (2760 HEU plates)   (b) Case 4-F (2520 HEU plates)

(c) Case 5-D (2280 HEU plates)   (d) Case 6-D (2040 HEU plates)

Fig. A4.9 Top view of EE1 ADS core with 100 MeV protons (Pb-Bi target) (Ref. [1])

A4.4.2 Reaction Rate Distributions

See (Figs. A4.10 and A4.11).

Fig. A4.10 Comparison between measured ^{115}In(n, γ)^{116m}In reaction rate distributions normalized by ^{93}Nb(n, 2n)^{92m}Nb along (16-17, L-A′) region shown in Fig. A4.7 (Ref. [1])

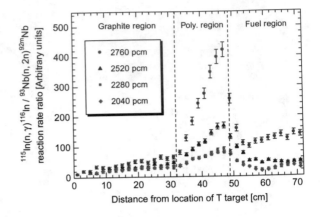

Fig. A4.11 Comparison between measured ^{115}In(n, γ)^{116m}In reaction rate distributions normalized by ^{115}In(n, n′)^{115m}In along (13-14, A-Q) region shown in Fig. A4.9 (Ref. [1])

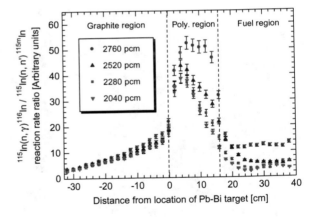

A4.4.3 Reaction Rates of Activation Foils

See (Tables A4.55, A4.56 and A4.57).

Table A4.55 Main characteristics and description of activation foils (Ref. [1])

Reaction	Dimension [mm]	Threshold [MeV]	Half-life	γ-ray Energy [keV]	Emission rate (%)
^{197}Au$(n, \gamma)^{198}$Au (bare and Cd* covered)	8 mm diam. 0.05 mm thick	-	2.697 d	411.9	95.51
Cd plate	10 mm diam. 1 mm thick	-	-	-	-
^{115}In$(n, \gamma)^{116m}$In (wire)	1 mm diam. 680 mm long	-	54.12 m	416.9 1097.3 1293.54	32.4 55.7 85.0
^{115}In$(n, n')^{115m}$In (foil; Pb–Bi target)	$10 \times 10 \times 1$	0.3	4.486 h	336.2	45.08
^{58}Ni$(n, p)^{58}$Co	$10 \times 10 \times 1$	0.9	70.82 d	810.8	99.4
^{56}Fe$(n, p)^{56}$Mn	$10 \times 10 \times 1$	5.0	2.578 h	846.8 1810.7	98.9 27.2
^{27}Al$(n, \alpha)^{24}$Na	$10 \times 10 \times 1$	5.6	14.96 h	1368.6	100
^{93}Nb$(n, 2n)^{92m}$Nb (T target)	$10 \times 10 \times 1$	9.0522	10.15 d	934.4	99.2

Cd*: Au foil (Cd covered) was sandwiched between two Cd plates (10 mm diam. and 1 mm thick).

Table A4.56 Measured reaction rates in Cases 3-D to 6-D (Ref. [1])

Measured reaction rate [s^{-1} cm^{-3}]

Reaction	Case 3-D	Case 4-D	Case 5-D	Case 6-D
^{197}Au$(n, \gamma)^{198}$Au (bare)	(9.657 ± 0.382)E+04	(5.065 ± 0.242)E+04	(3.670 ± 0.150)E+04	(2.051 ± 0.083)E+04
^{197}Au$(n, \gamma)^{198}$Au (Cd)	(6.514 ± 0.310)E+04	(4.083 ± 0.169)E+04	(1.780 ± 0.092)E+04	(2.041 ± 0.087)E+04
^{115}In$(n, n')^{115m}$In	(1.778 ± 0.101)E+02	(2.087 ± 0.109)E+02	(1.331 ± 0.057)E+02	(1.389 ± 0.058)E+02
^{58}Ni$(n, p)^{58}$Co	–	–	–	–
^{56}Fe$(n, p)^{56}$Mn	(8.056 ± 0.665)E+00	(1.895 ± 0.266)E+00	–	–
^{27}Al$(n, \alpha)^{24}$Na	(5.507 ± 0.701)E+00	(2.676 ± 0.473)E+00	–	–
^{93}Nb$(n, 2n)^{92m}$Nb (T target)	(2.801 ± 0.114)E+02	(4.765 ± 0.176)E+02	(6.652 ± 0.297)E+02	(6.533 ± 0.260)E+02

Table A4.57 Measured reaction rates in Cases 3-F to 6-F (Ref. [1])

| Reaction | Measured reaction rate [s^{-1} cm^{-3}] | | | |
	Case 3-F	Case 4-F	Case 5-F	Case 6-F
^{197}Au$(n, \gamma)^{198}$Au (bare)	(5.443 ± 0.180)E+06	(2.715 ± 0.090)E+06	(1.585 ± 0.053)E+06	(1.104 ± 0.036)E+06
^{115}In$(n, n')^{115m}$In	(1.671 ± 0.060)E+04	(0.842 ± 0.030)E+04	(0.518 ± 0.019)E+04	(0.401 ± 0.014)E+04
^{58}Ni$(n, p)^{58}$Co	(1.582 ± 0.071)E+04	(0.777 ± 0.029)E+04	(0.506 ± 0.019)E+04	(0.380 ± 0.015)E+04
^{56}Fe$(n, p)^{56}$Mn	(3.415 ± 0.195)E+02	(2.027 ± 0.121)E+02	(1.326 ± 0.077)E+02	(1.921 ± 0.074)E+02
^{27}Al$(n, \alpha)^{24}$Na	(6.696 ± 0.408)E+02	(3.110 ± 0.163)E+02	(1.155 ± 0.099)E+02	(1.291 ± 0.053)E+02
^{115}In$(n, n')^{115m}$In (Pb–Bi target)	(1.900 ± 0.060)E+05	(1.947 ± 0.062)E+05	(2.194 ± 0.068)E+05	(2.160 ± 0.068)E+05

References

1. Pyeon CH (2020) Experimental benchmarks of medium-fast spectrum core (EE1 core) in accelerator-driven system at Kyoto University Critical Assembly. KURNS-EKR-007
2. Pyeon CH, Yamanaka M, Pyeon CH, Endo T et al (2020) Neutron generation time in highly-enriched uranium core at Kyoto University Critical Assembly. Nucl Sci Eng. https://doi.org/10.1080/00295639.2020.1774230

Appendix A5: ^{232}Th-Fueled ADS Core

A5.1 Core Configurations

See (Figs. A5.1, A5.2, A5.3, A5.4, A5.5, A5.6, A5.7 and A5.8, Table A5.1).

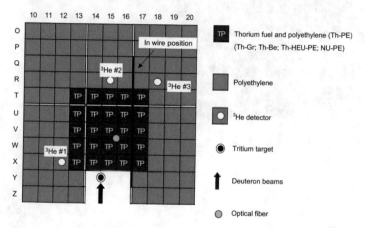

Fig. A5.1 Thorium-fueled ADS core configuration with 14 MeV neutrons (Th–PE, Th–Gr, Th–Be, Th–HEU–PE and NU–PE) (Refs. [1–4])

© The Editor(s) (if applicable) and The Author(s) 2021
C. H. Pyeon (ed.), *Accelerator-Driven System at Kyoto University Critical Assembly*,
https://doi.org/10.1007/978-981-16-0344-0

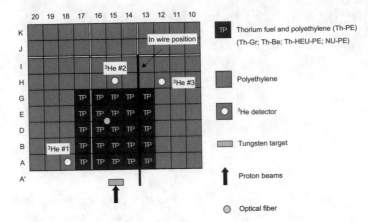

Fig. A5.2 Thorium-fueled ADS core configuration with 100 MeV protons (Tungsten target) (Th–PE, Th–Gr, Th–Be, Th–HEU–PE and NU–PE) (Refs. [1–4])

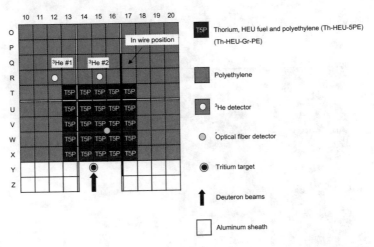

Fig. A5.3 Thorium-fueled ADS core configuration with 14 MeV neutrons (Th–HEU–5PE and Th–HEU–Gr–PE) (Refs. [1–4])

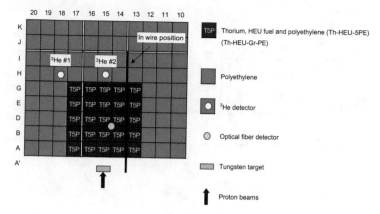

Fig. A5.4 Thorium-fueled ADS core configuration with 100 MeV protons (Tungsten target) (Th–HEU–5PE and Th–HEU–Gr–PE) (Refs. [1–2])

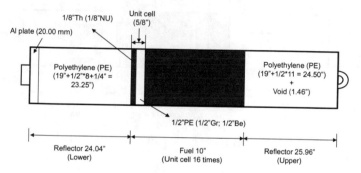

Fig. A5.5 Fall sideway view of fuel assembly Th–PE, Th–Gr, Th–Be and NU–PE (Ref. [1])

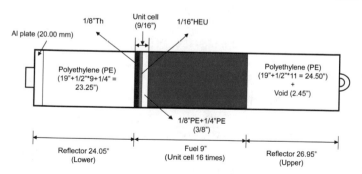

Fig. A5.6 Fall sideway view of fuel assembly Th–HEU–PE (Ref. [1])

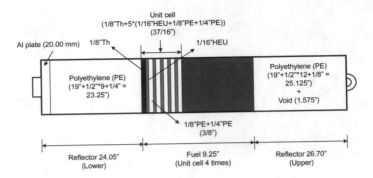

Fig. A5.7 Fall sideway view of fuel assembly Th–HEU–5PE (Ref. [1])

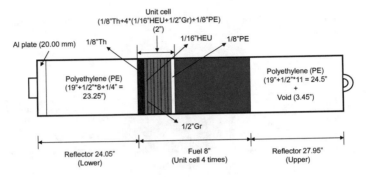

Fig. A5.8 Fall sideways view of fuel assembly Th–HEU–Gr–PE (Ref. [1])

Table A5.1 List of thorium-fueled ADS cores (Ref. [1])

Core	Cell pattern	k_{eff}
Th–PE	1/8″Th+1/2″PE	0.00613
Th–Gr	1/8″Th+1/2″Gr	0.00952
Th–Be	1/8″Th+1/2″Be	0.00765
Th–HEU–PE	1/8″Th+1/16″HEU+3/8″PE	0.58754
NU–PE	1/8″NU+1/2″PE	0.50867
Th–HEU–5PE	1/8″Th+5*(1/16″HEU+3/8″PE)	0.85121
Th–HEU–Gr–PE	1/8″Th+4*(1/16″HEU+1/2″Gr)+1/8″PE	0.35473

Th: Thorium; PE: Polyethylene; Gr: Graphite; Be: Beryllium;
HEU: Highly-enriched uranium; NU: Natural uranium

A5.2 Results of Experiments

A5.2.1 Reaction Rate Distributions

See (Figs. A5.9, A5.10, A5.11 and A5.12).

Fig. A5.9 Comparison between the measured reaction rates of indium wire obtained from the Th-fueled ADS experiment with 14 MeV neutrons for Th–PE, Th–Gr, Th–Be, Th–HEU–PE and NU–PE cores (Refs.[1–2])

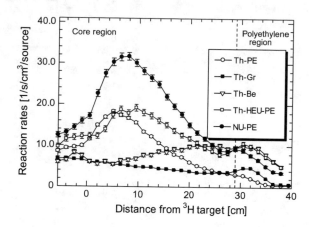

Fig. A5.10 Cparison between the measured reaction rates of indium wire obtained from the Th-fueled ADS experiment with 100 MeV protons for Th–PE, Th–Gr, Th–Be, Th–HEU–PE and NU–PE cores (Refs.[1–2])

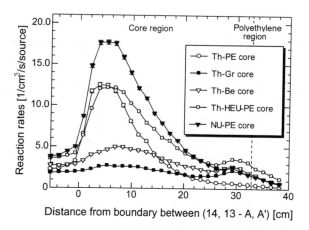

Fig. A5.11 Comparison between the measured reaction rates of indium wire obtained from the Th-fueled ADS experiment with 14 MeV neutrons for Th–HEU–5PE and Th–PE–Gr–PE cores (Refs. [1–3])

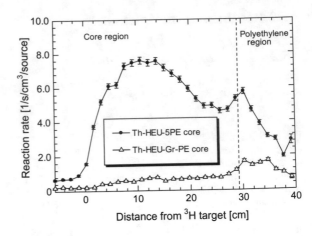

Fig. A5.12 Comparison between the measured reaction rates of indium wire obtained from the Th-fueled ADS experiment with 100 MeV protons for Th–HEU–5PE and Th–PE–Gr–PE cores (Refs. [1–3])

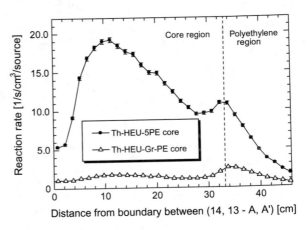

A5.2.2 PNS and Feynman-α Methods

See (Tables A5.2, A5.3 and A5.4).

Table A5.2 Beam characteristics of 14 MeV neutrons and 100 MeV protons in thorium-fueled cores (Ref. [1])

Core	14 MeV neutrons		100 MeV protons	
	Repetition (Hz)	Width (μs)	Repetition (Hz)	Width (ns)
Th–PE	100	10	–	–
Th–Gr	100	10	–	–
Th–Be	100	10	–	–
Th–HEU–PE	100	10	20	100
NU–PE	100	10	–	–
Th–HEU–5PE	10	10	20	100
Th–HEU–Gr–PE	10	10	20	100

Table A5.3 Measured results in prompt neutron decay constant deduced by least-squared fitting in PNS method (Refs. [1–2])

Core	Prompt neutron decay constant α [s^{-1}]		
	14 MeV neutrons		
	^{3}He #1	^{3}He #2	^{3}He #3
Th–PE	6642 ± 11	6224 ± 27	5751 ± 25
Th–Gr	6451 ± 12	5945 ± 15	5701 ± 17
Th–Be	6515 ± 8	6111 ± 17	5746 ± 20
Th–HEU–PE	5692 ± 11	5275 ± 7	5231 ± 9
NU–PE	5748 ± 11	6592 ± 15	5010 ± 11
Th–HEU–5PE	3110 ± 11	3104 ± 10	–
Th–HEU–Gr–PE	4980 ± 40	4939 ± 50	–
Core	Prompt neutron decay constant α [s^{-1}]		
	100 MeV protons		
	^{3}He #1	^{3}He #2	^{3}He #3
Th–PE	–	–	–
Th–Gr	–	–	–
Th–Be	–	–	–
Th–HEU–PE	5777 ± 11	5527 ± 35	5236 ± 37
NU–PE	–	–	–
Th–HEU–5PE	2776 ± 17	2917 ± 8	–
Th–HEU–Gr–PE	3186 ± 72	4656 ± 12	–

Table A5.4 Measured results in subcriticality in dollar units deduced by extrapolated area ratio method (Refs. [1–2])

Core	Subcriticality ρ [$] (by Area ratio method)		
	14 MeV neutrons		
	^{3}He #1	^{3}He #2	^{3}He #3
Th–PE	144.77 ± 1.56	1051.97 ± 28.70	2778.37 ± 135.26
Th–Gr	108.47 ± 18.06	400.34 ± 5.92	880.49 ± 20.41
Th–Be	33.42 ± 0.17	37.04 ± 0.16	105.35 ± 0.85
Th–HEU–PE	12.05 ± 0.03	29.75 ± 0.09	63.53 ± 0.35
NU–PE	13.40 ± 0.04	26.64 ± 0.12	82.76 ± 0.57
Th–HEU–5PE	16.61 ± 0.21	11.35 ± 0.09	–
Th–HEU–Gr–PE	63.76 ± 1.42	42.92 ± 0.66	–
Core	Subcriticality ρ [$] (by Area ratio method)		
	100 MeV protons		
	^{3}He #1	^{3}He #2	^{3}He #3
Th–PE	–	–	–
Th–Gr	–	–	–
Th–Be	–	–	–
Th–HEU–PE	32.30 ± 0.45	26.52 ± 0.29	44.27 ± 0.74
NU–PE	–	–	–
Th–HEU–5PE	3.77 ± 0.01	10.35 ± 0.05	–
Th–HEU–Gr–PE	3.83 ± 0.30	64.54 ± 11.13	–

References

1. Pyeon CH (2015) Experimental benchmarks on thorium-loaded accelerator-driven system (ADS) at Kyoto University Critical Assembly. KURRI-TR(CD)-48
2. Pyeon CH, Yagi T, Sukawa K et al (2014) Mockup experiments on the thorium-loaded accelerator-driven system in the Kyoto University Critical Assembly. Nucl Sci Eng 177:156
3. Yamanaka M, Pyeon CH, Yagi T et al (2016) Accuracy of reactor physics parameters in thorium-loaded accelerator-driven system experiments at Kyoto University Critical Assembly. Nucl Sci Eng 183:96
4. Yamanaka M, Pyeon CH, Misawa T (2016) Monte Carlo approach of effective delayed neutron fraction by k-ratio method with external neutron source. Nucl Sci Eng 184:551

Printed in the United States
by Baker & Taylor Publisher Services